D0142983

# Geometric Data Structures for Computer Graphics

# Geometric Data Structures for Computer Graphics

Elmar Langetepe
Gabriel Zachmann

A K Peters, Ltd.
Wellesley, Massachusetts

Editorial, Sales, and Customer Service Office

A K Peters, Ltd.
888 Worcester Street, Suite 230
Wellesley, MA 02482
www.akpeters.com



**Library of Congress Cataloging-in-Publication Data**

Langetepe, Elmar, 1967-
Geometric data structures for computer graphics / Elmar Langetepe, Gabriel Zachmann
p. cm.
Includes bibliographical references and index.
ISBN 13: 978-1-56881-235-9 (alk. paper)
ISBN 10: 1-56881-235-3 (alk. paper)
1. Computer graphics. 2. Data structures (Computer science) I. Zachmann, Gabriel, 1967- II. Title.

T385.L364 2005
006.6–dc22

2005051464

The frontispiece and quote are from Hartmut Böhme's *Albrecht Dürer, Melencolia I—Im Labyrinth der Deutung*, page 50, published by Fischer Taschenbuch Verlag, Frankfurt.

Printed in India
09 08 07 06 10 9 8 7 6 5 4 3 2 1

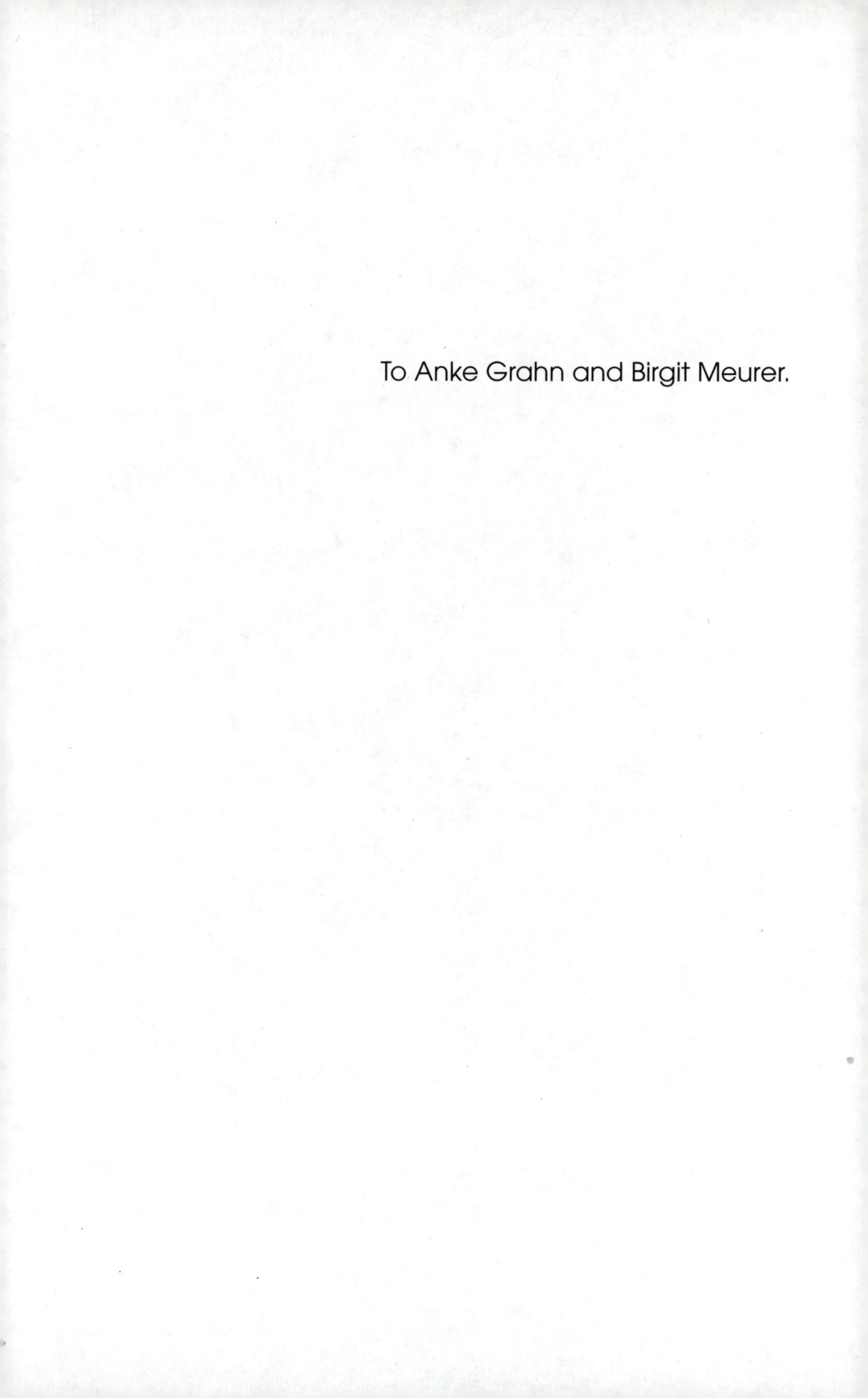

To Anke Grahn and Birgit Meurer.

# Contents

*Melencolia I,* Albrecht Dürer

Geometrie ist der Königsweg ästhetischer Erkenntnis:
sie offenbart "die kunst in der natur."
— Hartmut Böhme

# Preface

In recent years, methods from computational geometry have been widely adopted by the computer graphics community, yielding elegant and efficient algorithms. This book aims at endowing practitioners in the computer graphics field with a working knowledge of a wide range of geometric data structures from computational geometry. It will enable readers to recognize geometric problems and select the most suitable data structure when developing computer graphics algorithms.

The book will focus on algorithms and data structures that have proven to be versatile, efficient, fundamental, and easy to implement. Thus practitioners and researchers will benefit immediately from this book in their everyday work.

Our goal is to familiarize practitioners and researchers in computer graphics with some very versatile and ubiquitous geometric data structures, enable them to readily recognize geometric problems during their work, modify the algorithms to their needs, and hopefully make them curious about further powerful treasures to be discovered in the area of computational geometry.

In order to achieve these goals in an engaging yet sound manner, the general concept throughout the book is to present each geometric data structure in the following way: first, the data strucure will be defined and described in detail; then, some of its fundamental properties will be highlighted; after that, one or more computational geometry algorithms based on the data structure will be presented; and finally, a number of recent, representative, and practically relevant algorithms from computer graphics will be described in detail, showing the utilization of the data structure in a creative and enlightening way.

We do not try to provide an exhaustive survey of the topics touched upon here—this would be far beyond the scope of this book. Neither do we aspire to present the latest and greatest algorithms for a given problem, for two reasons. First, the focus is on geometric data structures, and we do not want to obstruct the view with complicated algorithms. Second, we feel that, for practical purposes, a good trade-off between simplicity and efficiency is important.

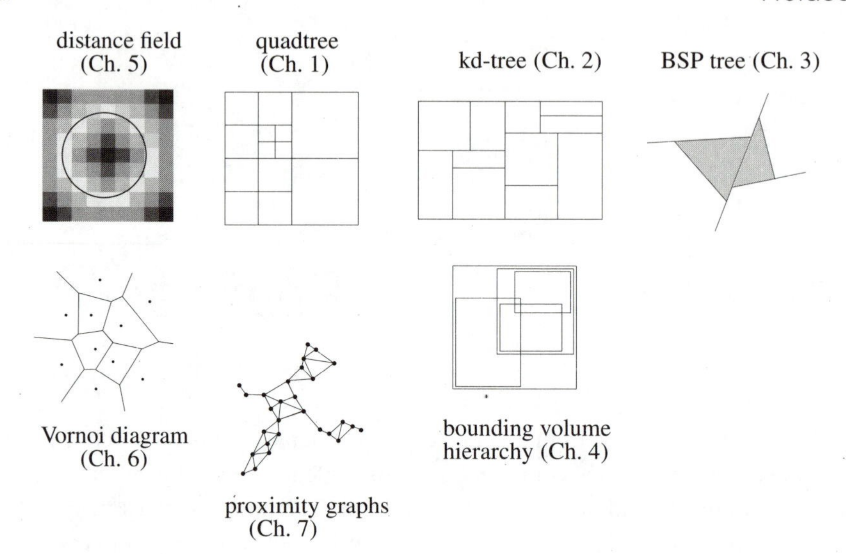


An overview of some of the data structures presented in this book.

The intended audience is practitioners working in 3D computer graphics (VR, CAD/CAM, entertainment, animation, etc.) and students of both computer graphics and computational geometry. Readers should be familiar with the basic principles of computer graphics and the type of problems in the area.

We have arranged the chapters in roughly increasing degree of difficulty. The hierarchical data structures are ordered by increasing flexibility, while the non-hierarchical chapters build on each other. Finally, the last three chapters present generic techniques for making geometric data structures kinetic, robust, and dynamic.

Figure provides an overview of some of the data structures that will be discussed in the following chapters. Chapter 1 presents quadtrees and octrees, which are, arguably, the most popular data structure in computer graphics. Lifting one of the restrictions, thus making the data structure more flexible, we arrive at kd-trees, which are presented in Chapter 2. We can make this even more flexible, yielding BSP trees, discussed in Chapter 3. From kd-trees, we can also derive bounding volume hierarchies, which are described in Chapter 4. Starting with a quadtree (or even a grid), we can store more information about the object(s), thus arriving at distance fields (Chapter 5). In a sense, distance fields are a discretized version of Voronoi diagrams, which are presented in Chapter 6. In Chapter 7, we discuss the general class of geometric proximity graphs, one kind of which, the Delaunay graph, we have already seen in Chapter 6. The general concept of specialized spatial data structures for moving objects is introduced in Chapter 8 and discussed by an example. In Chapter 9, we consider the problems of degeneracy and robustness in geometric computing. finally, in Chapter 10, we introduce a simple generic scheme for dynamization.

## Acknowledgments

We would like to thank Prof. Dr. Reinhard Klein and Prof. Dr. Rolf Klein for their encouragement and advice during the work. Thanks goes also to Ansgar Grüne, Tom Kamphans, Adalbert Prokop, Manuel Wedemeier, and Michael Bazanski for reading some parts of the manuscript. In addition, many thanks go to Jan Klein for the great collaboration. We are grateful to Alice and Klaus Peters for initiating this book. And we would like to thank Kevin Jackson-Mead for managing the project and for his great patience. Part of Zachmann's work was funded by DFG's grant ZA292/1.

# 1

# Quadtrees and Octrees

In this chapter, we introduce the quadtree and octree structures. Their definition and complexity, the recursive construction scheme, and a standard application are presented. Quadtrees and octrees have applications in mesh generation, as shown in Sections 1.3, 1.4, and 1.5.

## 1.1. Definition

A *quadtree* is a tree rooted so that every internal node has four children. Every node in the tree corresponds to a square. If a node $v$ has children, their corresponding squares are the four quadrants, as shown in Figure 1.1.

Quadtrees can store many kinds of data. We will describe the variant that stores a set of points and suggest a recursive definition. A simple recursive splitting of squares is continued until there is only one point in a square. Let $P$ be a set of points.

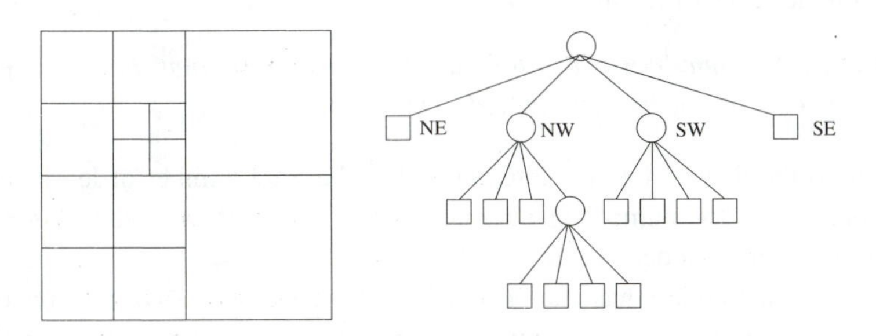


**Figure 1.1.** An example of a quadtree.

The definition of a quadtree for a set of points in a square $Q = [x1_Q : x2_Q] \times [y1_Q : y2_Q]$ is as follows:

- If $|P| \leq 1$, then the quadtree is a single leaf where $Q$ and $P$ are stored.
- Otherwise, let $Q_{NE}$, $Q_{NW}$, $Q_{SW}$, and $Q_{SE}$ denote the four quadrants. Let $x_{mid} := (x1_Q + x2_Q)/2$ and $y_{mid} := (y1_Q + y2_Q)/2$, and define

$$\begin{aligned} P_{NE} &:= \{p \in P : p_x > x_{mid} \wedge p_y > y_{mid}\}, \\ P_{NW} &:= \{p \in P : p_x \leq x_{mid} \wedge p_y > y_{mid}\}, \\ P_{SW} &:= \{p \in P : p_x \leq x_{mid} \wedge p_y \leq y_{mid}\}, \\ P_{SE} &:= \{p \in P : p_x > x_{mid} \wedge p_y \leq y_{mid}\}. \end{aligned}$$

The quadtree consists of a root node $v$, and $Q$ is stored at $v$. In the following, let $Q(v)$ denote the square stored at $v$. Furthermore, $v$ has four children: the $X$-child is the root of the quadtree of the set $P_X$, where $X$ is an element of the set $\{NE, NW, SW, SE\}$.

## 1.2. Complexity and Construction

The recursive definition implies a recursive construction algorithm. Only the starting square must be chosen adequately. If the split operation cannot be performed well, the quadtree is unbalanced. Despite this effect, the depth of the tree is related to the distance between the points.

**Theorem 1.1.** *The depth of a quadtree for a set P of points in the plane is at most $\log(s/c) + \frac{3}{2}$, where c is the smallest distance between any two points in P, and s is the side length of the initial square.*

The cost of the recursive construction and the complexity of the quadtree depends on the depth of the tree.

**Theorem 1.2.** *A quadtree of depth d that stores a set of n points has $O((d+1)n)$ nodes and can be constructed in $O((d+1)n)$ time.*

Due to the degree 4 of the internal nodes, the total number of leaves is one plus three times the number of internal nodes. Hence, it suffices to bound the number of internal nodes.

Any internal node $v$ has one or more points inside $Q(v)$. The squares of the node of a single depth cover the initial square. Therefore, at every depth, we have at most $n$ internal nodes, which gives the node bound.

The most time-consuming task in one step of the recursive approach is the distribution of the points. The amount of time spent is linear only in the number of points and the $O((d+1)n)$ time bound holds.

The 3D equivalents of quadtrees are *octrees*. The quadtree construction can be easily extended to octrees in 3D. The internal nodes of octrees have eight children, and the children correspond to boxes instead of squares.

Neighbor finding. A common task when utilizing quadtrees is *neighbor finding* (also called *navigation*), i.e., given a node $v$ and a direction north, east, south or west, find a node $v'$ so that $Q(v)$ is adjacent to $Q(v')$. Normally, $v$ is a leaf and $v'$ should be a leaf as well, but this is not necessarily unique. Obviously, one square may have many such neighbors, as shown in Figure 1.2.

For convenience, we extend the neighbor search. The given node can also be internal; that is, $v$ and $v'$ should be adjacent corresponding to the given direction *and* should also have the same depth. If there is no such node, we want to find the deepest node whose square is adjacent.

The algorithm works as follows. Suppose we want to find the north neighbor of $v$. If $v$ happens to be the $SE$ or $SW$ child of its parent, then its north neighbor is easy to find—it is the $NE$ or $NW$ child of its parent, respectively. If $v$ itself is the $NE$ or $NW$ child of its parent, then we proceed as follows. Recursively find the north neighbor of $\mu$ of the parent of $v$. If $\mu$ is an internal node, then the north neighbor of $v$ is a child of $\mu$; if $\mu$ is a leaf, the north neighbor we seek is $\mu$ itself.

This simple procedure runs in time $O(d+1)$.

**Theorem 1.3.** *Let $T$ be a quadtree of depth $d$. The neighbor of a given node $v$ in $T$ a given direction, as defined above, can be found in $O(d+1)$ time.*

Furthermore, there is also a simple procedure that constructs a balanced quadtree out of a given quadtree $T$. This can be done in time $O(d+1)m$ and $O(m)$ space if $T$ has $m$ nodes. For details, see also [de Berg et al. 00].

Similar results hold for octrees as well.

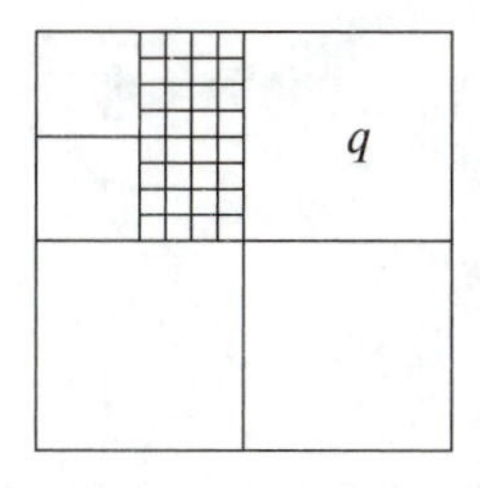


**Figure 1.2.** The square $q$ has many *west* neighbors.

## 1.3. Height Field Visualization

A special area in 3D visualization is the rendering of large terrains, or, more generally, of height fields. A height field is usually given as a uniformly gridded square array $h : [0, N-1]^2 \rightarrow \mathbb{R},\ N \in \mathbb{I}$, of height values, where $N$ is typically in the order of 16,384 up to several millions. In practice, such a raw height field is often stored in some image file format, such as GIF. A regular grid is, for instance, one of the standard forms in which the US Geological Survey publishes its data, known as the Digital Elevation Model (DEM) [Elassal and Caruso 84].

Obviously, a regular grid is not a very efficient data structure, both with regards to memory and rendering. So, alternatively, height fields can be stored as *triangular irregular networks* (TINs) (see Figure 1.3). They can adapt much better to the detail and features (or lack thereof) in the height field, so they can approximate any surface at any desired level of accuracy with fewer polygons than any other representation [Lindstrom et al. 96]. However, due to their much more complex structure, TINs do not lend themselves to interactive visualization as well as more regular representations.

The problem in terrain visualization is that, if the user looks at it from a low viewpoint directed at the horizon, there are a few parts of the terrain that are very close, while the majority of the visible terrain is at a greater distance. Close parts of the terrain should be rendered with high detail, while distant parts should be rendered with very little detail to maintain a high frame rate.

To solve this problem, a data structure is needed that allows us to quickly determine the desired level of detail in each part of the terrain. Quadtrees are such a data structure, particularly because they seem to be a good compromise between the simplicity of non-hierarchical grids and the good adaptivity of TINs. The general idea is to construct a quadtree over the grid, and then traverse this quadtree top-down in order to render it. At each node, we decide whether the detail offered by rendering it is enough, or if we have to go down further.

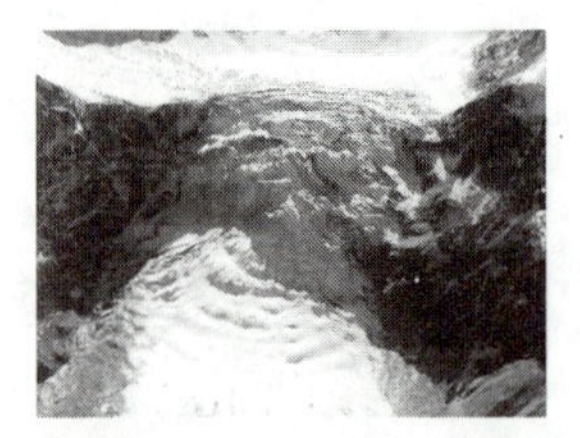
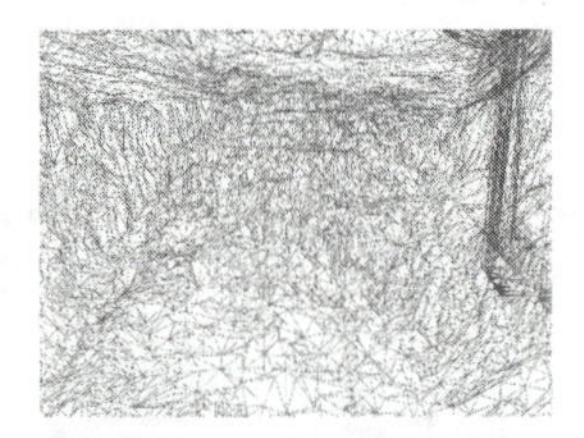

**Figure 1.3.** A terrain (left), a TIN of its height field (middle), and a superposition (right) (Wahl et al. 04). (See Color Plate I.)

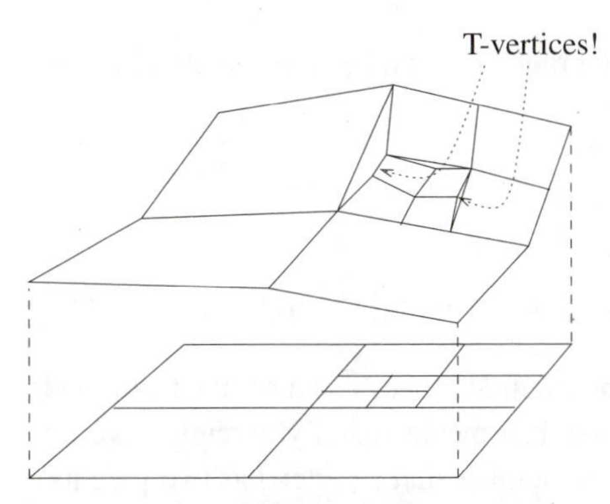


**Figure 1.4.** In order to use a quadtree for defining a height field mesh, it should be balanced. (Courtesy Prof. R. Klein and R. Wahl, Bonn University.)

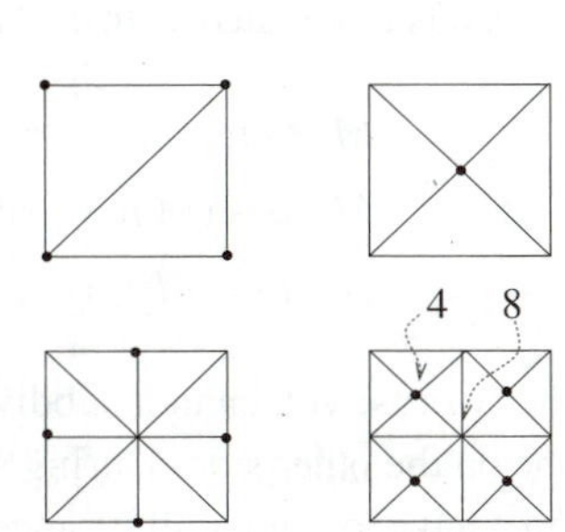


**Figure 1.5.** A quadtree defines a recursive subdivision scheme yielding a 4-8 mesh. The dots denote the newly added vertices. Some vertices have degree 4, and some have 8 (hence the name).

One problem with quadtrees (and quadrangle-based data structures in general) is that nodes are not quite independent of each other. Assume we have constructed a quadtree over some terrain, as depicted in Figure 1.4. If we render that as-is, there will be a gap (a *crack*) between the top-left square and the fine-detail squares inside the top-right square. The vertices causing this problem are called *T-vertices*. Triangulating them would help in theory, but in practice, this leads to long and thin triangles that have problems on their own. The solution is to triangulate each node.

Thus, a quadtree offers a recursive subdivision scheme to define a triangulated regular grid (see Figure 1.5): start with a square subdivided into two right-angle triangles; with each recursion step, subdivide the longest side of all triangles (the hypotenuse), yielding two new right-angle triangles each (hence, this scheme is sometimes referred to as "longest edge bisection") [Lindstrom and Pascucci 01]. This yields a mesh where all vertices have degree 4 or 8 (except the border vertices), which is why such a mesh is often called a 4-8 mesh.

This subdivision scheme induces a directed acyclic graph (DAG) on the set of vertices: vertex $j$ is a child of $i$ if it is created by a split of a right angle at vertex $i$. This will be denoted by an edge $(i, j)$. Note that almost all vertices are created twice (see Figure 1.5), so all nodes in the graph have four children and two parents (except the border vertices).

During rendering, we will choose cells of the subdivision at different levels. Let $M^0$ be the fully subdivided mesh (which corresponds to the original grid) and $M$ be the current, incompletely subdivided mesh. $M$ corresponds to a subset of the DAG of $M^0$. The condition of being crack-free can be reformulated in terms of the DAGs associated with $M^0$ and $M$:

$$
\begin{aligned}
&M \text{ is crack-free } \Leftrightarrow \\
&M \text{ does not have any T-vertices } \Leftrightarrow \\
&\forall (i,j) \in M : \ (i',j) \in M, \text{ where parent}(j) = \{i, i'\}. \qquad (1.1)
\end{aligned}
$$

In other words: you cannot subdivide one triangle alone; you also must subdivide the one on the other side. During rendering, this means that if you render a vertex, you also have to render all its ancestors (remember that a vertex has two parents).

Rendering such a mesh generates (conceptually) a single, long list of vertices that are then fed into the graphics pipeline as a single triangle strip. The pseudocode for the algorithm looks like this (simplified):

**submesh**($i, j$)
**if** error($i$) $< \tau$ **then**
  **return**
**end if**
**if** $B_i$ outside viewing frustum **then**
  **return**
**end if**
submesh( $j, c_l$ )
$V$ += $p_i$
submesh( $j, c_r$ )

where error($i$) is some error measure for vertex $i$, and $B_i$ is the sphere around vertex $i$ that completely encloses all descendant triangles.

Note that this algorithm can produce the same vertex multiple times consecutively; this is easy to check, of course. In order to produce one strip, the algorithm has to copy older vertices to the current front of the list at places where it makes a "turn"; again, this is easy to detect, and the interested reader is referred to [Lindstrom and Pascucci 01].

One can speed up the culling a bit by noticing that if $B_i$ is completely inside the frustum, we do not need to test the child vertices anymore.

We still need to think about the way we store our terrain subdivision mesh. Eventually, we will want to store it as a single linear array for two reasons:

- The tree is complete, so it really would not make sense to store it using pointers.

- We want to map the file that holds the tree into memory as-is (for instance, with the Unix `mmap` function), so pointers would not work at all.

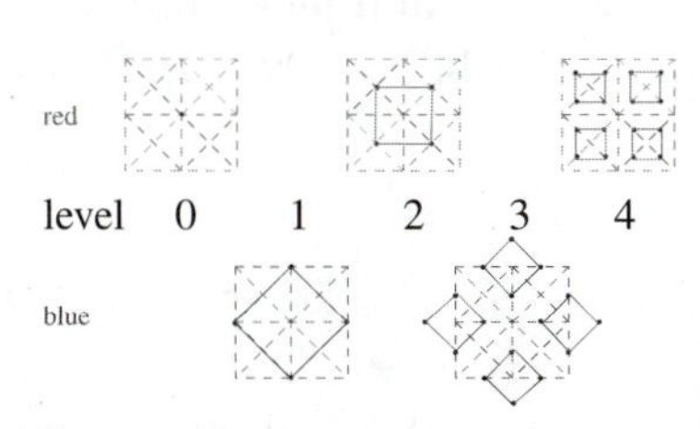


**Figure 1.6.** The 4-8 subdivision can be generated by two interleaved quadtrees. The solid lines connect siblings that share a common parent. (See Color Plate II.)

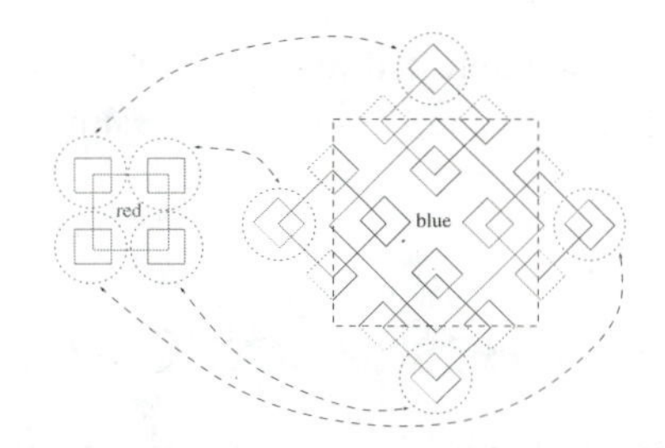


**Figure 1.7.** The red quadtree can be stored in the unused "ghost" nodes of the blue quadtree. (See Color Plate III.)

We should keep in mind, however, that with current architectures, every memory access that cannot be satisfied by the cache is extremely expensive (this is even more so with disk accesses, of course).

The simplest way to organize the terrain vertices is a matrix layout. The disadvantage is that there is no cache locality at all across the major index. To improve this, people often introduce some kind of blocking, where each block is stored in a matrix and all blocks are arranged in matrix order, too. Unfortunately, Lindstrom and Pascucci [Lindstrom and Pascucci 01] report that this is, at least for terrain visualization, worse than the simple matrix layout by a factor 10!

Enter quadtrees. They offer the advantage that vertices on the same level are stored fairly close in memory. The 4-8 subdivision scheme can be viewed as two quadtrees that are interleaved (see Figure 1.6): we start with the first level of the "red" quadtree that contains just the one vertex in the middle of the grid, which is the one that is generated by the 4-8 subdivision with the first step. Next comes the first level of the "blue" quadtree that contains four vertices, which are the vertices generated by the second step of the 4-8 subdivision scheme. This process repeats logically. Note that the blue quadtree is exactly like the red one, except it is rotated by 45°. When you overlay the red and the blue quadtree, you get exactly the 4-8 mesh.

Notice that the blue quadtree contains nodes that are outside the terrain grid; we will call these nodes "ghost nodes." The nice thing about them is that we can store the red quadtree in place of these ghost nodes (see Figure 1.7). This reduces the number of unused elements in the final linear array down to 33%.

During rendering, we need to calculate the indices of the child vertices, given the three vertices of a triangle. It turns out that by cleverly choosing the indices of the top-level vertices, this can be done as efficiently as with a matrix layout.

The interested reader can find more about this topic in [Lindstrom et al. 96, Lindstrom and Pascucci 01, Balmelli et al. 01, Balmelli et al. 99], and many others.

## 1.4. Isosurface Generation

One technique (among many others) of visualizing a 3D volume is to extract isosurfaces and render those as a regular polygonal surface. It can be used to extract the surfaces of bones or organs in medical scans, such as MRIs and CTs.

Assume for the moment that we are given a scalar field $f : \mathbb{R}^3 \to \mathbb{R}$. Then the task of finding an isosurface would "just" be to find all solutions (i.e., all roots) of the equation $f(\vec{x}) = t$.

Since we live in a discrete world (at least in computer graphics), the scalar field is usually given in the form of a *curvilinear grid*: the vertices of the *cells* are called *nodes*, and we have one scalar and a 3D point stored at each node (see Figure 1.8). Such a curvilinear grid is usually stored as a 3D array, which can be conceived as a regular 3D grid (here, the cells are often called *voxels*).

The task of finding an isosurface for a given value $t$ in a curvilinear grid amounts to finding all cells of which at least one node (i.e., corner) has a value less than $t$ and one node has a value greater than $t$. Such cells are then triangulated according to a look-up table (see Figure 1.9). So, a simple algorithm works as follows [Lorensen and Cline 87]: compute the sign for all nodes ($\oplus \triangleq > t$, $\ominus \triangleq < t$), and then considering each cell in turn, use the eight signs as an index into the look-up table, and triangulate it (if at all).

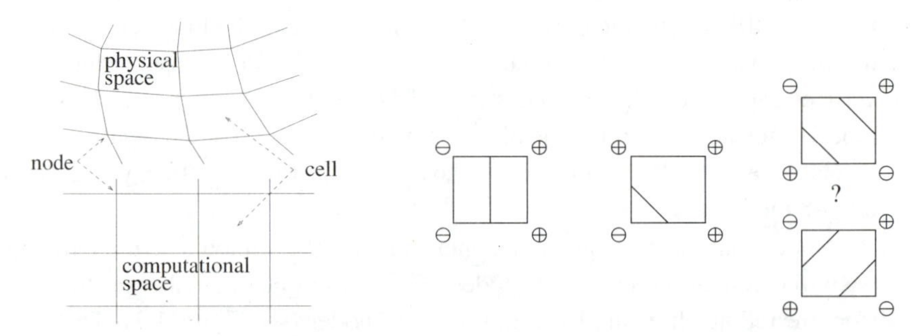


**Figure 1.8.** A scalar field is often given in the form of a curvilinear grid. By doing all calculations in computational space, we can usually save a lot of computational effort.

**Figure 1.9.** Cells straddling the isosurface are triangulated according to a look-up table. In some cases, several triangulations are possible, which must be resolved by heuristics.

Notice that in this algorithm, we have used only the 3D array; we have not used the information about exactly *where* in space the nodes are (except when actually producing the triangles). We have, in fact, made a transition from *computational space* (i.e., the curvilinear grid) to *computational space* (i.e., the 3D array). So in the following, we can, without loss of generality, restrict ourselves to consider only regular grids, that is, 3D arrays.

The question is: how can we improve the exhaustive algorithm? One problem is that we must not miss any little part of the isosurface. So, we need a data structure that allows us to discard large parts of the volume where the isosurface is guaranteed to *not* be. This calls for octrees.

The idea is to construct a complete octree over the cells of the grid [Wilhelms and Gelder 90] (for the sake of simplicity, we will assume that the grid's size is a power of two). The leaves point to the lower-left node of their associated cell (see Figure 1.10). Each leaf $v$ stores the minimum $v_{\min}$ and the maximum $v_{\max}$ of the eight nodes of the cell. Similarly, each inner node of the octree stores the minimum/maximum of its eight children.

Observe that an isosurface intersects the volume associated with a node $v$ (inner or leaf node) if and only if $v_{\min} \leq t \leq v_{\max}$. This already suggests how the algorithm works: start with the root and visit recursively all the children where the condition holds. At the leaves, construct the triangles as usual.

This can be accelerated further by noticing that if the isosurface crosses an edge of a cell, that edge will be visited exactly four times during the complete procedure. Therefore, when we visit an edge for the first time, we compute the vertex of the isosurface on that edge, and store the edge together with the vertex in a hash table. So, whenever we need a vertex on an edge, we first try to look up that edge in the hash table. Our observation also allows us to keep the size of the hash table fairly low. When an edge has been visited for the fourth time, we know that it cannot be visited anymore; therefore, we remove it from the hash table.

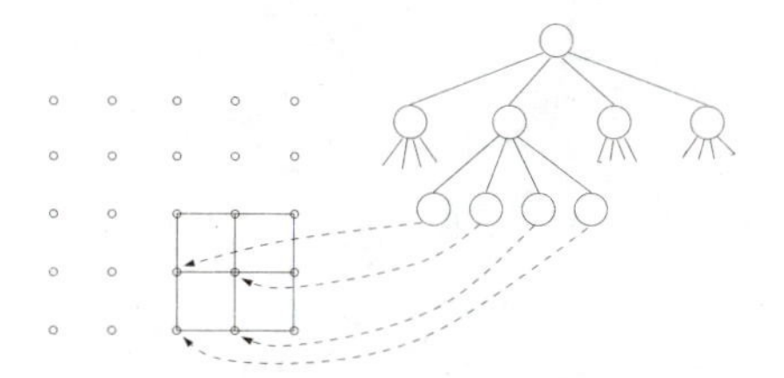

**Figure 1.10.** Octrees offer a simple way to compute isosurfaces efficiently..

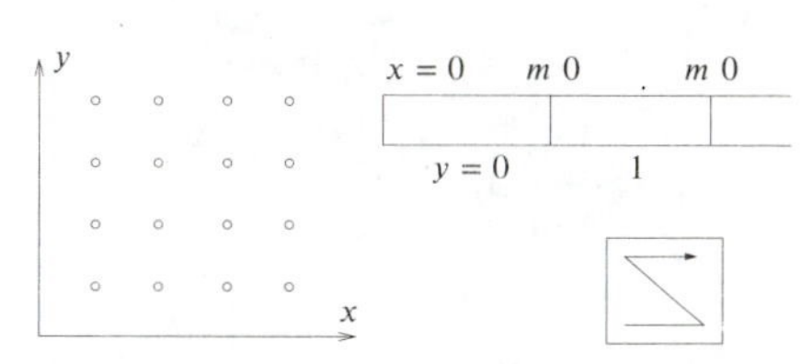


**Figure 1.11.** Volume data layout should match the order of traversal of the octree.

# 1.5. Ray Shooting

Ray shooting is an elementary task that frequently arises in ray tracing, volume visualization, and games for collision detection or terrain following. The task is, basically, to find the earliest hit of a given ray when following that ray through a scene composed of polygons or other objects.

A simple way to avoid checking the ray against all objects is to partition the universe into a regular grid (see Figure 1.12). With each cell, we store a list of objects that occupy that cell (at least partially). Then we just walk along the ray from cell to cell, and check the ray against all those objects that are stored with that cell.

In this scheme (and others), we need a technique called *mailboxes* that prevents us from checking the ray twice against the same object [Glassner 89]. Every ray gets a unique ID (we just increment a global variable holding that ID whenever we start with a new ray); during traversal, we store the ray's ID with the object whenever we have performed an intersection test with it. But before doing an intersection test with an object, we look into its mailbox to see whether the current ray's ID is already there; if so, we know that we have already performed the intersection test in an earlier cell.

In the following, we will present two methods that use octrees to further reduce the number of objects considered.

# 1.6. 3D Octree

A canonical way to improve any grid-based method is to construct an octree (see Figure 1.13). Here, the octree leaves store lists of objects (or, rather, pointers to

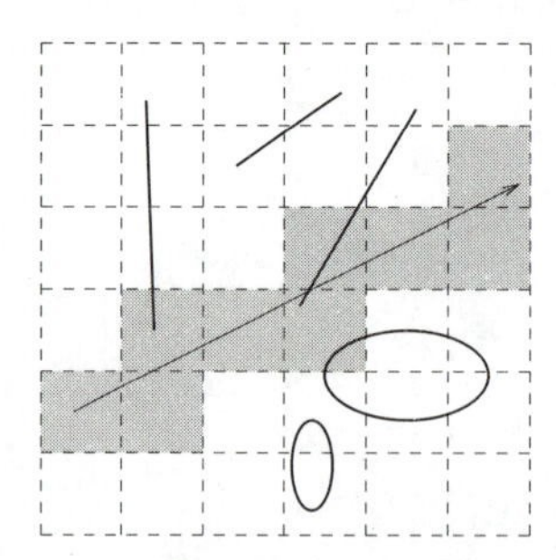

**Figure 1.12.** Ray shooting can be implemented efficiently with a grid.

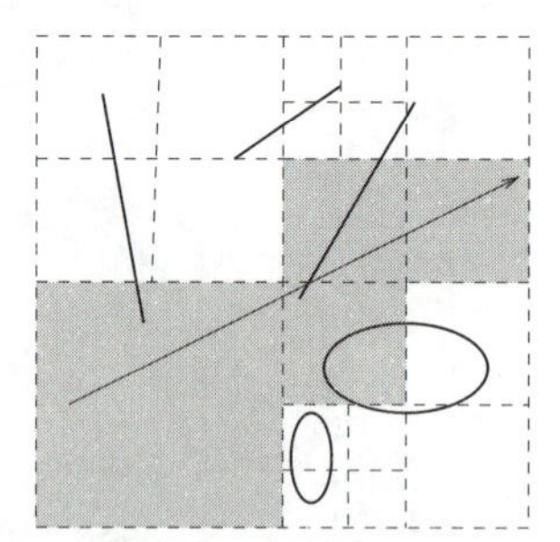

**Figure 1.13.** The same scenario utilizing an octree.

objects). Since we are dealing now with polygons and other graphical objects, the leaf rule for the octree construction process must be changed slightly: maximum depth reached, or only one polygon/object occupies the cell. We can try to better approximate the geometry of the scene by changing the rule to stop only when there are no objects in the cell (or the maximum depth is reached).

How do we traverse an octree along a given ray? As in the case of a grid, we have to make "horizontal" steps, which actually advance along the ray. With octrees, though, we also need to make "vertical" steps, which traverse the octree up or down.

All algorithms for ray shooting with octrees can be classified into two classes:

- Bottom-up: this method starts at that leaf in the octree that contains the origin of the ray; from there, it tries to find that neighbor cell that is stabbed next by the ray, etc.

- Top-down: this method starts at the root of the octree and tries to recurse down into exactly those nodes and leaves that are stabbed by the ray.

Here, we will describe a top-down method [Revelles et al. 00]. The idea is to work only with the ray parameter in order to decide which children of a node must be visited.

Let the ray be given by

$$\vec{x} = \vec{p} + t\vec{d}$$

and a voxel $v$ by

$$[x_l, x_h] \times [y_l, y_h] \times [z_l, z_h].$$

In the following, we will describe the algorithm assuming that all $d_i > 0$; later, we will show that the algorithm works also for all other cases.

First, observe that if we already have the line parameters of the intersection of the ray with the borders of a cell, it is trivial to compute the line intervals halfway in between (see Figure 1.14):

$$t_\alpha^m = \frac{1}{2}(t_\alpha^l + t_\alpha^h)\ ,\ \alpha \in \{x, y, z\}. \qquad (1.2)$$

So, for eight children of a cell, we need to compute only three new line parameters. Clearly, the line intersects a cell if and only if $\max\{t_i^l\} < \min\{t_j^h\}$. The algorithm can be outlined as follows:

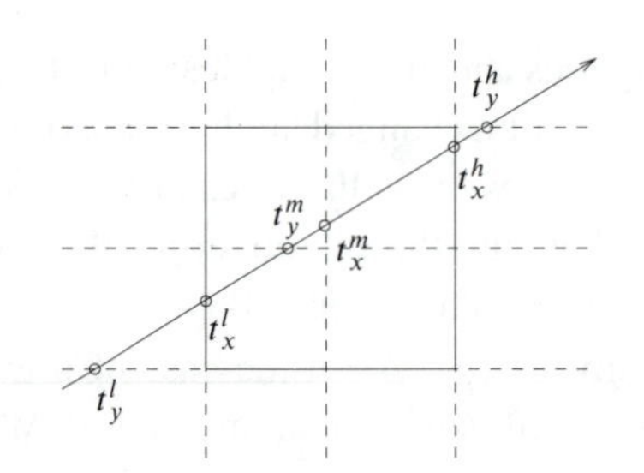


Figure 1.14. Line parameters are trivial to compute for children of a node.

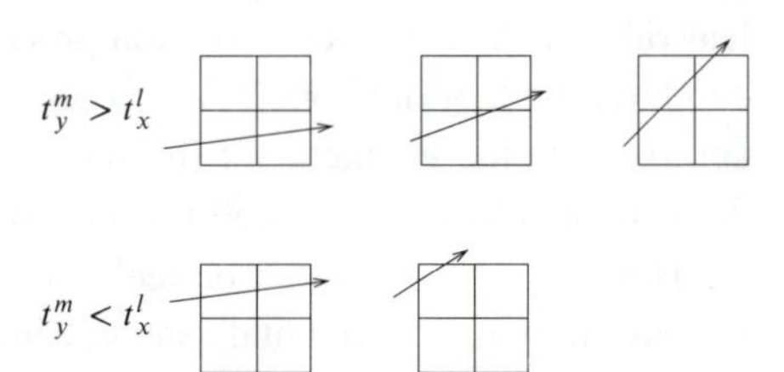


Figure 1.15. The sub-cell that must be traversed first can be found by simple comparisons. Here, only the case $t_x^l > t_y^l$ is depicted.

**traverse(** $v, t^l, t^h$ **)**
compute $t^m$
determine order in which sub-cells are hit by the ray
**for all** sub-cells $v_i$ that are hit **do**
    traverse( $v_i$, $t^l|t^m$, $t^m|t^h$ )
**end for**

where $t^l|t^m$ means that we construct the lower boundary for the respective cell by passing the appropriate components from $t^l$ and $t^m$.

To determine the order in which sub-cells should be traversed, we first need to determine which sub-cell is being hit first by the ray. In 2D, this is accomplished by two comparisons (see Figure 1.15). Then the comparison of $t_x^m$ with $t_y^m$ tells us which cell is next.

In 3D, this takes a little bit more work, but is essentially the same. First, we determine on which side the ray has been entering the current cell by Table 1.1.

Next, we determine the first sub-cell to be visited by Table 1.2 (see Figure 1.16 for the numbering scheme). The first column is the entering side determined in the first step. The third column yields the index of the first sub-cell to be visited: start with an index of zero; if one or both of the conditions of the second column

| $\max\{t_i^l\}$ | Side |
|---|---|
| $t_x^l$ | YZ |
| $t_y^l$ | XZ |
| $t_z^l$ | XY |

Table 1.1. Determines the entering side.

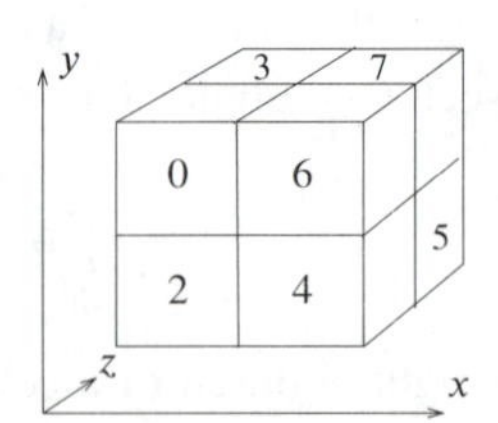


Figure 1.16. Sub-cells are numbered according to this scheme.

| Side | condition | index bits |
|---|---|---|
| XY | $t_z^m < t_x^l$ | 0 |
| | $t_y^m < t_x^l$ | 1 |
| XZ | $t_x^m < t_y^l$ | 0 |
| | $t_z^m < t_y^l$ | 2 |
| YZ | $t_y^m < t_x^l$ | 1 |
| | $t_z^m < t_x^l$ | 2 |

**Table 1.2.** Determines the first sub-cell.

| current | exit side | | |
|---|---|---|---|
| sub-cell | YZ | XZ | XY |
| 0 | 4 | 2 | 1 |
| 1 | 5 | 3 | ex |
| 2 | 6 | ex | 3 |
| 3 | 7 | ex | ex |
| 4 | ex | 6 | 5 |
| 5 | ex | 7 | ex |
| 6 | ex | ex | 7 |
| 7 | ex | ex | ex |

**Table 1.3.** Determines the traversal order of the sub-cells.

hold, then the corresponding bit in the index as indicated by the third column should be set.

Finally, we can traverse all sub-cells according to Table 1.3, where "ex" means the exit side of the ray for the *current* sub-cell.

If the ray direction contains a negative component(s), then we just have to mirror all tables along the respective axis (axes) conceptually. This can be implemented efficiently by an XOR operation.

## 1.7. 5D Octree

In the previous, simple algorithm, we still walk along a ray every time we shoot it into the scene. However, rays are essentially static objects, just like the geometry of the scene! This is the basic observation behind the following algorithm [Arvo and Kirk 87, Arvo and Kirk 89]. Again, it makes use of octrees to adaptively decompose the problem.

The underlying technique is a discretization of rays, which are 5D objects. Consider a cube enclosing the unit sphere of all directions. We can identify any ray's direction with a point on that cube; hence, it is called a *direction cube* (see Figure 1.17). The nice thing about it is that we can now perform any hierarchical partitioning scheme that works in the plane, such as an octree: we just apply the scheme individually on each side.

Using the direction cube, we can establish a one-to-one mapping between direction vectors and points on all six sides of the cube, i.e.:

$$S^2 \quad \leftrightarrow \quad [-1,+1]^2 \times \{+x,-x,+y,-y,+z,-z\}.$$

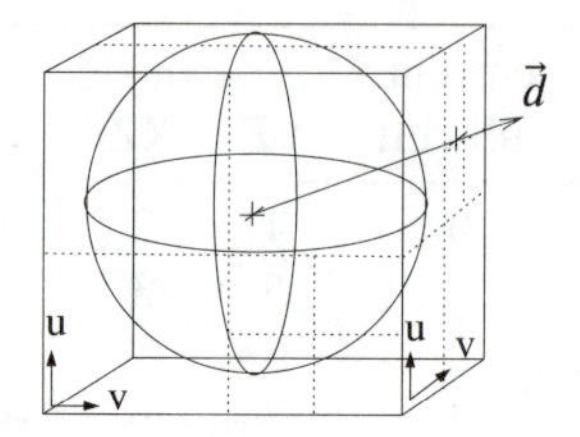


Figure 1.17. With the direction cube, we can discretize directions and organize them with any hierarchical partitioning scheme.

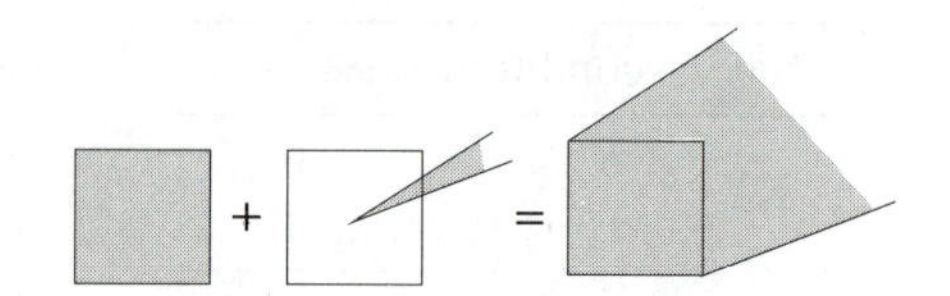


Figure 1.18. A $uv$ interval on the direction cube plus a $xyz$ interval in 3-space yield a beam.

We will denote the coordinates on the cube's side by $u$ and $v$. Here, $\{+x, -x, \ldots\}$ are just some kind of IDs that are used to identify the side of the cube, on which a point $(u, v)$ lives (we could have used $\{1, \ldots, 6\}$ just as well).

Within a given universe $B = [0, 1]^3$ (we assume it is a box), we can represent all possibly occurring rays by points in

$$R = B \times [-1, +1]^2 \times \{+x, -x, +y, -y, +z, -z\}, \tag{1.3}$$

which can be implemented conveniently by six copies of 5D boxes.

Returning to our goal, we now build six 5D octrees as follows. Associate (conceptually) all objects with the root. Partition a node in the octree if there are too many objects associated with it *and* the node's cell is too large. If a node is partitioned, we must also partition its set of objects and assign each subset to one of the children.

Observe that each node in the 5D octree defines a beam in 3-space: the $xyz$-interval of the first three coordinates of the cell define a box in 3-space, and the remaining two $uv$-intervals define a cone in 3-space. Together (more precisely, taking their Minkowski sum), they define a beam in 3-space that starts at the cell's box and extends in the general direction of the cone (see Figure 1.18).

Since we have now defined what a 5D cell of the octree represents, it is almost trivial to define how objects are assigned to sub-cells: we just compare the bounding volume of each object against the sub-cells' 3D beam. Note that an object can be assigned to several sub-cells (just like in regular 3D octrees). The test whether or not an object intersects a beam could be simplified further by enclosing a beam with a cone and then checking the objects' bounding sphere against that cone. This just increases the number of false positives a bit.

Having computed the six 5D octrees for a given scene, ray tracing through that octree is almost trivial: map the ray onto a 5D point via the direction cube;

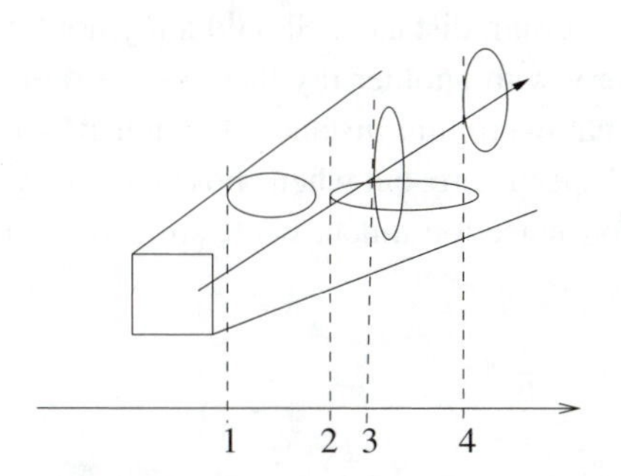


**Figure 1.19.** By sorting objects within each 5D leaf, we can often stop checking ray intersection quite early.

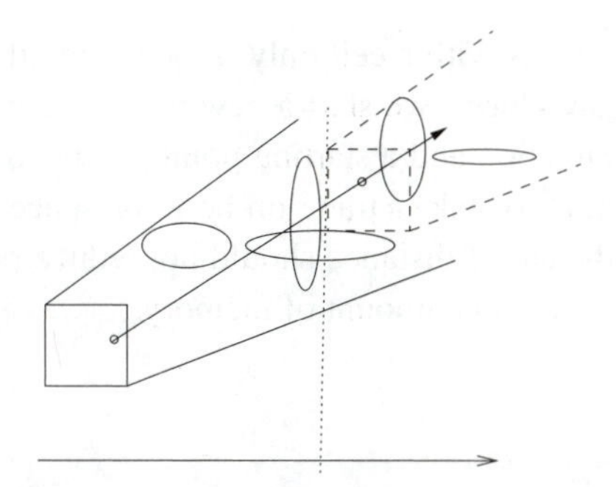

**Figure 1.20.** By truncating the beam (or rather, the list of objects), we can save a lot of memory usage of a 5D octree, while reducing performance only insignificantly.

start with the root of the octree that is associated with the side of the direction cube onto which the ray was mapped; find the leaf in that octree that contains the 5D point (i.e., the ray); and check the ray against all objects associated with that leaf.

By locating a leaf in one of the six 5D octrees, we have discarded all objects that do *not* lie in the general direction of the ray. But we can optimize the algorithm even further.

First of all, we sort all objects associated with a leaf along the dominant axis of the beam by their minimum (see Figure 1.19). If the minimum coordinate of an object along the dominant axis is greater than the current intersection point, then we can stop—all other possible intersection points are farther away.

Second, we can utilize ray coherence as follows. We maintain a cache for each level in the ray tree that stores the leaves of the 5D octrees that were visited last time. When following a new ray, we first look into the octree leaf in the cache to see whether it is contained therein before we start searching for it from the root.

Another trick (which works with other ray acceleration schemes as well) is to exploit the fact that we do not need to know the *first* occluder between a point on a surface and a light source. Any occluder suffices to assert that the point is in shadow. So, we also keep a cache with each light source, which stores the object (or a small set) that was an occluder last time.

Finally, we would like to mention a memory optimization technique for 5D octrees, because they can occupy a lot of memory. It is based on the observation that within a beam defined by a leaf of the octree, the objects at the back (almost) never intersect with a ray emanating from that cell (see Figure 1.20). So, we store objects with a cell only if they are within a certain distance. Should a ray not hit any object, we start a new intersection query with another ray that has the same direction and a starting point just behind that maximum distance. Obviously, we have to make a trade-off between space and speed here, but when chosen properly, the cutoff distance should not reduce performance too much, while still saving a significant amount of memory.

# 2

# Orthogonal Windowing and Stabbing Queries

In this chapter, we introduce some tree-based geometric data structures for answering windowing and stabbing queries. Such queries are useful in many computer graphics algorithms.

A *stabbing query* reports all objects that are *stabbed* by a single object. For a set of segments $S$, a typical stabbing query reports all segments that are stabbed by a single query line $l$. On the other hand, a *windowing query* reports all objects that lie inside a window. For a set of points $S$, a typical windowing query reports all points of $S$ inside query box $B$.

We start with some simple queries and data structures, and then progress to more sophisticated queries. Furthermore, we present time and space bounds for construction and queries.

All data structures are considered to be *static*; that is, we assume that the set of objects will not change over time and we do not have to consider Insert and Delete operations. Such operations might be very costly or complicated. For example, the hierarchical structure of the balanced kd-tree in Section 2.4 requires a lot of reconstruction effort if new elements are inserted or old elements are deleted.

For a dynamization, we use the simple and efficient generic dynamization techniques presented in Chapter 10. We consider simple WeakDelete operations for all data structures in order to apply the dynamization approach adequately. A WeakDelete operation marks an object as deleted; the object is not removed from the memory. After a set of efficient WeakDelete operations, it is necessary to reconstruct the complete structure; see Chapter 10 for details. The corresponding running times can be found in Section 10.5.

The presented pseudocode algorithms make use of straightforward and self-explanatory operations. For example, in Algorithm 2.2, the operation $L$. ListInsert($s$) indicates that the element $s$ is inserted into the list $L$.

For each data structure, first, we consider the query operation. Then we give a sketch of the construction and query operations. Additionally, the corresponding algorithms are represented in pseudocode. We also give short proofs for the complexity of construction and query operations. A query is denoted by the following:

- the dimension of the space,
- a tuple (object/query object) of the corresponding objects,
- the type of the query operation.

For example, in Section 2.1, the corresponding query is denoted as a one-dimensional (interval/point) stabbing query.

## 2.1. Interval Trees

The *interval tree* is used for answering the following one-dimensional (interval/point) stabbing query efficiently:

- Input: a set $S$ of closed intervals on the line;
- Query: a single value $x_q \in \mathbb{R}$;
- Output: all intervals $I \in S$ with $x_q \in I$.

For example, in Figure 2.1, there are seven intervals $s_1, s_2, \ldots, s_7$ and a query value $x_q$. For convenience, the intervals are illustrated by horizontal line segments $s_i$ in 2D and the query value is illustrated as a horizontal line $x_q$. The (interval/point) stabbing query should report the segments $s_2$, $s_5$, and $s_6$. We can assume that a segment $s_i$ is represented by the $x$-coordinate of its left and right endpoints $l_i$ and $r_i$, respectively.

We construct a data structure that answers the above query efficiently. The information of the intervals are stored in a binary tree. Let $S$ denote a set of $n$ intervals $[l_i, r_i]$ for $i = 1, \ldots, n$. The binary interval tree is constructed recursively, as shown in the construction sketch. A node of the tree is dedicated to a median value of the endpoint of all segments. For all segments that contain this value, there are two lists for checking wether a query value is also covered by

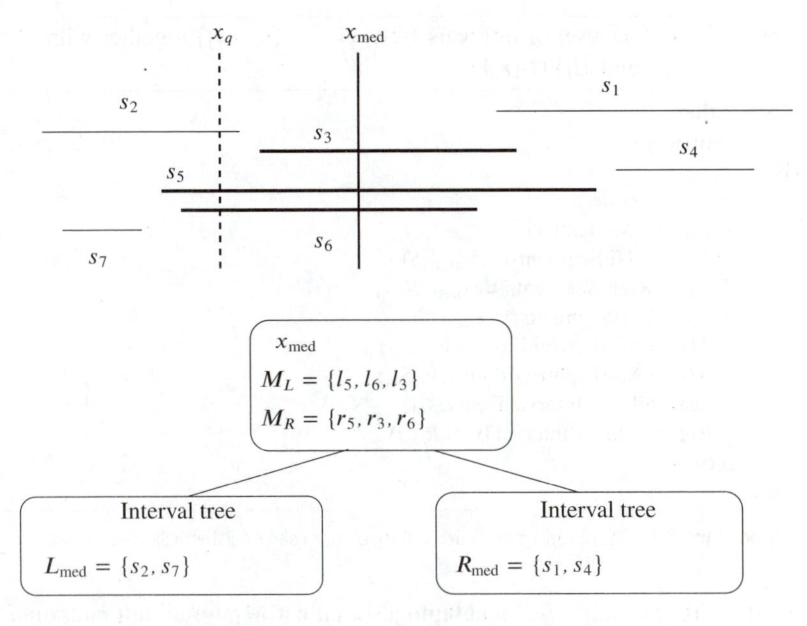


**Figure 2.1.** An example of the node information in an interval tree. For convenience, the intervals are represented by 2D line segments. The left and right endpoints $l_i$ and $r_i$ of the segments $s_i$ hit by the median $x_{med}$ are represented in sorted lists $M_L$ and $M_R$, respectively.

those intervals; see Figure 2.1 for an example of the node information. The query operation will be explained in detail later in this chapter.

The following is the interval tree construction algorithm sketch (see also Algorithm 2.1):

- Input: given a set of intervals $S$ represented by $[l_i, r_i]$ for $i = 1, \ldots, n$;
- If $S$ is empty, then the interval tree is an empty leaf. Otherwise, allocate a node $v$ with two children;
- For the node $v$, compute the median $x_{med}$ of $\{l_1, \ldots, l_n, r_1, \ldots, r_n\}$. This means that half of the interval endpoints lie to the left and half of the interval endpoints lie to the right of $x_{med}$. Note that the median normally will not be equal to a value $l_i$ or $r_i$;
- Let $L_{med}$ denote the set of intervals to the left of $x_{med}$, let $S_{med}$ denote the set of intervals that contain $x_{med}$, and let $R_{med}$ denote the set of intervals to the right of $x_{med}$;

IntervalTree($S$) ($S$ is a set of intervals $\{[l_1, r_1], \dots, [l_n, r_n]\}$ together with sorted lists for $\{l_i\}$, $\{r_i\}$, and $\{l_i\} \cup \{r_i\}$)

```
if S = ∅ then
        return Nil
else
        v := new node
        v.x_med := Median(S)
        v.S_med := HitSegments(v.x_med, S)
        R_med := RightSegments(v.x_med, S)
        L_med := LeftSegments(v.x_med, S)
        v.M_L := SortLeftEndPoints(v.S_med)
        v.M_R := SortRightEndPoints(v.S_med)
        v.LeftChild := IntervalTree(L_med)
        v.RightChild := IntervalTree(R_med)
        return v
end if
```

**Algorithm 2.1.** Computing an interval tree for a set of intervals, recursively.

- At the root $v$, store $x_{\text{med}}$ and build a sorted list $M_L$ for all left endpoints of $S_{\text{med}}$ and a sorted list $M_R$ for all right endpoints of $S_{\text{med}}$;
- Build the interval trees for $L_{\text{med}}$ and $R_{\text{med}}$ recursively for the children of $v$.

Algorithm 2.1 computes the interval tree efficiently, as shown in Lemma 2.1.

**Lemma 2.1.** *The interval tree for a set $S$ of $n$ intervals has size $O(n)$ and can be constructed in $O(n \log n)$ time.*

*Proof:* In a preliminary step, we sort the interval endpoints by $l_i$ order, by $r_i$ order, and by total ($l_i$ and $r_i$) order. Obviously, the depth of the constructed tree is in $O(\log n)$. Let $n_v$ denote the number of intervals at node $v$. Computing the median takes $O(n_v)$ time using the total order of the interval endpoints. Let $m_v = |S_{\text{med}}|$ for the set of intervals $S_{\text{med}}$ at $v$. Computing $M_L$ and $M_R$ from the $l_i$ and $r_i$ order takes at most $O(m_v)$ time. Since all occurring sets $S_{\text{med}}$ are distinct, this gives $O(n)$ altogether. Maintaining the $l_i$, $r_i$, and total order for the recursive steps takes $O(n_v)$ time. Altogether, with the recursive steps, we have the recursive cost function $T(n_v) \leq 2T(\frac{n_v}{2}) + O(n_v)$. Together with the depth $O(\log n)$ of the tree, we conclude the given result. □

Next, we consider the recursive stabbing query operation for a value $x_q \in \mathbb{R}$ and the root $v$ of the interval tree. The median $x_{\text{med}}$ of node $v$ is used for binary branching. The node information is used for answering the stabbing query for the intervals that contain the median.

IntervalStabbing($v, x_q$) ($v$ is the root of an interval tree, $x_q \in \mathbb{R}$)

```
D := new list
if v.x_med < x_q then
        L := v.M_L
        f := L.First
        while f ≠ Nil and f < x_q do
                D.ListInsert(Seg(f))
                f := L.Next
        end while
        D_1 := IntervalStabbing(v.LeftChild, x_q)
else if v.x_med ≥ x_q then
        R := M_R(v)
        l := R.Last
        while l ≠ Nil and l > x_q do
                D.ListInsert(Seg(l))
                l := R.Prev
        end while
        D_1 := IntervalStabbing(v.RightChild, x_q)
end if
D := D.ListAdd(D_1)
return D
```

**Algorithm 2.2.** Answering a stabbing query for an interval tree $v$ and a value $x_q$, recursively.

The following is the stabbing query operation sketch:

- Input: given the root $v$ of an interval tree and the query point $x_q \in \mathbb{R}$;
- If $x_q < x_{\text{med}}$ then
  - Scan the sorted list $M_L$ of the left endpoints in increasing order and report all stabbed segments. Stop if $x_q$ is smaller than the current left endpoint;
  - Recursively, continue with the interval tree of $L_{\text{med}}$;
- If $x_q > x_{\text{med}}$ then
  - Scan the sorted list $M_R$ of the right endpoints in decreasing order and report all stabbed segments. Stop if $x_q$ is bigger than the current right endpoint;
  - Recursively, continue with the interval tree of $R_{\text{med}}$.

For example, assume that there is a stabbing query at a node $v$ as pointed out in Figure 2.1. Starting from the left with $M_L$, the segments $s_5$ and $s_6$ are reported since $x_q$ lies to the right of $l_5$ and $l_6$. Then the stabbing query recursively goes on with $L_{\text{med}}$ and will find $s_2$.

**Lemma 2.2.** *An (interval/point) stabbing query with an interval tree for a set $S$ of $n$ intervals and a value $x_q \in \mathbb{R}$ report all $k$ intervals $I$ of $S$ with $x_q \in I$ in $O(k + \log n)$ time.*

*Proof:* Scanning through $M_R$ or $M_L$ at node $v$ takes time proportional to the number of reported stabbed intervals because we stop as soon as we find an interval that does not contain the point $x_q$. The sets $S_{\text{med}}$ of all $v$ are distinct, which gives $O(k)$ for all scannings. The size of the considered subtree is divided by at least two at every level, which gives $O(\log n)$ for the path length. Altogether, the given bound holds. □

The interval tree was considered in [Edelsbrunner 80] and [McCreight 80]. Note that the interval tree has no direct generalization to higher dimensions, but can support combined queries; see Section 2.6.

A WeakDelete operation for a segment $s$ (see Chapter 10) can be done in $O(\log n)$ time. We find the corresponding node with $s \in S_{\text{med}}$ and mark the endpoints as deleted in the sorted lists $M_L$ and $M_R$.

**Lemma 2.3.** *A* WeakDelete *operation for a segment $s$ in an interval tree of $n$ intervals is performed in $O(\log n)$ time.*

## 2.2. Segment Trees

A segment tree is constructed for answering the following 2D (line segment/line) stabbing query efficiently:

- Input: a set $S$ of segments in the plane;
- Query: a vertical[1] line $l$;
- Output: all segments $s \in S$ crossed by $l$.

Let $S = \{s_1, s_2, \ldots, s_n\}$ be the set of segments and let $E$ be the sorted set of $x$-coordinates of the endpoints. We assume general position; that is, $E =$

[1] Arbitrary query lines $l$ can be handled by application of a transformation matrix; see Section 9.5.3.

$\{e_1, e_2, \ldots, e_{2n}\}$ with $e_i < e_j$ for $i < j$ (Section 9.5). For the construction of the tree, we can split $E$ into $2n + 1$ atomic intervals

$$[-\infty, e_1], [e_1, e_2], \ldots, [e_{2n-1}, e_{2n}], [e_{2n}, \infty],$$

which represent the leaves of the segment tree.

The segment tree is a balanced binary tree. Each internal node $v$ represents an elementary interval $I$, which is split into intervals $I_l$ and $I_r$ for the two children of $v$. The intervals are split with respect to the endpoints of the segments; see Figure 2.2. In the first place, the segment tree is a one-dimensional search tree for the $x$-coordinates of the endpoints of the segments. The one-dimensional search tree contains search keys in every node; the data points are represented in the leaves of the tree. Every node $v$ additionally represents an interval $I_v$ so that all points in the (sub)tree represented by root node $v$ are in $I_v$. In Section 2.5, we generalize the one-dimensional search tree to arbitrary dimension $d$. An example of a one-dimensional search tree is given in Figure 2.7.

The following is the one-dimensional search tree sketch (see also Algorithm 2.3):

- Input: given a sorted list $S$ of $n$ elements $x_1, x_2, \ldots, x_n$;
- Output: the root node of a one-dimensional search tree for $S$;
- If $|S| = 0$, then set $v :=$ Nil and return $v$;
- Otherwise, if $|S| \geq 1$ then

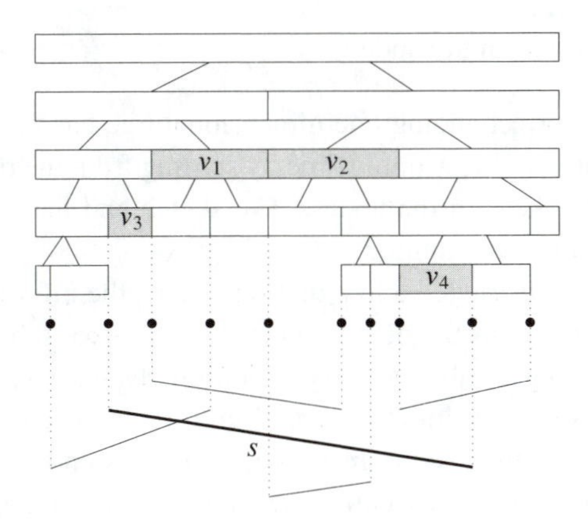


**Figure 2.2.** A segment tree for a set of segments. The segment $s$ is minimally covered by the nodes $v_1$, $v_2$, $v_3$, and $v_4$.

SearchTree($S$) ($S$ is a sorted list $\{x_1, \dots, x_n\}$)

```
if S = ∅ then
        v := Nil
else
        v := new node
        n := |S|
        m := ⌈n/2⌉
        (L, R) := S.Split(m)
        v.Key := S.Get(m)
        v.I := [S.First, S.Last])
        /* Alternatively: v.I := [x_1, ..., x_n] */
        v.LeftChild := SearchTree(L)
        v.RightChild := SearchTree(R)
end if
return v
```

**Algorithm 2.3.** Construction of a simple balanced one-dimensional search tree.

- Allocate a node $v$ with two children for the root of the tree;
- Choose the element $x_m$ of $S$ with $m := \lceil \frac{n}{2} \rceil$ and insert the branch key $x_m$ at $v$. Insert the interval $I = [x_1, x_n]$ at $v$. If necessary, $I$ also contains the data points of the interval;
- Compute a one-dimensional search tree $v_l$ for $x_1, \dots, x_m$ and compute a one-dimensional search tree $v_r$ for $x_{m+1}, \dots, x_n$;
- Insert $v_r$ as the right child of $v$ and $v_l$ as the left child of $v$;
- Finally, return the node $v$.

Obviously, the corresponding one-dimensional tree has height $O(\log n)$. We can find an element $x_i$ in logarithmic time by starting from the root and branching towards $x_i$ using the keys in the nodes. The sketch and the pseudocode of this simple query procedure are omitted.

So far, we have constructed a tree that represents the information of the endpoints of the segments plus the dummy nodes $\infty$ and $-\infty$. Obviously, this is not sufficient for answering a stabbing query. Additionally, we have to incorporate the information of the segments into the tree. Therefore, let us consider a single segment $s$ represented by the $x$-coordinates $(e_j, e_k)$ of its endpoints. The segment $s$ can be represented by a consecutive set of elementary intervals. We choose a minimal set of elementary intervals (or the corresponding nodes) in the one-dimensional search tree so that $s$ is fully covered. Due to Algorithm 2.3, every node represents an elementary interval $v.I$ and, if necessary, its data points. *Minimal* means that the nodes are chosen as close as possible to the root of the tree; see Figure 2.2. We store $s$ in each of the associated nodes. More precisely, let $v$.pred be the predecessor of $v$ in the interval tree. A segment $s = (e_j, e_k)$ represented by the $x$-coordinates of the endpoints is stored in a list $v.L$ at $v$ iff $v.I \subseteq [e_j, e_k]$ and $(v.\text{pred}).I \not\subseteq [e_j, e_k]$. Thus, each node represents an elementary interval $v.I$ and a list of segments $v.L$; for an example, see Figure 2.3.

The one-dimensional search tree with intervals $v.I$ is built up in $O(n \log n)$ time, as was shown previously. However, we need to show how the segment partitions are inserted into the tree. This is done by the following recursive insert procedure, which has to run for every segment $s = (e_j, e_k)$.

The following is the sketch of the segment insertion in the one-dimensional search tree:

- Input: given a line segment $s = (e_j, e_k)$ and the root $v$ of the segment tree;
- If the interval $v.I$ is already a subset of the interval $[e_j, e_k]$, then insert $s$ into the segment set $v.L$ and stop;
- Otherwise, let $v$. LeftChild and $v$. RightChild be the children of $v$. This means that the interval $v.I$ equals $(v.\text{LeftChild}).I \cup (v.\text{RightChild}).I$;
  - If the interval $[e_j, e_k]$ and the set $(v.\text{LeftChild}).I$ overlap, then run the insert procedure recursively with $s$ and the segment tree $v$. LeftChild;
  - If the interval $[e_j, e_k]$ and the set $(v.\text{RightChild}).I$ overlap, then run the insert procedure recursively with $s$ and the segment tree $v$. RightChild.

By inserting every segment into the one-dimensional search tree of the line segment's endpoints, we will finally obtain the segment tree. Altogether, Algorithm 2.4 constructs the segment tree by using Algorithm 2.5 as a subroutine. Note that $S_x$. Extend extends $S_x$ by the dummy elements $\infty$ and $-\infty$.

**Lemma 2.4.** *The segment tree for $n$ segments can be built in $O(n \log n)$ and uses $O(n \log n)$ space.*

*Proof:* The binary tree has depth $O(logn)$ and $O(n)$ nodes. At each level of the segment tree $T$, a segment $s = (e_j, e_k)$ is stored at most twice. To prove this fact, let us assume that three nodes $u_l$, $u_m$, and $u_r$ of the same level in $T$ contain the

SegmentTree($S$) ($S$ is a set of line segments given by endpoints)

```
S_x := S.SortX
S_x := S_x.Extend
v := SearchTree(S_x)
while s ≠ ∅ do
        s := S.First
        SegmentInsertion(v, s)
        S.DeleteFirst
end while
```

**Algorithm 2.4.** Building a segment tree.

segment $s$. The intervals of their parents do not overlap. This is because they can overlap only if a pair of $u_l$, $u_m$, and $u_r$ has a common predecessor. In this case, the segment has to be inserted for the predecessor. If the intervals of the parents of $u_l$, $u_m$, and $u_r$ do not overlap, the intermediate node of $u_l$, $u_m$, and $u_r$ must have a parent whose interval fully contains $[e_j, e_k]$; therefore, $s$ is not inserted at $u_m$ which gives a contradiction. Summing up over all segments and levels, the segment tree has space $O(n \log n)$.

With a similar argument, one can prove that, during the insertion operation, only four nodes on every level could be visited. Thus a single segment is inserted in $O(\log n)$ time, which gives the overall time bound. □

For a stabbing query with a vertical line $l$, we will proceed as follows. We start with the $x$-coordinate $l_x$ of the vertical line $l$ and the root node $v$ of a segment tree.

SegmentInsertion($v$, $s$) ($v$ is one-dimensional search tree of the $x$-coordinates of the endpoints of segment set $S$, $s \in S$.)

```
e_j := s.LeftXCoord
e_k := s.RightXCoord
if v.I ⊂ [e_j, e_k] then
        (v.L).ListAdd(s)
else
        if (v.LeftChild).I ∩ [e_j, e_k] ≠ ∅ then
                SegmentInsertion(v.LeftChild, s)
        end if
        if (v.RightChild).I ∩ [e_j, e_k] ≠ ∅ then
                SegmentInsertion(v.RightChild, s)
        end if
end if
```

**Algorithm 2.5.** Insertion of a segment $s$ in the segment tree.

StabbingQuery($v, q$) ($v$ is the root node of a segment tree and $l$ a vertical line segment)

```
L := Nil
if v ≠ Nil and l_x ∈ v.I then
    L := v.L
    L_l := StabbingQuery(v.LeftChild, l)
    L_r := StabbingQuery(v.RightChild, l)
    L.ListAdd(L_r)
    L.ListAdd(L_l)
end if
return L
```

**Algorithm 2.6.** The stabbing query for a segment tree.

The following is the sketch of the stabbing query with a segment tree:

- Input: given the $x$-coordinate $l_x$ of the vertical line $l$ and the root node $v$ of a segment tree;
- If $l_x$ is inside the interval $v.I$, then report all segments in $v.L$;
- If $v$ is not a leaf, then for the children $v$. LeftChild and $v$. RightChild of $v$, we know that the interval $v.I$ equals $(v.\text{LeftChild}).I \cup (v.\text{RightChild}).I$ and we proceed as follows:
  - If $l_x$ lies inside $(v.\text{LeftChild}).I$, then start a query with segment tree $v$. LeftChild and $l_x$;
  - Else we have $l_x \in (v.\text{RightChild}).I$, and we start a query with segment tree $v$.RightChild and $l_x$.

Figure 2.3 shows an example for a segment query. The shaded intervals indicate the query path from the root to the leaf.

**Lemma 2.5.** *A vertical line stabbing query for n segments in the plane can be answered by a segment tree in time $O(k + \log n)$, where k stands for the number of reported segments.*

*Proof:* Obviously, the tree is traversed with $O(\log n)$ steps. The running time $O(k + \log n)$ stems from the cardinality $k$ of the corresponding sets $v.L$. □

The segment tree was first considered in [Bentley 77] and extended to higher dimensions by several authors; see, for example, [Edelsbrunner and Maurer 81] and [Vaishnavi and Wood 82].

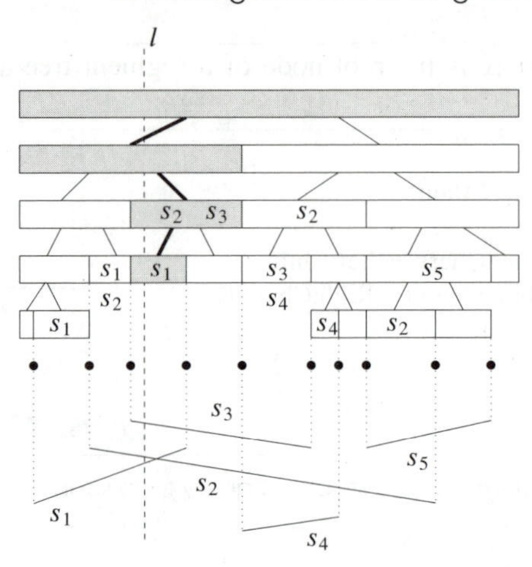


**Figure 2.3.** The query procedure for a segment tree and a query line $l$. The shaded intervals indicate the query path from the root to the leaf. All segments stored at the corresponding nodes are reported.

A WeakDelete operation can be performed in $O(\log n)$ time (see Chapter 10). We have to mark a segment $s$ in every list $v.L$ with $s \in v.L$. By construction, a segment $s$ is stored at most twice in every level of the segment tree and occurs at most $O(\log n)$ times. We proceed as if we would like to insert the segment $s$, which can be done in $O(\log n)$ time; see also the proof of Lemma 2.4.

**Lemma 2.6.** *A* WeakDelete *operation for a segment tree of n segments in the plane can be performed in time* $O(\log n)$.

# 2.3. Multi-Level Segment Trees

In the preceding sections, we considered segment trees for 2D objects. Now, we generalize the approach. Namely, we consider the following multi-dimensional stabbing problem and make use of a set of inductively defined segment trees. Multi-level segment trees support (axis-parallel box/point) stabbing queries as follows:

- Input: a set $S$ of $n$ axis-parallel boxes in $\mathbb{R}^d$;
- Query: a query point $q$ in $\mathbb{R}^d$;
- Output: all boxes $B \in S$ with $q \in B$.

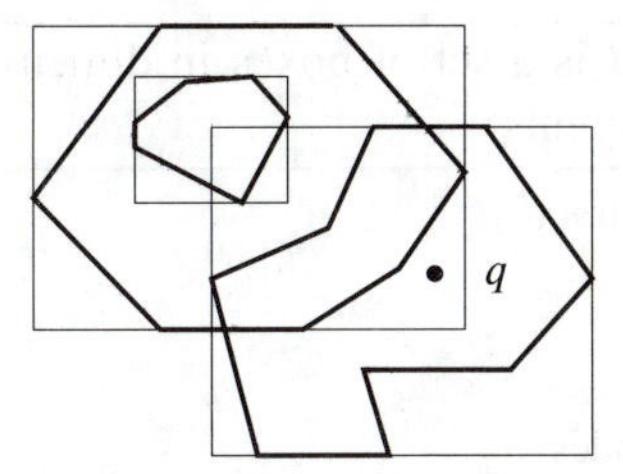


**Figure 2.4.** A stabbing query for a set of bounding boxes can report all polygonal objects near $q$.

This stabbing query has many applications in computer graphics. For example, objects in the sphere may be approximated by their axis-parallel bounding box. For every point $q$, the (bounding box/point) stabbing query will report all objects that are near $q$. Figure 2.4 shows a simple example in 2D.

A single $d$-dimensional box $B_i$ can be defined by a set of $d$ axis-parallel segments $\{S_{i_1}, S_{i_2}, \ldots, S_{i_d}\}$ starting at a unique corner $C_i = (C_{i_1}, C_{i_2}, \ldots, C_{i_d})$; see Figure 2.5 for a 2D example. A multi-level segment tree $T^d$ with respect to $(x_1, x_2, \ldots, x_d)$, where $x_i$ denotes the $i$th coordinate, is inductively defined.

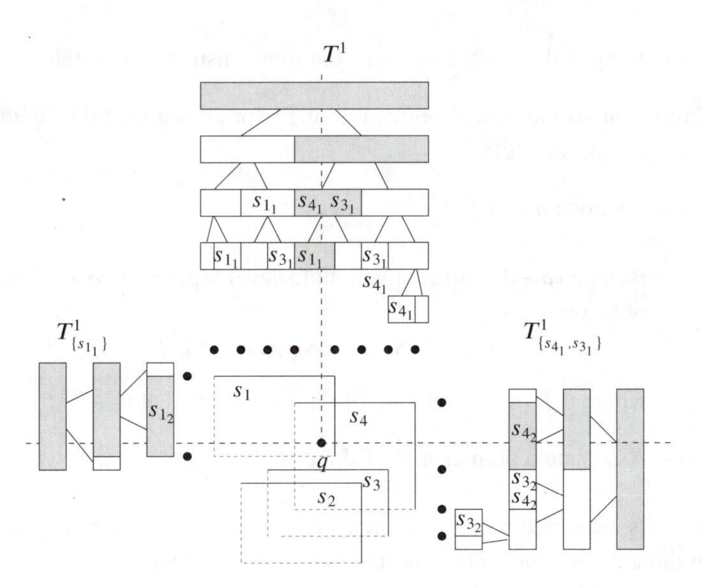


**Figure 2.5.** Some parts of the multi-level segment tree of four boxes in 2D. The segment tree $T^1$ and the relevant trees for the nodes $u_1$ and $u_2$ with $L_{u_1} = \{s_{1_1}\}$ and $L_{u_2} = \{s_{3_1}, s_{4_1}\}$ for the query point $q$ are shown.

MLSegmentTree($B, d$) ($B$ is a set of boxes in dimension $d$; each box is represented by a set of line segments)

```
S := B.FirstSegmentsExtract
T := SegmentTree(S)
T.Dim := d
if d > 1 then
        N := T.GetAllNodes
        while N ≠ ∅ do
                u := N.First   N.DeleteFirst
                L := u.L
                B := new list
                while L ≠ ∅ do
                        s := L.First   L.DeleteFirst
                        B := B.ListAdd( s.Box(d − 1) )
                end while
                u.Pointer := MLSegmentTree(B, d − 1)
                B.Deallocate
        end while
end if
return T
```

**Algorithm 2.7.** The inductive construction algorithm of a multi-level segment tree.

The following is the multi-level segment tree construction sketch:

- Build a one-dimensional segment tree $T^1$ for all segments $S_{i_1}$ with respect to the $x_1$ coordinate;
- For every node $u$ of $T^1$:
  - Build a $(d-1)$-dimensional multi-level segment tree $T_u^{d-1}$ for the set of boxes
  $$\{(S_{j_2}, \ldots, S_{j_d}) : S_{j_1} \in u.L\},$$
  where $u.L$ denotes the set of segments stored at $u$ in $T^1$;
  - Associate a pointer in $T^1$ from $u$ to $T_u^{d-1}$.

A query is answered recursively in a straightforward manner. If a set of segments in dimension $j$ has to be reported, we recursively check the corresponding tree with dimension $j-1$. The corresponding box is reported only if the query succeeds in every dimension. This means that finally we get an answer for a tree $T_u^1$.

MLSegmentTreeQuery($T, q, d$) ($q \in \mathbb{R}^d$, $T$ is the root node of a $d$-dimensional multi-level segment tree)

```
if d = 1 then
        L := StabbingQuery(T, q)
        A := L.GetBoxes
else
        A := new list
        L := SearchTreeQuery(T, q.First)
        while L ≠ do
                t := (L.First).Pointer
                B := MLSegmentTreeQuery(t, q.Rest, d − 1)
                A.ListAdd(B)
                L.DeleteFirst
        end while
        return A
end if
```

**Algorithm 2.8.** The inductive query operation for a multi-level segment tree.

More precisely, a query for a query point $q = (q_1, q_2, \ldots, q_d)$ and a multi-level segment tree $T^d$ for a set of segments $S_1, \ldots, S_n$ is given in the following sketch.

The following is the multi-level segment tree query sketch:

- If $d = 1$, answer the question with the corresponding segment tree $T^1$; see Algorithm 2.6 and return the boxes associated to the segments reported;
- Otherwise, traverse $T^1$ and find the nodes $u_1, \ldots, u_l$ so that $q_1$ lies in the intervals $(u_i).I$ of the nodes $u_i$ of $T^1$ for $i = 1$ to $l$;
- For $1 \leq i \leq l$, answer recursively the query $(q_2, \ldots, q_d)$ for the tree $T_{u_i}^{(d-1)}$ of the set of segments $(u_i).L$ stored at $u_i$, respectively.

The following time and space bounds hold.

**Theorem 2.7.** *A multi-level segment tree for a set of $n$ axis-parallel boxes in dimension $d$ can be built up in $O(n \log^d n)$ time and needs $O(n \log^d n)$ space. A (axis-parallel box/point) stabbing query can be answered in $O(k + \log^d n)$ time.*

*Proof:* First, we sort the segment endpoints with respect to every dimension. The first segment tree $T^1$ is built up in $O(n \log n)$ time, occupies space $O(n \log n)$, and has $O(n)$ nodes. Next, we construct $(d-1)$-dimensional segment trees for the set $v.L$ of every node $v$ in $T^1$. By induction, we can assume that the $(d-1)$-dimensional segment tree for a node $v \in T^1$ with $|v.L|$ segments is computed in

$O(|v.L| \log^{(d-1)} |v.L|)$. Additionally, as already proven, every segment $s_i$ occurs in $O(\log n)$ lists $v.L$, i.e., in $O(\log n)$ $(d-1)$-dimensional segment trees. This means that $\sum_{v \in T^1} |v.L|$ is in $O(n \log n)$, and the overall running time is given by

$$\sum_{v \in T^1} |v.L| \log^{(d-1)} |v.L| \leq D\, n \log^d n\,.$$

The required space can be measured as follows. The whole structure $T^d$ consists of one-dimensional segment trees $T^1$ only. In how many trees will a part of the box $B_i$ appear? In the first tree $T^1$, the segment $s_{i_1}$ of the box $B_i$ appears in $O(\log n)$ nodes. Therefore, it appears in $O(\log n)$ trees $T_v^{(d-1)}$. In each of these trees, the segment $s_{i_2}$ of the box $B_i$ appears again in $O(\log n)$ nodes. Altogether, segments of $B_i$ appear in

$$\sum_{i=1}^{d} \log^i \in O(\log^d n)$$

nodes. Summing up over all $n$ segments gives $O(n \log^d n)$ entries. By the same argument, the number of trees is bounded by $O(\log^d n)$ also, so that the number of empty nodes is in $O(n \log^d n)$.

Analogously, a query traverses along $O(\log^d n)$ nodes. For the first tree $T^1$, the query traverses $O(\log n)$ nodes and therefore $O(\log n)$ new queries have to be considered. Recursively, $O(\log n)$ nodes in $O(\log n)$ new trees may be traversed. In the *final* segment trees, $k$ boxes may also be reported. Altogether

$$\sum_{i=1}^{d} \log^i \in O(\log^d n)$$

nodes are visited, and the running time for a query is $O(k + \log^d n)$. □

A WeakDelete operation (see Chapter 10) for a segment $s$ in the multi-level segment tree can be performed in $O(\log^d n)$ time due to the number of nodes where $s$ appears and due to the time for visiting all these nodes. Technically, we can *reinsert* the segment $s$ recursively into the multi-level segment tree.

**Lemma 2.8.** *A* WeakDelete *operation for a $d$-dimensional multi-level segment tree for a set of $n$ axis-parallel boxes can be performed in time $O(\log^d n)$.*

Next, we turn to the subject of orthogonal windowing queries.

## 2.4. Kd-Trees

In general, a windowing query considers a set of objects and a multi-dimensional window, such as a box. All objects that intersect with the window should be reported. Thus, one can easily decide which objects are located within a certain area, one of the classic problems arising in computer graphics.

The kd-tree is a natural generalization of the one-dimensional search tree considered in the previous section. It can be viewed also as a generalization of quadtrees. With the help of a kd-tree, one can efficiently answer (point/axis-parallel box) windowing queries as follows:

- Input: a set of points $S$ in $\mathbb{R}^d$;
- Query: an axis-parallel $d$-dimensional box $B$;
- Output: all points $p \in S$ with $p \in B$.

Let $D$ be a set of $n$ points in $\mathbb{R}^k$. For convenience, we assume that for a while $k = 2$, and that all $x$- and $y$-coordinates are different. If this is not the case, we will have a so-called *degenerate* situation, and we can apply the techniques presented in Section 9.5. Let us further assume that we have decided to split $D$ along the $x$-axis, and that we have determined the *split value* $s$ of the $x$-coordinates. Then we split $D$ by the *split line* $x = s$ into subsets

$$\begin{aligned} D_{<s} &= \{(x, y) \in D | x < s\}, \\ D_{>s} &= \{(x, y) \in D | x > s\}. \end{aligned}$$

We repeat the process recursively with the constructed subsets. For every split operation, we have to determine the split axis and the split value. The simplest strategy is to choose the axes in a cyclic manner (i.e., $x$-, $y$-, $z$-axis, etc.)—this is called a *cyclic* kd-tree. More precisely, a kd-tree for dimension $d$ is constructed as follows.

The following is the inductive construction of a kd-tree sketch:

- Input: given a set of points $D$ in dimension $d$ and the split coordinate $x_i$;
- If $D$ is empty, then return an empty node $v$;
- Otherwise, allocate a node $v$ for the root of the kd-tree with two children $v$.LeftChild and $v$.RightChild. Choose a split value $s_i$ with respect to the chosen coordinate $x_i$. Split the set $D$ into subsets

$$\begin{aligned} D_{<s_i} &= \{(x_1, \ldots, x_i, \ldots, x_n) \in D | x_i < s\}, \\ D_{>s_i} &= \{(x_1, \ldots, x_i, \ldots, x_n) \in D | x_i > s\}; \end{aligned}$$

KdTreeConstr($D, i$) ($D$ is a set of points in $\mathbb{R}^d$, $i \in \{1, \ldots, d\}$)

```
if D = ∅ then
    v := Nil
else
    v := new node
    if |D| = 1 then
        v.Element := D.Element
        v.LeftChild := Nil  v.RightChild := Nil
    else
        s := D.SplitValue(i)
        v.Split := s  v.Dim := i
        D_<s := D.Left(i, s)
        D_>s := D.Right(i, s)
        j := (i mod d) + 1
        v.LeftChild := KdTreeConstr(D_<s, j)
        v.RightChild := KdTreeConstr(D_>s, j)
    end if
end if
return v
```

**Algorithm 2.9.** Recursive construction of a kd-tree.

- Recursively, build the kd-trees $v$.LeftChild and $v$.RightChild for the set $D_{<s}$ and $D_{>s}$ with respect to the next coordinate $x_j$, $j = (i \mod d) + 1$, respectively. Finally, return the node $v$.

The tree will be built up by KdTreeConstr($D$, 1). Thus, we simply obtain a binary tree. The balance of the tree depends on the choice of the split value in procedure SplitValue. In case of $d = 2$, we obtain a 2D kd-tree[2] of the point set $D$; see Figure 2.6. Each internal node of the tree corresponds to a split line. For every node $v$ of the kd-tree, we define the rectangle $R(v)$, the *region* of $v$, which is the intersection of half-planes corresponding to the path from the root to $v$. For the root $r$, $R(r)$ is the plane itself; the children of $r$, say $\lambda$ and $\rho$, correspond to two half-planes $R(\lambda)$ and $R(\rho)$, and so on. The set of rectangles $\{R(l) | l \text{ is a leaf}\}$ gives a non-overlapping partition of the plane into rectangles. Every $R(l)$ has exactly one point of $D$ inside. We do not have to store the rectangles explicitly; they are given by the path from the root to the corresponding node; see Figure 2.6.

For simplicity, we again consider the 2D case. The kd-tree in 2D efficiently supports range queries of axis-parallel rectangles. If $Q$ is an axis-parallel rectangle, the set of sites $v \in D$ with $v \in Q$ can be computed as follows. We have to

[2]According to [de Berg et al. 00], the term *k-d tree* was meant as a template to be specialized like *2-d tree*, *3-d tree*, etc. Today, it is customary to specialize it like *2D kd-tree*, etc.

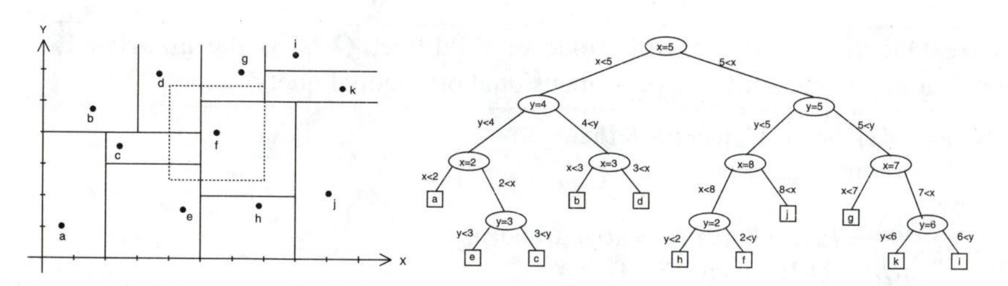


**Figure 2.6.** A 2D kd-tree and a rectangular range query. Each node corresponds to a split line. Additionally, each node represents a unique rectangular range $R(v)$ according to the path from the root to the node.

compute all nodes $v$ with:

$$R(v) \cap Q \neq \varnothing .$$

If the condition holds for node $v$, it will hold also for the predecessor $u$ of $v$ in the kd-tree since $R(v) \subset R(u)$. Thus, we can start searching from the root to the leaves. Finally, if we reach a leaf of the tree with the given property, we still have to check whether the data point of the leaf is inside $Q$.

For general dimension $d$, every node $v$ implicitly represents an orthogonal box in dimension $d$. Analogously, the box is given by the path from the root to $v$. For the query operation, we store the current orthogonal box explicitly during the query process. We obtain the following query procedure.

The following is the $d$-dimensional axis-parallel query sketch:

- Input: the root $r$ of a kd-tree in dimension $d$ and a $d$-dimensional orthogonal range $R$. The $d$-dimensional orthogonal box $Q(v)$ defines the range associated with node $v$. In the beginning, $Q(r)$ represents the full $d$-dimensional space for the root $r$;

- Let $v$ be the current node;

- If $v$ is a leaf, then check whether the element $v.\text{Element}$ stored in $v$ lies in $R$, and if so, report the element. Stop the procedure;

- Otherwise, the given $Q(v)$ is split into the regions $Q(v.\text{LeftChild})$ and $Q(v.\text{RightChild})$ for the left and the right child of $v$ by using the split line $v.\text{Split}$;

- If $Q(v.\text{LeftChild})$ or $Q(v.\text{RightChild})$, respectively, is fully contained in $R$, then report *all* points in $v.\text{LeftChild}$ or $v.\text{RightChild}$, respectively;

KdTreeQuery($v, Q, R$) ($v$ is the node of a kd-tree, $Q$ is its the associated $d$-dimensional range, and $R$ is a $d$-dimensional orthogonal query range)

```
if v is a leaf and v.Element ∈ R then
        return v.Element
else
        v_l := LeftChild(v) v_r := RightChild(v)
        Q_l := Q.LeftPart(v.Split)
        Q_r := Q.RightPart(v.Split)
        if Q_l ⊂ R then
                (v.LeftChild).Report
        else if Q_l ∩ R ≠ ∅ then
                KdTreeQuery(v.LeftChild, Q_l, R)
        end if
        if Q_r ⊂ R then
                (v.RightChild).Report
        else if Q_r ∩ R ≠ ∅ then
                KdTreeQuery(v.RightChild, Q_r, R)
        end if
end if
```

**Algorithm 2.10.** The query procedure of the $d$-dimensional kd-tree.

- If $Q(v.\text{LeftChild})$ or $Q(v.\text{RightChild})$, respectively, intersect with $R$, then rerun the procedure with $v.$LeftChild and $Q(v.\text{LeftChild})$ or $v.$RightChild and $Q(v.\text{RightChild})$, respectively.

For convenience, we discuss time and space complexity for the 2D case. It can be shown that in 2D, the number of nodes that fulfill $R(v) \cap Q \neq \emptyset$ is restricted to $O(2^{\frac{h}{2}} + k)$, where $h$ denotes the depth of the tree and $k$ denotes the number of points in $Q$, i.e., the size of the answer; see [Klein 05] or [de Berg et al. 00]. Altogether, the efficiency of the kd-tree with respect to range queries depends on the depth of the tree. A balanced kd-tree can be easily constructed. We sort the points with respect to the $x$- and $y$-coordinates. With this order, we recursively split the set into subsets of equal size in time $O(\log n)$. The construction runs in time $O(n \log n)$, and the tree has depth $O(\log n)$. Altogether, the following theorem can be proven.

**Theorem 2.9.** *A balanced kd-tree for n points in the plane can be constructed in $O(n \log n)$ and needs $O(n)$ space. A range query with an axis-parallel rectangle can be answered in time $O(\sqrt{n} + a)$, where a denotes the size of the answer.*

As mentioned earlier, the 2D kd-tree can be easily generalized to arbitrary dimension $d$, splitting the points successively with respect to the given axes in a balanced way. Fortunately, the depth of the balanced tree still is bounded by $O(\log n)$, and the tree can be built up in time $O(n \log n)$ and space $O(n)$ for fixed dimension $d$. Therefore, the kd-tree is optimal in space. A rectangular range query can be answered in time $O(n^{(1-\frac{1}{d})} + k)$.

**Theorem 2.10.** *A balanced $d$-dimensional kd-tree for $n$ points in $\mathbb{R}^d$ can be constructed in $O(n \log n)$ and needs $O(n)$ space. A range query with an axis-parallel box in $\mathbb{R}^d$ can be answered in time $O(n^{(1-\frac{1}{d})} + k)$, where $k$ denotes the size of the answer.*

The main advantage of the $d$-dimensional kd-tree lies in its small size. In the following section, we will see that rectangular range queries can be answered more efficiently with the help of range trees. A WeakDelete operation in the kd-tree can be done in $O(\log n)$. We simply traverse the tree up to the corresponding node and mark the node as deleted.

**Lemma 2.11.** *A* WeakDelete *operation for a balanced $d$-dimensional kd-tree for $n$ points in $\mathbb{R}^d$ can be performed in $O(\log n)$ time.*

# 2.5. Range Trees

Range trees are defined for arbitrary dimension $d$ and support exactly the same windowing query as the kd-tree. A range tree can answer the corresponding query more efficiently. On the negative side, a range tree requires more space than the kd-tree.

With the help of a $d$-dimensional range tree, one can efficiently answer (point/axis-parallel box) windowing queries as follows:

- Input: a set of points $S$ in $\mathbb{R}^d$;
- Query: an axis-parallel $d$-dimensional box $B$;
- Output: all points $p \in S$ with $p \in B$.

The range tree is defined similar to the multi-level segment tree (see Section 2.3). The main difference is that we do not need to represent segments in the tree. Let us first consider a simple balanced one-dimensional search tree for a set $S$ of points on the $x$-axis. As mentioned earlier, each node $v$ represents a uniquely determined interval $v.I$ of points in $S$. In turn, for a query interval $I$, the points of $S$ inside $I$ are covered by a minimal set of intervals $v.I$; see Figure 2.7. The construction of the one-dimensional search tree was already discussed in Section 2.2.

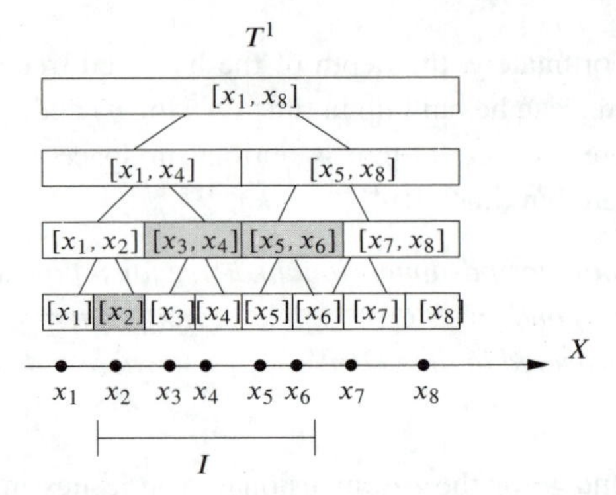


**Figure 2.7.** Every node $v$ in the one-dimensional search tree represents a unique interval $v.I$. A query interval $I$ is uniquely covered by a minimal set of intervals $v.I$.

For the inductive construction of a $d$-dimensional range tree, let $x_i$ denote the $i$th coordinate. A point $x_i \in S$ is given by $\{x_{i_1}, x_{i_2}, \ldots, x_{i_d}\}$; see Figure 2.8 for a 2D example. A $d$-dimensional range tree with respect to $(X_1, X_2, \ldots, X_d)$ is inductively defined as follows.

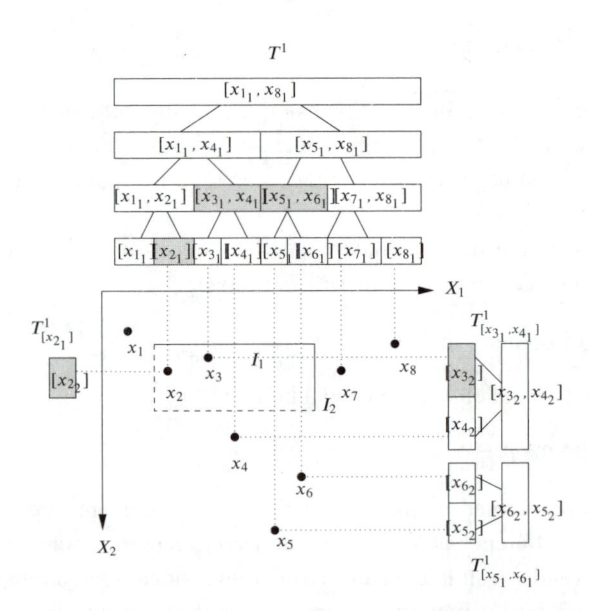


**Figure 2.8.** Some parts of the 2D range tree of eight points in 2D. The range tree $T^1$ and the relevant trees $T^1_{u_i}$ for the nodes $u_1, u_2$, and $u_3$ with $I_{u_1} = [x_{2_1}]$, $I_{u_2} = [x_{3_1}, x_{4_1}]$, and $Iu_3 = [x_{5_1}, x_{6_1}]$ for the query rectangle $R = I_1 \times I_2$ with intervals $I_1 \subset X$ and $I_2 \subset Y$.

```
RangeTreeConstr(D, d) (S is a set of points in ℝ^d)

  S_f := S.FirstCoordElements
  S_f.Sort
  T := S_f.SearchTree
  T.Dim := d
  if d > 1 then
        N := T.GetAllNodes
        while N ≠ ∅ do
              u := N.First  N.DeleteFirst
              L := u.I
              S := new list
              while L ≠ ∅ do
                    x := L.First  L.DeleteFirst
                    S := S.ListAdd(x.Point(d − 1))
              end while
              Pointer(u) := RangeTreeConstr(S, d − 1)
        end while
  end if
  return T
```

**Algorithm 2.11.** The inductive construction algorithm of a $d$-dimensional range tree.

The following is the $d$-dimensional range tree construction sketch:

- Build a one-dimensional balanced search tree $T^1$ for all points $x_i$ with respect to $X_1$;
- For every node $u$ of $T_1$:
  - Build a $(d-1)$-dimensional range tree $T_u^{d-1}$ for the set of points

    $$\{(x_{j_2}, \ldots, x_{j_d}) : x_{j_1} \in u.I\},$$

    where $u.I$ represents the interval at node $u$ in $T_1$;
  - Associate a pointer in $T^1$ from $u$ to $T_u^{d-1}$.

The time and space complexity analysis for range a trees is very similar to the multi-level segment tree analysis; see the proof of Theorem 2.7. Since we do not have to insert segments into the trees, we can neglect a log factor.

The query operation for an axis-parallel query box $q = I_1 \times I_2 \times \cdots \times I_d \subset \mathbb{R}^d$ and a $d$-dimensional range tree $T^d$ of a set of points $p_1, \ldots, p_n$ $T^d$ is explained in the following sketch; see Figure 2.8 for an example.

The following is the range tree query sketch:

RangeTreeQuery($T, B, d$) ($T$ is the root of a $d$-dimensional range tree, $B$ is a $d$-dimensional box)

```
if d = 1 then
        L := SearchTreeQuery(T, B)
        A := L.GetPoints
else
        A := new list
        L := MinimalCoveringIntervals(T, B.First)
        while L ≠ do
                t := (L.First).Pointer
                D := RangeTreeQuery(t, B.Rest, d − 1)
                A.ListAdd(D)
                L.DeleteFirst
        end while
        return A
end if
```

**Algorithm 2.12.** The inductive query operation for a $d$-dimensional range tree.

- If $d = 1$, answer the question with the corresponding one-dimensional search tree $T^1$;
- Otherwise, traverse $T^1$ and find the nodes $u_1, \ldots, u_l$ so that the intervals of the nodes $(u_i).I$ minimally cover $I_1$;
- For $1 \leq i \leq l$, answer the query $I_2 \times \cdots \times I_d \subset \mathbb{R}^{(d-1)}$ recursively with the trees $T_{u_i}^{(d-1)}$ of the data point sets of $(u_i).I$.

**Theorem 2.12.** *A $d$-dimensional range tree for a set of $n$ points in dimension $d$ can be built up in $O(n \log^{(d-1)} n)$ time and needs $O(n \log^{(d-1)} n)$ space. An axis-parallel box windowing query can be answered in $O(k + \log^d n)$ time.*

*Proof:* First, we sort the segments endpoints with respect to every dimension. The first range tree $T^1$ is built up in $O(n)$, has space $O(n \log n)$, and has $O(n)$ nodes. Next, we construct $(d-1)$-dimensional segment trees for the set of points in the interval $v.I$ of every node $v$ in $T^1$. Let $|v.I|$ denote the number of data points in $v.I$. By induction, we can assume that the $(d-1)$-dimensional segment tree for node $v \in T^1$ with all points in $v.I$ is computed in $O(|v.I| \log^{(d-2)} |v.I|)$. Additionally, every data point $p$ occurs in $O(\log n)$ intervals $v.I$; i.e., in only $O(\log n)$ $(d-1)$-dimensional segment trees. Therefore, the overall running time is given by

$$\sum_{v \in T^1} |v.I| \log^{(d-2)} |v.I| \leq D\, n \log^{(d-2)} n\,.$$

The required space is measured as follows. The structure $T^d$ consists only of one-dimensional range trees $T^1$. In how many trees will a data point $p_i$ appear? Since $p$ lies inside $O(\log n)$ intervals of $T^1$, the point $p$ appears in $O(\log n)$ trees $T_u^{(d-1)}$. In each of these trees, the data point $p$ again lies inside $O(\log n)$ intervals and will be a data point in the corresponding trees. Altogether, a data point $p$ appears in

$$\sum_{i=1}^{(d-1)} \log^i n \in O(\log^{(d-1)} n)$$

nodes. Summing up over all $n$ data points gives $O(n \log^{(d-1)} n)$ entries. By the same argument, the number of trees is also in $O(\log^{(d-1)} n)$, so that the number of empty nodes is bounded by $O(n \log^{(d-1)} n)$.

On the other hand, a query traverses along $O(\log^d n)$ nodes. For the first tree $T^1$, the query traverses $O(\log n)$ nodes and $O(\log n)$ new queries must be considered. Recursively, $O(\log n)$ nodes of $O(\log n)$ new trees may be traversed. In the *final* range trees, $k$ boxes may also be reported. Altogether,

$$\sum_{i=1}^{d} \log^i n \in O(\log^d n)$$

nodes are visited, and the running time for a query is $O(k + \log^d n)$. □

A WeakDelete operation (see Chapter 10) for a point $p$ in the $d$-dimensional range tree can be performed in $O(\log^d n)$ time due to the number of nodes that must be traversed to find all entries belonging to $p$.

**Lemma 2.13.** *A* WeakDelete *operation for a balanced $d$-dimensional range tree for $n$ points in $\mathbb{R}^d$ can be performed in $O(\log^d n)$ time.*

Finally, with the help of the given results, we will answer a more general windowing query.

## 2.6. The (Axis-Parallel Box/Axis-Parallel Box) Windowing Problem

We consider the problem of answering the following (axis-parallel box/axis-parallel box) windowing query. For a set of rectangular boxes, we want to find all boxes that are intersected by a query box $B$:

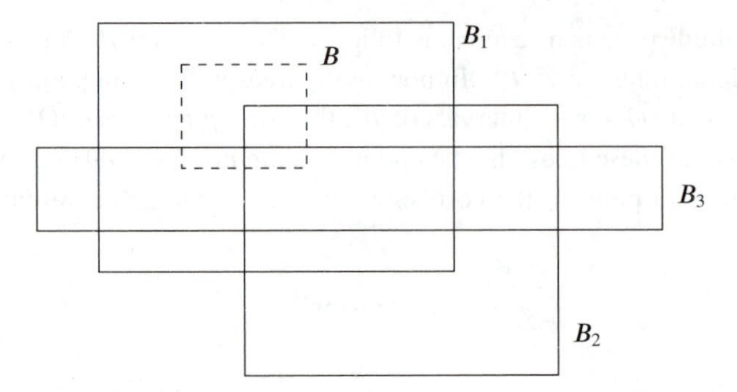


**Figure 2.9.** The box $B_i$ reported by a (axis-parallel box/axis-parallel box) windowing query for $B$ contains $B$ for $i = 1$, has a vertex inside $B$ for $i = 2$, and has a segment that crosses $B$ for $i = 3$.

- Input: a set $S$ of rectangular boxes in 2D;
- Query: a rectangular box $B$ in 2D;
- Output: all boxes of $S$ that intersect with $B$.

Let us assume that a set $S$ of $n$ rectangular boxes $B_1, B_2, \ldots, B_n$ is given. A rectangular box $B_i$ has points inside $B$ in the following three cases (see Figure 2.9):

- The query $B$ is fully contained in $B_i$;
- An endpoint of $B_i$ lies inside $B$;
- A segment of $B_i$ crosses $B$, and no endpoint of $B_i$ is inside $B$.

For the first case, we can use a 2D segment tree of the boxes of $S$; see Section 2.3. We answer (axis-parallel box/point) stabbing queries for the four endpoints of $B$. These operations require $O(k_1 + \log^2 n)$ time and $O(n + \log^2 n)$ space, where $k_1$ denotes the number of boxes in $S$ that fully contain $B$.

In the second case, we can use a range tree in 2D and answer a rectangular (point/axis-parallel box) windowing query for the endpoints of the $B_i$s with respect to $B$; see Section 2.5. This can be performed in time $O(k_2 + \log^2 n)$ and $O(n \log n)$ space, where $k_2$ denotes the number boxes $B_i$ with endpoints in $B$.

The third case cannot be solved directly with the former approaches. We have to answer the following (horizontal line segment/vertical line segment) stabbing query for the vertical line segments of box $B$ and the horizontal line segments of all boxes $B_i$ for $i = 1, \ldots, n$.

- Input: a set $S$ of horizontal line segments in 2D;
- Query: a vertical line segment $s$ in 2D;
- Output: all segments of $S$ intersected by $s$.

Of course, we have to answer also a (vertical line segment/horizontal line segment) stabbing query. This is done by a simple rotation of 90 degrees.

A subset of horizontal line segments of $S$ is stabbed by the line passing through the vertical line segment $s$. Obviously, among these line segments, we must report all line segments whose left endpoints lie in a rectangular range grounded in $s$ and going to $\infty$; see Figure 2.10. If $s$ is given by $(x, y_l)$ and $(x, y_u)$, the corresponding box is represented by $[-\infty, x] \times [y_l, y_u]$.

Now, we have a stabbing query and a range query, and we adapt an interval tree accordingly. The horizontal segments of $S$ are stored in an interval tree for the $x$-coordinates of the segments. We have to report the horizontal segments $h$ stabbed by the $x$-coordinate $x$ of $s$. For the left endpoints of such a segment $h$, we have to take care that the $y$-coordinate of $h$ lies also in the interval $[y_l, y_u]$. A node $v$ of a *regular* interval tree contains the list $S_{\text{med}}$ of the segments hit by $x_{\text{med}}$; see Section 2.1. We replace the sorted list $M_L$ and $M_R$ of the left and right endpoints of $S_{\text{med}}$ by two range trees. A 2D range tree $M_L$ for the left endpoints of the segments in $S_{\text{med}}$ and a 2D range tree $M_R$ for the right endpoints of the segments in $S_{\text{med}}$.

In Algorithm 2.1, we simply replace

$$v.\,M_L := \text{SortLeft}(v.\,S_{\text{med}})\quad v.\,M_R) := \text{SortRight}(v.\,S_{\text{med}})$$

with

$$v.\,M_L := \text{RangeTreeConstr}((v.\,S_{\text{med}}).\,\text{LeftPoints}, 2)$$
$$v.\,M_R := \text{RangeTreeConstr}((v.\,S_{\text{med}}).\,\text{RightPoints}, 2)\,.$$

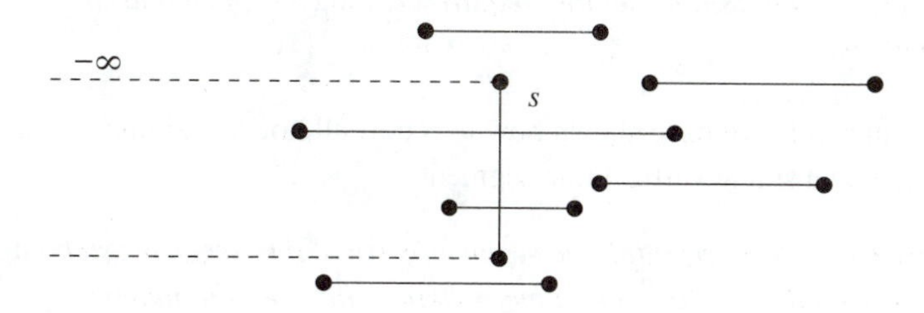


**Figure 2.10.** A rectangular range query by an infinite box grounded in $s$ will answer the windowing query in the third case.

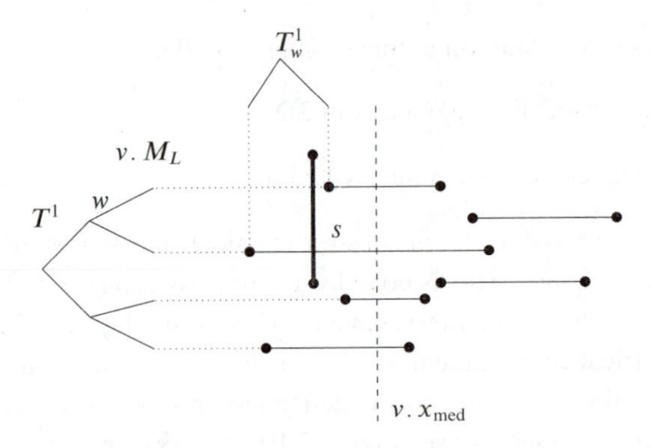


**Figure 2.11.** At the node $v$ with a median $v.x_{\text{med}}$ in an interval tree, there is also a 2D range tree $v.M_L$ of the left endpoints of the segments in $v.S_{\text{med}}$. $v.M_L$ answers the query $[y_l, y_u] \times [-\infty, x]$ for a segment $s = [(x, y_l), (x, y_u)]$.

This gives an additional log factor for the space of the structure. For an interval query and a node $v$, it suffices to check the corresponding 2D range trees. Note that all segments in $S_{\text{med}}$ are crossed by the line $X = x_{\text{med}}$. For example, if $x < x_{\text{med}}$, we use the range tree $M_L$ of the left endpoints of $S_{\text{med}}$; see Figure 2.11. We answer the range query for $[y_l, y_u] \times [-\infty, x]$. Algorithm 2.2 is extended in a straightforward manner.

In time $O(\log n + k_v)$, we can report the $k_v$ segments of $S_{\text{med}}$ at node $v$ that hit $s$ and go on with the left children and $L_{\text{med}}$ in the interval tree query. Altogether, we traverse $\log n$ nodes and report all $k$ intersecting segments, which gives $O(k + \log^2 n)$ query time.

With the results from the previous section, we can construct the extended interval tree in $O(n \log n)$ with $O(n \log n)$ space.

**Theorem 2.14.** *A 2D (axis-parallel box/axis parallel box) windowing query can be answered in $O(\log^2 + k)$ time where k denotes the number of reported boxes. The corresponding data structures require $O(n \log n)$ space and are built up in $O(n \log n)$ time.*

As a subresult, we have shown how to report all horizontal line segments that are crossed by a single vertical line segment.

**Theorem 2.15.** *A (horizontal line segment/vertical line segment) stabbing query can be answered in $O(\log^2 + k)$ time, where k denotes the number of reported line segments. The corresponding data structures require $O(n \log n)$ space and are built up in $O(n \log n)$ time.*

A WeakDelete operation for the combined interval range tree can be done in $O(\log^2 n)$ time, since we additionally have to make a WeakDelete in the 2D range tree.

**Lemma 2.16.** A WeakDelete *operation for the combined interval range tree for n boxes in 2D can be performed in* $O(\log^2 n)$ *time.*

## 2.7. Texture Synthesis

Textures are visual details of the rendered geometry. Textures have become very important in the past few years because the cost of rendering with texture is the same as the cost without texture. Virtually all real-world objects have texture, so it is extremely important to render them in synthetic worlds, too.

Texture synthesis generally tries to synthesize new textures, from given images, from a mathematical description, or from a physical model. Mathematical descriptions can be as simple as a number of sine waves to generate water ripples. Physical models try to describe the physical or biological effects and phenomena that lead to some texture (such as patina or fur). In all of these "model-based" methods, the knowledge about the texture is in the model and the algorithm. The other class of methods starts with one or more images, tries to find some statistical or stochastic description (explicitly or implicitly) of these, and finally generates a new texture from the statistic.

Basically, textures are images with the following properties:

- Stationary: if a window with the proper size is moved about the image, the portion inside the window always appears the same;

- Local: each pixel's color in the image depends on only a relatively small neighborhood.

Of course, images not satisfying these criteria can be used as textures as well (such as façades), but if you want to synthesize such images, a statistical or stochastic approach is probably not feasible.

In the following, we will describe a stochastic algorithm that is very simple and efficient, and works remarkably well [Wei and Levoy 00]. Given a sample image, it does not, like most other methods, try to compute explicitly the stochastic model. Instead, it uses the sample image itself, which implicitly contains that model already.

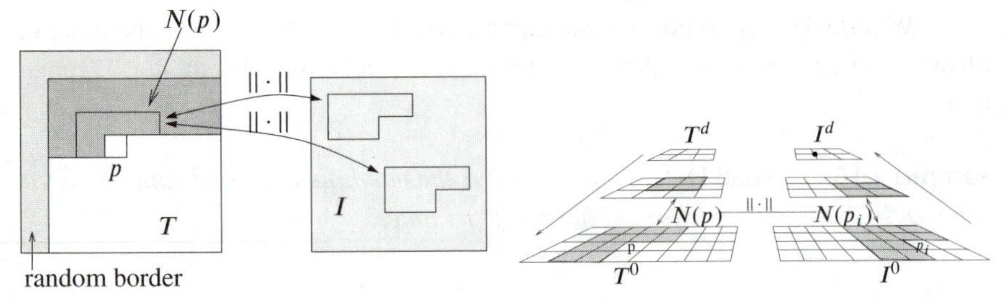


**Figure 2.12.** The texture synthesis algorithm proceeds in scan line order through the texture and considers only the neighborhood around the current pixel as shown.

**Figure 2.13.** Using an image pyramid, the texture synthesis process becomes fairly robust against different scales of detail in the sample images.

We will use the following terminology:

$$\begin{aligned} I &= \text{original (sample) image}, \\ T &= \text{new texture image}, \\ p_i &= \text{pixel from } I, \\ p &= \text{pixel from } T \text{ to be generated next}, \\ N(p) &= \text{neighborhood of } p \text{ (see Figure 2.12)}. \end{aligned}$$

Initially, $T$ is cleared to black. The algorithm starts by adding a suitably sized border at the left and the top, filled with random pixels (this will be thrown away again at the end). Then it performs the following simple loop in scan line order (see Figure 2.12):

1: **for all** $p \in T$ **do**
2: find the $p_i \in I$ that minimizes $|N(p) - N(p_i)|^2$
3: $p := p_i$
4: **end for**

The search in line 2 is exactly a nearest-neighbor search! This can be performed efficiently with the algorithm presented in Section 6.4.1: if $N(p)$ contains $k$ pixels, then the points are just $3k$-dimensional vectors of RGB values, and the distance is just the Euclidean distance.

Obviously, all pixels of the new texture are deterministically defined, once the random border has been filled. The shape of the neighborhood $N(p)$ can be chosen arbitrarily; it must just be chosen such that all pixels but the current pixel are already computed. Likewise, other "scans" of the texture are possible and sensible (for instance, a spiral scan order); they must just match the shape of $N(p)$.

The quality of the texture depends on the size of the neighborhood $N(p)$. However, the optimal size itself depends on the "granularity" in the sample image. In order to make the algorithm independent, we can synthesize an image pyramid (see Figure 2.13). First, we generate a pyramid $I^0, I^1, \ldots, I^d$ for the sample image $I^0$. Then we synthesize the texture pyramid $T^0, T^1, \ldots, T^d$ level by level with the above algorithm, starting at the coarsest level. The only difference is that we extend the neighborhood $N(p)$ of a pixel $p$ over $k$ levels, as depicted by Figure 2.13. Consequently, we have to build a nearest-neighbor search structure for each level, because as we proceed downwards in the texture pyramid, the size of the neighborhood grows.

Of course, now we have replaced the parameter of the best size of the neighborhood by the parameter of the best size per level and the best number of levels to consider for the neighborhood. However, as Wei and Levoy [Wei and Levoy 00] report, a neighborhood of $9 \times 9$ (at the finest level) across two levels seems to be sufficient in almost all cases.

Figure 2.14 shows two examples of the results that can be achieved with this method.

**Figure 2.14.** Some results of the texture synthesis algorithm [Wei and Levoy 00]. In each pair, the image on the left is the original one, and the one on the right is the (partly) synthesized one. (See Color Plate IV.) (Courtesy of L.-Y. Wei and M. Levoy, and ACM.)

## 2.8. Shape Matching

As the availability of 3D models on the Internet and in databases increases, searching for such models becomes an interesting problem. Such a functionality is needed, for instance, in medical image databases or CAD databases. One question is how to specify a query. Usually, most researchers pursue the "query by content" approach, where a query is specified by providing a (possibly crude) shape, for which the database is to return best matches.[3] The fundamental step here is the matching of shapes; i.e., the calculation of a *similarity* measure.

Almost all approaches perform the following steps:

1. Define a transformation function that takes a shape and computes a so-called *feature vector* in some high-dimensional space, which (hopefully) captures the shape in its essence. Naturally, those transformation functions that are invariant under rotation and/or translation and tessellation are preferred;

2. Define a *similarity measure* $d$ on the feature vectors, such that if $d(f_1, f_2)$ is large, then the associated shapes $s_1, s_2$ do not *look* similar. Obviously, this is (partly) a human factor issue. In almost all algorithms, $d$ is just the Euclidean distance;

3. Compute a feature vector for each shape in the database and store the vectors in a data structure that allows for fast nearest-neighbor search;

4. Given a query (a shape), compute its feature vector and retrieve the nearest neighbor from the database. Usually, the system also retrieves all $k$ nearest neighbors. Often, you are not interested in the exact $k$ nearest neighbors but only in *approximate* nearest neighbors (because the feature vector is an approximation of the shape anyway).

The main difference among most shape-matching algorithms is, therefore, the transformation from shape to feature vector.

So, fast shape retrieval essentially requires a fast (approximate) nearest-neighbor search. We could stop our discussion of shape matching here, but for sake of completeness, we will describe a very simple algorithm (from the plethora of others) to compute a feature vector [Osada et al. 01].

The general idea is to define some *shape function* $f(P_1, \ldots, P_n) \rightarrow \mathbb{R}$, which computes some geometrical property of a number of points, and then evaluate this

[3]This idea seems to originate from image database retrieval, where it was called QBIC, for query by image content.

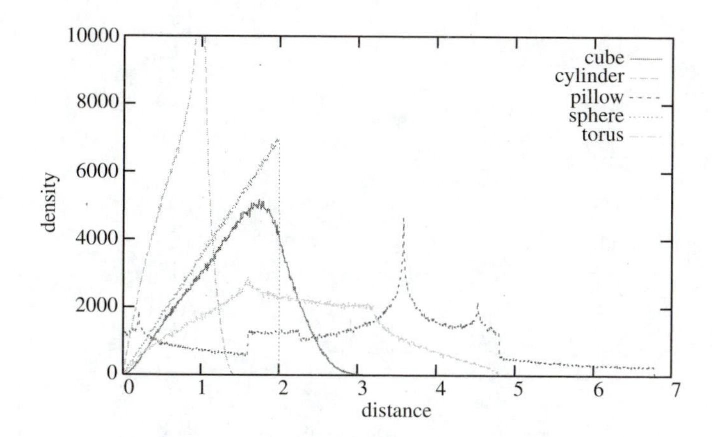


**Figure 2.15.** The shape distributions of a number of different simple objects. (See Color Plate XX.)

function for a large number of random points that lie on the surface of the shape. The resulting distribution of $f$ is called a *shape distribution.*

For the shape function, there are a lot of possibilities (your imagination is the limit). There are some examples:

- $f(P_1, P_2) = |P_1 - P_2|$,
- $f(P_1) = |P_1 - P_0|$, where $P_0$ is a fixed point, such as the bounding box center,
- $f(P_1, P_2, P_3) = \angle(\overline{P_1 P_2}, \overline{P_1 P_3})$,
- $f(P_1, P_2, P_3, P_4) =$ volume of the tetrahedron between the four points.

Figure 2.15 shows the shape distributions of a few simple objects with the distance between two points as shape function.

# 3

# BSP Trees

*Binary space partitioning (BSP) trees* can be viewed as a generalization of kd-trees. Like kd-trees, BSP trees are binary trees, but now the orientation and position of a splitting plane can be chosen arbitrarily. To get a feeling for a BSP tree, Figure 3.1 shows an example for a set of objects.

The definition of a BSP (short for BSP tree) is fairly straightforward. Here, we will present a recursive definition. Let $h$ denote a plane in $\mathbb{R}^d$, and $h^+$ and $h^-$ denote the positive and negative half-space, respectively.

**Definition 3.1. (BSP tree.)** Let $S$ be a set of objects (points, polygons, groups of polygons, or other spatial objects), and let $S(v)$ denote the set of objects associated with a node $v$. Then the BSP $T(S)$ is defined by the following.

1. If $|S| \leq 1$, then $T$ is a leaf $v$ which stores $S(v) := S$.
2. If $|S| > 1$, then the root of $T$ is a node $v$; $v$ stores a plane $h_v$ and a set $S(v) := \{x \in S | x \subseteq h_v\}$ (this is the set of objects that lie completely inside $h_v$; in 3D, these can be only polygons, edges, or points). $v$ also has two children $T^-$ and $T^+$; $T^-$ is the BSP for the set of objects $S^- := \{x \cap h_v^- | x \in S\}$, and $T^+$ is the BSP for the set of objects $S^+ := \{x \cap h_v^+ | x \in S\}$.

This can readily be turned into a general algorithm for constructing BSPs. Note that a splitting step (i. e., the construction of an inner node) requires us to split each object into two disjoint *fragments* if it straddles the splitting plane of that node. In some applications though (such as ray shooting), this is not really necessary; instead, we can just put those objects into both subsets.

Note that a convex cell (which is possibly unbounded) is associated with each node of the BSP. The "cell" associated with the root is the whole space, which is convex. Splitting a convex region into two parts yields two convex regions. In Figure 3.1, the convex region of one of the leaves has been highlighted as an example.

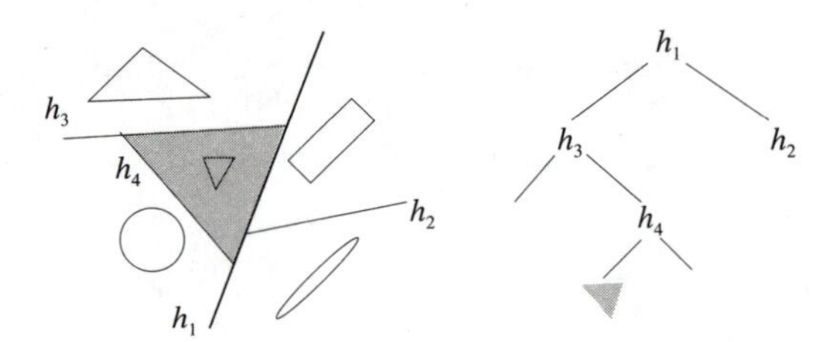


**Figure 3.1.** An example of a BSP tree for a set of objects.

With BSPs, we have much more freedom to place the splitting planes than with kd-trees. However, this also makes that decision much harder (as almost always in life). If our input is a set of polygons, then a very common approach is to choose one of the polygons from the input set and use this as the splitting plane. This is called an *auto-partition* (see Figure 3.2).

While an auto-partition can have $\Omega(n^2)$ fragments, it is possible to show the following in 2D [de Berg et al. 00, Paterson and Yao 90].

**Lemma 3.2.** *Given a set S of n line segments in the plane, the expected number of fragments in an auto-partition* $T(S)$ *is in* $O(n \log n)$. *It can be constructed in time* $O(n^2 \log n)$.

In higher dimensions, it is not possible to show a similar result. In fact, one can construct sets of polygons such that any BSP (not just auto-partitions) must have $\Omega(n^2)$ many fragments (see Figure 3.2 for a "bad" example for auto-partitions).

However, all of these examples producing quadratic BSPs violate the *principle of locality*: polygons are small compared to the extent of the whole set. In practice, no BSPs have been observed that exhibit the worst-case quadratic behavior [Naylor 96].

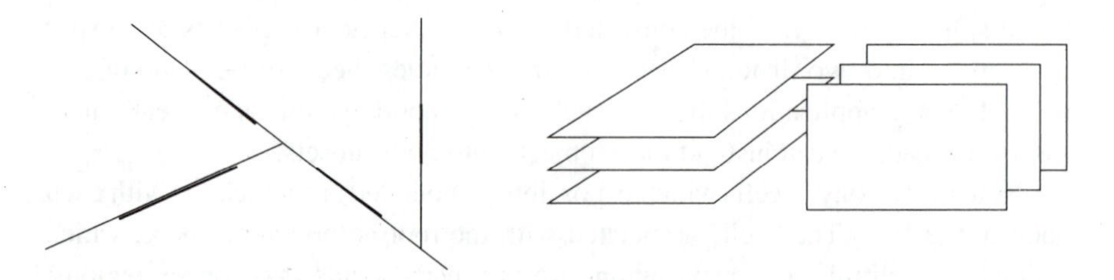

**Figure 3.2.** Left: an auto-partition. Right: an example of a configuration in which any auto-partition must have quadratic size.

## 3.1. Rendering without a Z-Buffer

BSP trees were introduced to computer graphics in [Fuchs et al. 80]. At the time, *hidden-surface removal* was still a major obstacle to interactive computer graphics, because a Z-buffer was just too costly in terms of memory.

In this section, we will describe how to solve this problem, not so much because the application itself is relevant today, but because it nicely exhibits one of the fundamental "features" of BSP trees: they enable efficient enumeration of all polygons in *visibility order* from any point in any direction.[1]

A simple algorithm to render a set of polygons with correct hidden-surface removal, and without a Z-buffer, is the *painter's algorithm*: render the scene from back to front as seen from the current viewpoint. Front polygons will just overwrite the contents of the frame buffer, thus effectively hiding the polygons in the back. There are polygon configurations where this kind of sorting is not always possible, but we will deal with that later.

How can we efficiently obtain such a *visibility order* of all polygons? Using BSP trees, this is almost trivial: starting from the root, first traverse the branch that does *not* contain the viewpoint, then render the polygon stored with the node, and then traverse the other branch containing the viewpoint (see Figure 3.3).

For the sake of completeness, we would like to mention a few strategies to optimize this algorithm. First, we should make use of the viewing direction by skipping BSP branches that lie completely behind the viewpoint.

Furthermore, we can perform back-face culling as usual (which does not cause any extra costs). We can also perform view-frustum culling by testing all vertices of the frustum against the plane of a BSP node.

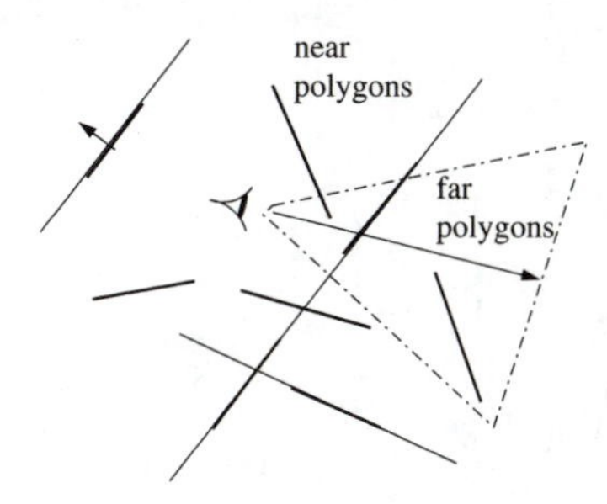


**Figure 3.3.** BSP trees are an efficient data structure encoding visibility order of a set of polygons.

[1]Actually, the first version of Doom used exactly this algorithm to achieve its fantastic frame rate (at the time) on PCs, even without any graphics accelerator.

Another problem with the simple algorithm is that, potentially, a pixel gets overwritten many times (this is exactly the *pixel complexity*), although only the last write "survives." To remedy this, we must traverse the BSP from front to back. But in order to actually save work, we also need to maintain a 2D BSP for the screen that allows us to quickly discard those parts of a polygon that fall onto a screen area that is already occupied. In that 2D screen BSP, we mark all cells as either "free" or "occupied."

Initially, each BSP consists only of a "free" root node. When a new polygon is to be rendered, it is first run through the screen BSP, splitting it into smaller and smaller convex parts until it reaches the leaves. If a part reaches a leaf that is already occupied, nothing happens; if it reaches a free leaf, then it is inserted beneath that leaf, and this part is drawn on the screen.

## 3.2. Representing Objects with BSPs

BSPs offer a nice way to represent volumetric polygonal objects, i. e., closed objects. In other words, their border is represented by polygons and they have an "inside" and an "outside." Such a BSP representation of an object is just like an ordinary BSP for the set of polygons (we can, for instance, build an autopartition), except that here we stop the construction process (see Definition 3.1) only when the set is empty. These leaves represent homogeneous convex cells of the space partitioning—they are completely "in" our "out."

Figure 3.4 shows an example of such a BSP representation. In this section, we will follow the convention that normals point to the "outside," and that the right child of a BSP node lies in the positive half-space and the left child in the negative

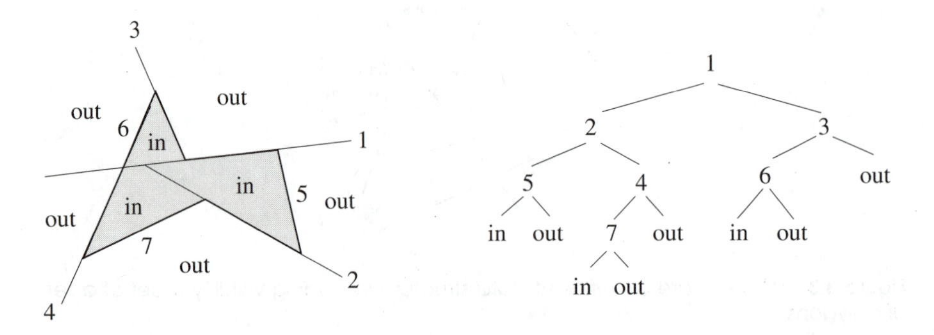


**Figure 3.4.** Each leaf cell of the BSP representation of an object is completely inside or completely outside.

half-space. So, in a real implementation that adheres to these conventions, we can still stop the construction when only one polygon is left, because we know that the left child of such a pseudo-leaf will be "in" and the right one will be "out."

Given such a representation, it is very easy and efficient, for instance, to determine whether a given a point is inside an object. We just filter the point down through the BSP until we reach a leaf.

In the next section, we will describe an algorithm for solving a slightly more difficult problem.

## 3.3. Boolean Operations

In solid modeling, a very frequent task is to compute the intersection or union of a pair of objects. More generally, given two objects A and B, we want to compute $C := A \text{ op } B$, where op $\in \{\cup, \cap, \setminus, \ominus\}$ (see Figure 3.5). This can be computed efficiently using the BSP representation of objects [Naylor et al. 90, Naylor 96]. Furthermore, the algorithm is almost the same for all of these operations. Only the elementary step that processes two leaves of the BSPs is different.

We will present the algorithm for Boolean operations bottom-up in three steps. The first step is a subprocedure for computing the following simple operation: given a BSP $T$ and a plane $H$, construct a new BSP $\hat{T}$ whose root is $H$, such that $\hat{T}^- \triangleq T \cap H^-$, $\hat{T}^+ \triangleq T \cap H^+$ (see Figure 3.6). This basically splits a BSP tree by a plane and then puts that plane at the root of the two halves. Since we will not need the new tree $\hat{T}$ explicitly, we will describe only the splitting procedure (which is the bulk of the work anyway).

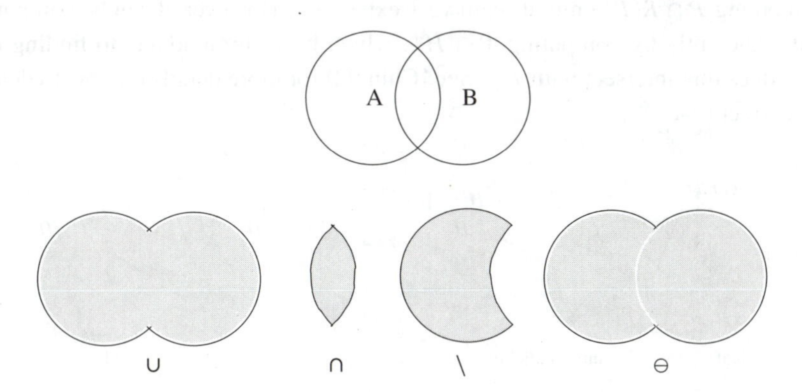


**Figure 3.5.** Using BSPs, we can efficiently compute these Boolean operations on solids.

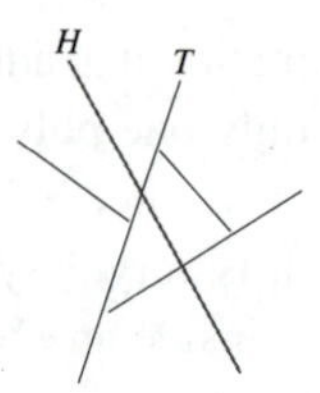


**Figure 3.6.** The fundamental step of the construction is this simple operation, which merges a BSP and a plane.

First, we need to define some nomenclature:

$$T^-, T^+ = \text{left and right child of } T\text{, respectively,}$$
$$R(T) = \text{region of the cell of node } T \text{ (which is convex),}$$
$$T^\oplus, T^\ominus = \text{portion of } T \text{ on the positive/negative side of } H\text{, respectively.}$$

Finally, we would like to define a node $T$ by the tuple $(H_T, p_T, T^-, T^+)$, where $H$ is the splitting plane and $p$ is the polygon associated with $T$ (with $p \subset H$).

Algorithm 3.1 shows the pseudocode for this first step. It is organized into eight cases, four of which are illustrated in Figure 3.7. This might look a little confusing at first sight, but it is really pretty simple. If $T$ is a leaf, then everything is trivial. The case where $T$'s splitting plane $H_T$ and $H$ are coincident (i. e., same plane and same or opposite direction) is trivial, too. The only case, where it actually needs to do some work is the "mixed" case, in which $H \cap R(T)$ intersects with $H_T \cap R(T)$.

The polygon $P$ is needed only to find the case applying at each recursion. Computing $P \cap R(T^+)$ might seem very expensive. However, it can be computed quite efficiently by computing $P \cap H_T^+$, which basically amounts to finding the two edges that intersect with $H_T$. See [Chin 92] for more details on how to detect the correct case.

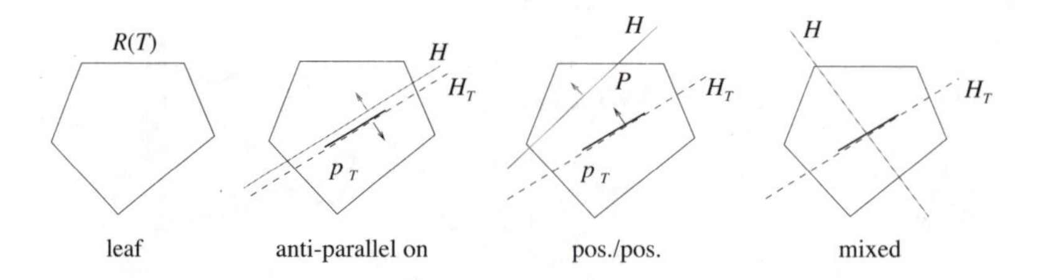


**Figure 3.7.** The main building block of the algorithm consists of these four cases (plus analogous ones).

**split-tree**( $T, H, P$ ) $\rightarrow (T^{\ominus}, T^{\oplus})$
$\{P = H \cap R(T)\}$
**case** $T$ is a leaf :
    **return** $(T^{\ominus}, T^{\oplus}) \leftarrow (T, T)$
**case** "anti-parallel" and "on" :
    **return** $(T^{\ominus}, T^{\oplus}) \leftarrow (T^{+}, T^{-})$
**case** "pos./pos." :
    $(T^{+\ominus}, T^{+\oplus}) \leftarrow$ split-tree$(T^{+}, H)$
    $T^{\ominus} \leftarrow (H_T, p_T, T^{-}, T^{+\ominus})$
    $T^{\oplus} \leftarrow T^{+\oplus}$
**case** "mixed" :
    $(T^{+\ominus}, T^{+\oplus}) \leftarrow$ split-tree$(T^{+}, H, P \cap R(T^{+}))$
    $(T^{-\ominus}, T^{-\oplus}) \leftarrow$ split-tree$(T^{-}, H, P \cap R(T^{-}))$
    $T^{\ominus} \leftarrow (H_T, p_T \cap H^{-}, T^{-\ominus}, T^{+\ominus})$
    $T^{\oplus} \leftarrow (H_T, p_T \cap H^{+}, T^{-\oplus}, T^{+\oplus})$
    **return** $(T^{\ominus}, T^{\oplus})$
**end case**

**Algorithm 3.1.** The first building block of the algorithm for Boolean operations is the procedure that splits a BSP.

It might seem surprising at first sight that Algorithm 3.1 does almost no work—it just traverses the BSP tree, classifies the case found at each recursion, and computes $p \cap H^{+}$ and $p \cap H^{-}$.

The previous algorithm is already the main building block of the overall Boolean operation algorithm. The next step towards that end is an algorithm that performs a so-called *merge* operation on two BSP trees $T_1$ and $T_2$. Let $C_i$ denote the set of elementary cells of a BSP, i. e., all regions $R(L_j)$ of tree $T_i$ where $L_j$ are all the leaves. Then the merge of $T_1, T_2$ yields a new BSP tree $T_3$ such that $C_3 = \{c_1 \cap c_2 | c_1 \in C_1, c_2 \in C_2, c_1 \cap c_2 \neq \varnothing\}$ (see Figure 3.8).

The merge operation consists of two cases. The first, almost trivial, case occurs when one of the two operands is a leaf: then at least one of the two regions is homogenous, i. e., completely inside or outside. In the other case, both trees are not homogenous over the same region of space: then, we just split one of the

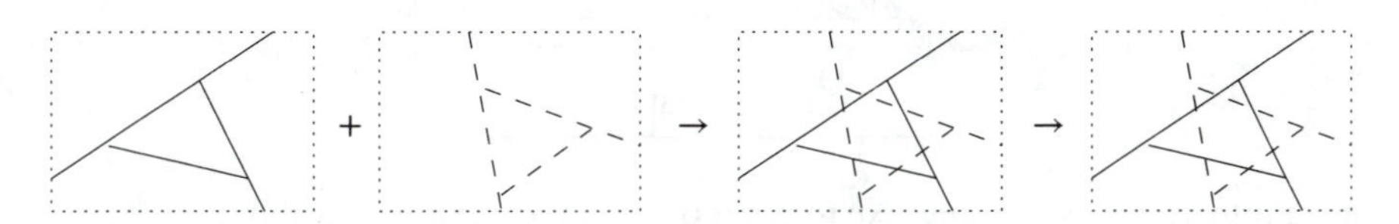

**Figure 3.8.** Computation of Boolean operations is based on a general merge operation.

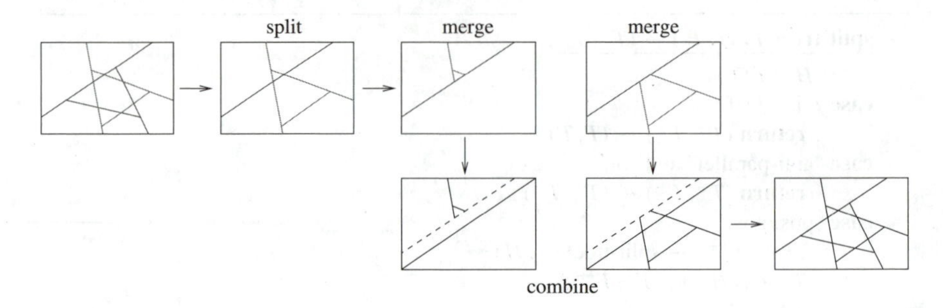


**Figure 3.9.** A graphical depiction of the merge step in the algorithm for Boolean operations on objects represented by BSP trees.

two with the splitting plane from the root of the other, and we obtain two pairs of BSPs, which are smaller and still cover the same regions in space. Those two pairs can be merged recursively (see Figure 3.9). This recursive procedure is described more formally by the pseudocode in Algorithm 3.2.

The third, and last, step is the subprocedure `cell-op`, which is called at the base of the recursion in Algorithm 3.2. This is the only place where the *semantic* of the general merge operation is defined, i. e., whether it should perform a union, an intersection, or anything else. When we have reached that point, then we know that one of the two cells is homogeneous, so we can just replace it by the other node's subtree suitably modified according to the Boolean operation. The following table lists the details of this function (assuming that $T_1$ is the leaf):

| Operation | $T_1$ | Result |
|---|---|---|
| $\cup$ | in | $T_1$ |
| | out | $T_2$ |
| $\cap$ | in | $T_2$ |
| | out | $T_1$ |
| $\setminus$ | in | $T_2^c$ |
| | out | $T_1$ |
| $\ominus$ | in | $T_2^c$ |
| | out | $T_2$ |

Furthermore, we would like to point out that the `merge` algorithm is symmetric: it does not matter whether we partition $T_2$ with $H_1$ or, the other way around, $T_1$ with $H_2$—the result will be the same.

```
merge( T1, T2 ) → T3
if T1 or T2 is a leaf then
    perform the cell-op as required by the Boolean operation to be constructed (see
        below)
else
    (T2⊖, T2⊕) ← split-tree(T2, H1, ...)
    T3− ← merge(T1−, T2⊖)
    T3+ ← merge(T1+, T2⊕)
    T3 ← (H1, T3−, T3+)
end if
```

**Algorithm 3.2.** The second building block for Boolean operations merges two BSPs.

## 3.4. Construction Heuristics

One can prove that in general, auto-partitions have the same complexity as other partitionings [Paterson and Yao 90, de Berg et al. 00]. In addition, it has been proven that "fat" objects (i. e., objects with bounded aspect ratios) and "uncluttered" scenes allow a BSP with linear size [de Berg 95, de Berg 00].

However, for practical applications, the "hidden" constants do matter. So, the construction strategy should produce "good" BSPs. Depending on the application itself, however, the definition of exactly what a "good" BSP tree is can be quite different. Basically, there are two kinds of applications:

- Classification: these are applications where the BSP is used to determine the inside/outside status of a point, or whether or not a line intersects with an object. In this case, we try to optimize the balance of the BSP.

- Visibility: these are applications where the BSP is used to sort the polygons in "visibility order" such as rendering without a Z-buffer. Consequently, we try to minimize the number of splits, i. e., the size of the BSP.

### 3.4.1. Convex Objects

As an example, consider a convex object. In that case, an auto-partition has size $O(n)$, takes time $O(n^2)$ to construct, and is a linear tree. This does not seem to be a very smart BSP (although it is perfectly suitable for visibility ordering).

If we allow arbitrary splitting planes, then we can balance the BSP much better. The construction time will be

$$T(n) = n + 2T(\tfrac{n}{2} + \alpha n) \in O(n^{1+\delta}) \ , \ 0 < \alpha < \tfrac{n}{2},$$

where $\alpha$ is the fraction of polygons that are split at each node. The following table shows the actual complexities for some values of $\alpha$:

| $\alpha$ | 0.05 | 0.2 | 0.4 |
|---|---|---|---|
| | $n^{1.15}$ | $n^2$ | $n^7$ |

As we mentioned, the question now is how to choose the splitting planes. We propose the following simple heuristic:[2] compute a representative vertex for each polygon (barycenter, bounding box center, etc.). Determine a plane, such that about the same number of points are on both sides, and such that all points are far away from the plane (obviously, this is an optimization, which can, for instance, be solved by principal component analysis).

### 3.4.2. Cost-Driven Heuristic

In order to derive construction criteria, one needs to define the *quality* of a BSP. An abstract measure is the *cost* of a BSP tree $T$:

$$C(T) = 1 + P(T^-)C(T^-) + P(T^+)C(T^+), \tag{3.1}$$

where $P(T^-)$ is the probability that the left subtree $T^-$ will be visited under the condition that the tree $T$ has been visited (ditto for $P(T^+)$). This probability depends on the application the BSP is going to be used for. For instance, in the case of inside/outside queries,

$$P(T^-) = \frac{\mathrm{Vol}(T^-)}{\mathrm{Vol}(T)}.$$

Obviously, trying to optimize Equation (3.1) globally is prohibitively expensive. Therefore, a local heuristic must be employed to guide the splitting process. A simple heuristic is the following [Naylor 96]: estimate the cost of the subtrees by the number of polygons, and add a penalty for polygons that are split by the current node, so that

$$C(T) = 1 + |S^-|^\alpha + |S^+|^\alpha + \beta s, \tag{3.2}$$

where $S$ is the set of polygons associated with a node, $s$ is the set of polygons split by a node, and $\alpha$ and $\beta$ are two parameters that can be used to make the BSP more balanced or smaller. ([Naylor 96] reports that $\alpha = 0.8, \ldots, 0.95$ and

[2]For the case of convex objects, a heuristic has already been proposed in [Torres 90]. However, we believe that this heuristic has some flaws.

$\beta = \frac{1}{4}, \ldots, \frac{3}{4}$ are usually good values to start from.) Again, this is an optimization process, but now only a local one.

If the BSP is to be an auto-partition, then a very quick approximization of the local optimum yields very good results: just sort the polygons according to their size and evaluate Equation (3.2) for the first $k$ polygons. Then choose the one that produces the least costly BSP (subtree). The rationale is that the probability that a polygon will be split is proportional to its size, so we try to get rid of the polygons as early as possible.

An even simpler heuristic was proposed in [Fuchs et al. 80]: randomly choose $k$ polygons from $S$ and select the one that produces the least number of splits. They report that $k = 5$ yielded near-optimal BSPs (for visibility ordering).

### 3.4.3. Non-Uniform Queries

In the previous section, we assumed that all queries are uniformly distributed over a certain domain. This is a valid assumption if nothing is known about the distribution. On the other hand, if we know more about it, then we should make use of this and construct the BSP such that frequent queries can be answered most quickly.[3]

Indeed, quite often, we know more about the queries. For instance, in ray tracing, the starting points are usually not uniformly distributed in space; for instance, they usually do not emanate from the interior of objects. Also, the prominent polygons of an object are hit more frequently than those that are within cavities or completely inside.

According to [Ar et al. 00], we can estimate the costs of a query by

$$\begin{aligned} C(\text{query}) &= \#\,\text{nodes visited} \\ &\leq \text{depth(BSP)} \cdot \#\,\text{stabbed leaf cells}. \end{aligned}$$

So, according to this, we should minimize the number of stabbed leaf cells before a polygon hit occurs. At least two factors influencing the probability that a ray hits a polygon are:

- if the angle between a ray and a polygon is large, then the probability is large;

- if the polygon is large (relative to the total size of the object/universe), then the probability is large.

[3]The same principle underlies the Huffman encoding scheme.

Let $\omega(l)$ denote the density of all rays $l$ over some domain $D$; this could be measured, or it could be derived from the geometry. Let $S$ be the set of polygons over which the BSP is to be built. Assign a score $p$ to each polygon

$$\text{score}(p) = \int_D w(S, p, l)\omega(l)dl,$$

where the weight $w$ is defined as

$$w(S, p, l) = \sin^2(\mathbf{n}_p, \mathbf{r}_l)\frac{\text{Area}(p)}{\text{Area}(S)},$$

where $\mathbf{n}_p$ is the normal of $p$, and $\mathbf{r}_l$ is the direction of $l$.

So the algorithm for constructing a BSP adapted to a given distribution is the following randomized greedy strategy: sort all polygons according to score($p$); choose randomly one out of the "top $k$," and split the set $S$. Thus, polygons with a higher probability of getting hit by the ray tend to end up higher in the tree.

The BSP thus constructed has been named the *customized* BSP by [Ar et al. 00]. The authors report that it usually has about two times as many polygons as its "oblivious" version, but the queries have to visit only one-half to one-tenth as many polygons.

### 3.4.4. Deferred, Self-Organizing BSPs

Now, what should we do if we just don't know the query distribution, or if it is too expensive to measure it by experiments? The answer is to defer the complete construction of the BSP. In other words, we build only as much of the BSP as is absolutely necessary. In addition, we keep a history (in some form) of all the queries we have seen so far, so whenever we need to build a new part of the BSP, we base the construction on that history (as a best guess of the probability distribution of all queries to come) [Ar et al. 02].

As an example, let us consider the problem of detecting intersections between a 3D line segment and a set of polygons.[4]

Since a BSP is now no longer completely constructed before we use it, the nodes must store additional information:

- the polygon $P$ defining the splitting plane; or, if it is a preliminary leaf,
- a list $\mathcal{L} \subseteq S$ of polygons associated with it, for which the subtree has not been built yet.

[4]Sometimes, this is also called *collision detection*, in particular in the game programming industry, because we are interested only in a yes/no answer; in ray tracing, we want to determine the *earliest* intersection.

```
testray(R,v)
if v is a leaf then
        for all P ∈ L_v do
                if R intersects P then
                        return hit
                end if
        end for
else
        v1 ← child of v that contains startpoint of R
        v2 ← other child
        testray(R,v1)
        if no hit in v1 then
                testray(R,v2)
        end if
end if
```

**Algorithm 3.3.** Testing for ray-scene intersections with a deferred BSP.

Now, the algorithm for answering queries with a ray $R$ also triggers the construction of the BSP (see Algorithm 3.3). The initial BSP tree is just one node (the root) with $\mathcal{L} = S$, i.e., all polygons of the object or scene. Since the line segment is finite (in particular, if it is short), this algorithm can usually stop much earlier, but the details have been omitted here.

In the following, we will fill in the open issues, namely:

- when do we split a preliminary leaf? (and which one?)
- how do we split?

For the "when" question: we keep an access counter per node, which is incremented every time that node is traversed during a query. Whenever one of the counters is over a certain threshold, we split that node (leaf). The threshold can be an absolute value or relative to the sum of all counters.

For the "how" question: we keep a counter per polygon $P \in \mathcal{L}_v$, which is incremented every time an intersection with $P$ is found. We sort $\mathcal{L}_v$ according to this counter; this can be done incrementally whenever one of the counters changes. If a split is to be performed for $v$, then we use the first polygon from $\mathcal{L}_v$ as the splitting plane.

It turns out that many polygons are never hit by a line segment. Therefore, with this algorithm, a BSP subtree will never be "wasted" on these polygons, and they will be stored at the end of the lists at the leaves.

There are other ways to organize the polygon lists at the leaves: move the polygon currently being hit to the front of the list or swap it with the one in front of it. However, according to [Ar et al. 02], the strategy that sorts the list seems to work best.

According to the same authors, the performance gain in their experiments was a factor of about 2 to 20.

# 4 Bounding Volume Hierarchies

In Chapter 3, we obtained a new data structure (BSPs) by starting from kd-trees and lifting one restriction, namely that splitting planes be always perpendicular to one of the coordinate axes. Starting again from kd-trees, we can regard them as a way to hierarchically group objects (points, polygons, etc.), rather than a way to partition space. In the simple kd-tree, all nodes on the same level partition space, and their extents are implicitly given by the path from the root. Since we now abandon this property, we can as well explicitly store the smallest bounding box enclosing all objects associated with a node, which is usually smaller than the extent of the kd-tree cell of that node. Thus, we have arrived at one particular type of bounding volume (BV) hierarchy.

Often times, bounding volume hierarchies (BVHs) are described as the opposite of spatial partitioning schemes, such as quadtrees or BSP trees: instead of partitioning space, the idea is to partition the set of objects recursively until some leaf criterion is met. However, we hope it has become clear by now that we regard BVHs as just one class of geometric data structures within the whole spectrum of hierarchical data structures.

Like the previous hierarchical data structures, BVHs are mostly used to prevent performing an operation exhaustively on all objects. Almost all queries that can be implemented with other space partitioning schemes can also be answered using BVHs. Examples of queries and operations are ray shooting, frustum culling, occlusion culling, point location, nearest neighbor, and collision detection (the latter will be discussed in more detail later in this chapter).

**Definition 4.1. (BVH.)** Let $O = \{o_1, \ldots, o_n\}$ be a set of elementary objects. A BVH for $O$, BVH($O$), is defined by

1. if $|O| = e$, then BVH($O$) := a leaf node that stores $O$ and a bounding volume of $O$;

2. if $|O| > e$, then BVH($O$) := a node $v$ with $n(v)$ children $v_1, \ldots, v_n$, where each child $v_i$ is a BVH, BVH($O_i$), over a subset $O_i \subset O$, such that $\bigcup O_i = O$. In addition, $v$ stores a BV of $O$.

The definition mentions two parameters. The threshold $e$ is often set to 1, but depending on the application, the optimal $e$ can be much larger. Just as in sorting, when the set of objects is small, it is often cheaper to perform the operation iteratively on all of them, because recursive algorithms always incur some overhead.

Another parameter in the definition is the arity. Mostly, BVHs are constructed as binary trees, but again, the optimum can be larger. And what is more, as the definition suggests, the out-degree of nodes in a BVH does not necessarily have to be constant, although this often simplifies implementations considerably.

Effectively, these two parameters, $e$ and $n(v)$, control the balance between linear search/operation (which is exhaustive) and a maximally recursive algorithm.

There are more design choices possible according to the definition. For inner nodes, it requires only that $\bigcup O_i = O$; this means that the same object $o \in O$ could be associated with several children. Depending on the application, the type of BVs, and the construction process, this may not always be avoidable. But if possible, you should split the set of objects into disjoint subsets.

Finally, there is at least one more design choice: the type of BV used at each node. This does not necessarily mean that each node uses the same type of BV. Figure 4.1 shows a number of the most commonly used BVs. The difference

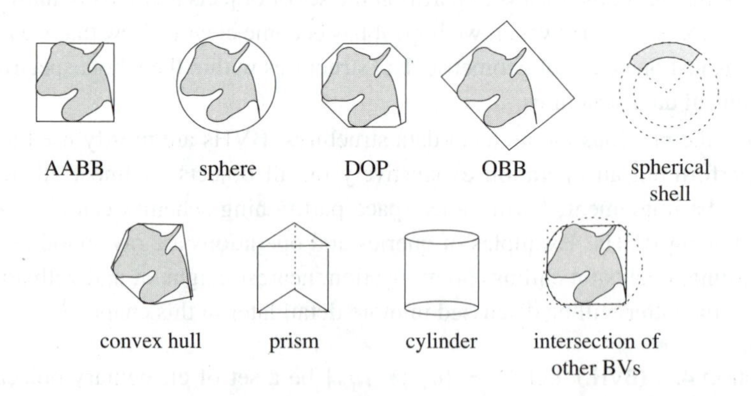


**Figure 4.1.** Some bounding volumes. (Courtesy of Blackwell Publishing.)

between OBBs [Arvo and Kirk 89] and AABBs is that OBBs can be oriented arbitrarily (hence *oriented bounding boxes*). DOPs [Zachmann 98, Klosowski et al. 98, Kay and Kajiya 86] are a generalization of AABBs; basically, they are the intersection of $k$ slabs. Prisms and cylinders have been proposed by [Barequet et al. 96] and [Weghorst et al. 84], but they seem to be too expensive computationally. A spherical shell is the intersection of a shell and a cone (the cone's apex coincides with the sphere's center), and a shell is the space between two concentric spheres. Finally, one can always take the intersection of two or more different types of BVs [Katayama and Satoh 97].

There are three characteristic properties of BVs: tightness, memory usage, and the number of operations needed to test the query object against a bounding volume.

Often, one must make a trade-off between these properties, because generally, the type of BV that offers better tightness also requires more operations per query and more memory.

Regarding the tightness, one can establish a theoretical advantage of OBBs. But first, we need to define tightness [Gottschalk et al. 96].

**Definition 4.2. (Tightness by Hausdorff distance.)** Let $B$ be a BV, and $G$ be some geometry bounded by $B$, i. e., $g \subset B$. Let

$$h(B, G) = \max_{b \in B} \min_{g \in G} d(b, g)$$

be the *directed Hausdorff distance*, i. e., the maximum distance of $B$ to the nearest point in $G$. (Here, $d$ is any metric, very often just the Euclidean distance.) Let

$$\text{diam}(G) = \max_{g, f \in G} d(g, f)$$

be the *diameter* of $G$.

Then we can define *tightness* as

$$\tau := \frac{h(B, G)}{\text{diam}(G)}.$$

See Figure 4.2 for an illustration.

Since the Hausdorff distance is very sensitive to outliers, one could also think of other definitions such as the following one.

**Definition 4.3. (Tightness by volume.)** Let $C(v)$ b the set of children of a node $v$ of the BVH. Let $\text{Vol}(v)$ be the volume of the BV stored with $v$.

Then we can define the tightness as

$$\tau := \frac{\mathrm{Vol}(v)}{\sum_{v' \in C(v)} \mathrm{Vol}(v')}.$$

Alternatively, we can define it as

$$\tau := \frac{\mathrm{Vol}(v)}{\sum_{v' \in L(v)} \mathrm{Vol}(v')},$$

where $L(v)$ is the set of leaves beneath $v$.

Getting back to the tightness definition based on the Hausdorff distance, we observe a fundamental difference between AABBs and OBBs [Gottschalk et al. 96] as follows.

- The tightness of AABBs depends on the orientation of the enclosed geometry. What is worse is that the tightness of the children of an AABB enclosing a surface of small curvature is almost the same as that of the parent.

  The worst case is depicted in Figure 4.3. The tightness of the parent is $\tau = h/d$, while the tightness of a child is $\tau' = \frac{h'}{d/2} = \frac{h/2}{d/2} = \tau$.

- The tightness of OBBs does not depend on the orientation of the enclosed geometry. Instead, it depends on its curvature, and it decreases approximately linearly with the depth in the hierarchy.

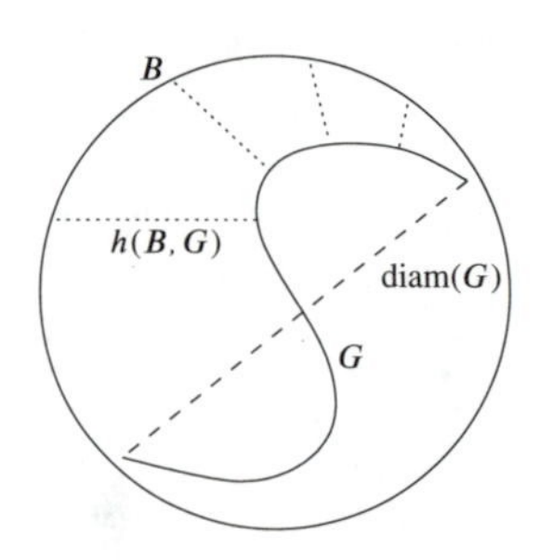


**Figure 4.2.** One way to define tightness is via the directed Hausdorff distance.

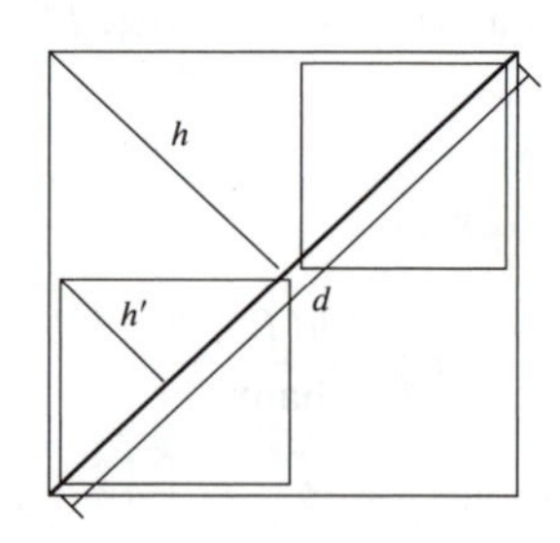


**Figure 4.3.** The tightness of an AABB remains more or less constant xhroughout the levels of an AABB hierarchy for surfaces of small curvature.

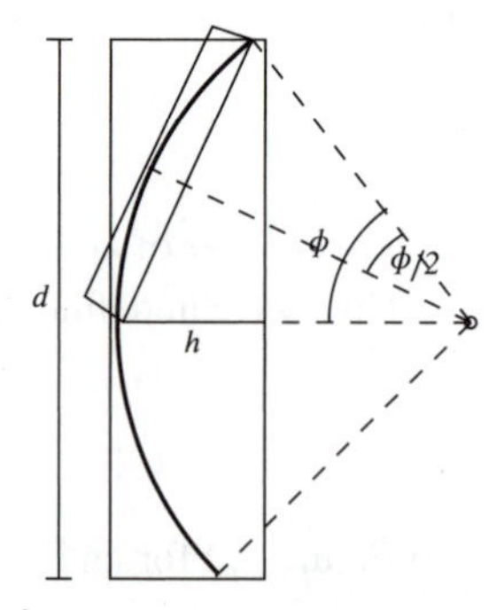


**Figure 4.4.** The tightness of an OBB decreases for deeper levels in an OBB hierarchy for small curvature surfaces

Figure 4.4 depicts the situation for a sphere. The Hausdorff distance from an OBB to an enclosed spherical arc is $h = r(1 - \cos\phi)$, while the diameter of the arc is $d = 2r\sin\phi$. Thus, the tightness for an OBB bounding a spherical arc of degree $\phi$ is $\tau = \frac{1-\cos\phi}{2\sin\phi}$, which approaches 0 linearly as $\phi \to 0$.

This makes OBBs seem much more attractive than AABBs. The price of the much improved tightness is the higher computational effort needed for most queries per node when traversing an OBB tree with a query.

## 4.1. Construction of BVHs

Essentially, there are three strategies to build BVHs: bottom-up, top-down, and insertion.

From a theoretical point of view, one could pursue a simple top-down strategy, which just splits the set of objects into two equally sized parts, where the objects are assigned randomly to either subset. Asymptotically, this usually yields the same query time as any other strategy. However, in practice, the query times offered by such a BVH are worse by a large factor.

During construction of a BVH, it is convenient to forget about the graphical objects or primitives, and instead deal with their bounding volumes and consider those as the atoms. Sometimes, another simplification is to just approximate each object by its center (barycenter or bounding box center), and then deal only with sets of points during the construction. Of course, when the BVs are finally computed for the nodes, the true extents of the primitives or objects must be considered.

In the following, we will describe algorithms for each construction strategy.

**Bottom-up.** In this class, we will actually describe two algorithms.

Let $B$ be the set of BVs on the topmost level of the BVH that has been constructed so far [Roussopoulos and Leifker 85]. For each $b_i \in B$, find the nearest neighbor $b'_i \in B$; let $d_i$ be the distance between $b_i$ and $b'_i$. Sort $B$ with respect to $d_i$. Then combine the first $k$ nodes in $B$ under a common parent; do the same with the next $k$ elements from $B$, etc. This yields a new set $B'$, and the process is repeated.

Note that this strategy does not necessarily produce bounding volumes with a small "dead space." In Figure 4.5, the strategy would choose to combine the left pair (distance = 0), while choosing the right pair would result in much less dead space.

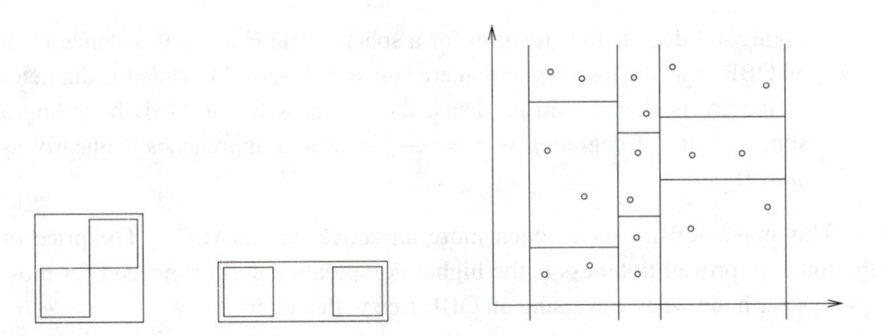

**Figure 4.5.** A simple greedy strategy can produce much "dead space."

**Figure 4.6.** A less greedy strategy combines bounding volumes by computing a "tiling."

The second strategy is less greedy in that it computes a tiling for each level. We will describe it first in 2D [Leutenegger et al. 97]. Again, let $B$ be the set of bounding volumes on the topmost level so far constructed, with $|B| = n$. The algorithm first computes the center $c^i$ for each $b^i \in B$. Then it sorts $B$ along the $x$-axis with respect to $c^i_x$. Now, the set $B$ is split into $\sqrt{n/k}$ vertical "slices" (again with respect to $c^i_x$). Now, each slice is sorted according to $c^i_y$ and subsequently split into $\sqrt{n/k}$ tiles, so that we end up with $k$ tiles (see Figure 4.6). Finally, all nodes in a tile are combined under one common parent, its bounding volume is combined, and the process repeats with a new set $B'$.

In $\mathbb{R}^d$, it works quite similarly: we just split each slice repeatedly by $\sqrt[d]{n/k}$ along all coordinate axes.

Insertion. This construction scheme starts with an empty tree. Let $B$ be the set of elementary bounding volumes. Algorithm 4.1 describes the general procedure.

```
1: while |B| > 0 do
2:     choose next b ∈ B
3:     v := root
4:     while v ≠ leaf do
5:         choose child v',
               so that insertion of b into v' causes minimal increase in the costs
             of the total tree
6:         v := v'
7:     end while
8: end while
```

**Algorithm 4.1.** Construction of BVHs by insertion.

All insertion algorithms vary only Step 2 and/or 5. Step 2 is important because a "bad" choice in the beginning can probably never be made right afterwards. Step 5 depends on the type of query that is to be performed on the BVH. See the next section for a few criteria.

Usually, algorithms in this class have complexity $O(n \log n)$.

**Top-down.** This scheme is the most popular one. It seems to produce very good hierarchies while still being very efficient, and usually it can be implemented easily.

The general idea is to start with the complete set of elementary bounding volumes (which will become the leaves), split that set into $k$ parts, and create a BVH for each part recursively. The splitting should be guided by some heuristic or criterion that produces good hierarchies.

### 4.1.1. Construction Criteria

In the literature, there is a vast number of criteria for guiding the splitting, insertion, or merging during BVH construction. (Often, the authors endow the BVH thus constructed with a new name, even though the bounding volumes utilized are well known.) Obviously, the criterion depends on the application for which the BVH is to be used. In the following, we will present a few of these criteria.

For ray tracing, if we can estimate the probability that a ray will hit a child box when it has hit the parent box, we know how likely it is that we need to visit the child node when we have visited the parent node. Let us assume that all rays emanate from the same origin (see Figure 4.7). Then we can observe that the probability that a ray $s$ hits a child box $v'$ under the condition that it has hit the father box $v$ is

$$P(s \text{ hits } v' | s \text{ hits } v) = \frac{\theta_{v'}}{\theta_v} \approx \frac{\text{Area}(v')}{\text{Area}(v)}, \tag{4.1}$$

where Area denotes the surface area of the BV, and $\theta$ denotes the solid angle subtended by the BV. This is because, for a convex object, the solid angle subtended by it, when seen from large distances, is approximately proportional to its surface area [Goldsmith and Salmon 87]. So, a simple strategy is to minimize the surface area of the bounding volumes of the children that are produced by a split.[1]

A more elaborate criterion tries to establish a *cost function* for a split and minimize that. For ray tracing, this cost function can be approximated by

$$C(v_1, v_2) = \frac{\text{Area}(v_1)}{\text{Area}(v)} C(v_1) + \frac{\text{Area}(v_2)}{\text{Area}(v)} C(v_2), \tag{4.2}$$

[1]For the insertion scheme, the strategy is to choose that child node whose area is increased least [Goldsmith and Salmon 87].

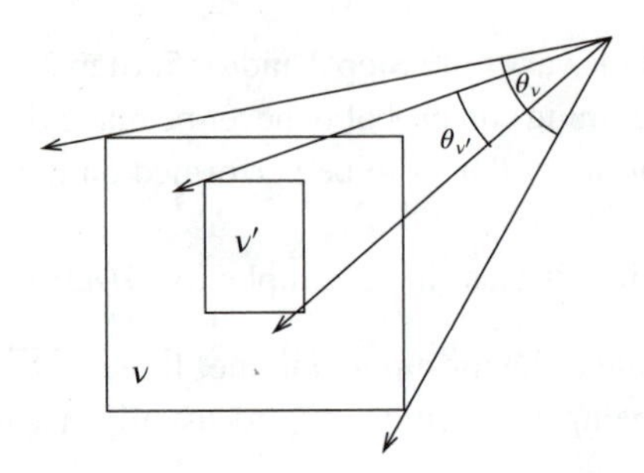


**Figure 4.7.** The probability of a ray hitting a child box can be estimated by the surface area.

where $v_1, v_2$ are the children of $v$. The optimal split $B = B_1 \cup B_2$ minimizes this cost function:

$$C(B_1, B_2) = \min_{B' \in \mathcal{P}(B)} C(B', B \setminus B'),$$

where $B_1, B_2$ are the subsets of elementary bounding volumes (or objects) assigned to the children. Here, we have assumed a binary tree, but this can be extended to other arities analogously.

Of course, such a minimization is too expensive in practice, in particular, because of the recursive definition of the cost function. So, [Fussell and Subramanian 88, Müller et al. 00, Beckmann et al. 90] have proposed the following approximation algorithm:

**for** $\alpha = x, y, z$ **do**
sort $B$ along axis $\alpha$ with respect to the bounding volume centers
find

$$k^\alpha = \min_{j=0\ldots n} \left\{ \frac{\text{Area}(b_1, \ldots, b_j)}{\text{Area}(B)} j + \frac{\text{Area}(b_{j+1}, \ldots, b_n)}{\text{Area}(B)} (n-j) \right\}$$

**end for**
choose the best $k^\alpha$

where $\text{Area}(b_1, \ldots, b_j)$ denotes the surface area of the bounding volume enclosing $b_1, \ldots, b_j$.

If the query is a point location query (e. g., is a given point inside or outside the object), then the volume instead of the surface area should be used. This is because the probability that a point is contained in a child BV, under the condition that it is contained in the parent BV, is proportional to the ratio of the two volumes.

For collision detection, the volume seems to be a good probability estimation, too, as you will see in the next section.

### 4.1.2. The Criterion for Collision Detection

In the following, we will describe a general criterion that can guide the splitting process of top-down BVH construction algorithms, such that the hierarchy produced is good in the sense of fast collision detection [Zachmann 02] (see Section 4.3).

Let $C(A, B)$ be the expected costs of a node pair $(A, B)$ under the condition that we have already determined during collision detection that we need to traverse the hierarchies further down. Assuming binary trees and unit costs for an overlap test, this can be expressed by

$$C(A, B) = 4 + \sum_{i,j=1,2} P(A_i, B_j) \cdot C(A_i, B_j), \tag{4.3}$$

where $A_i, B_j$ are the children of A and B, respectively, and $P(A_i, B_j)$ is the probability that this pair must be visited (under the condition that the pair $(A, B)$ has been visited).

An optimal construction algorithm would need to expand Equation (4.3) down to the leaves:

$$\begin{aligned} C(A, B) = & P(A_1, B_1) + P(A_1, B_1)P(A_{11}, B_{11}) \\ & + P(A_1, B_1)P(A_{12}, B_{11}) + \ldots + \\ & P(A_1, B_2) + P(A_1, B_2)P(A_{11}, B_{21}) \\ & + \ldots \end{aligned} \tag{4.4}$$

and then find the minimum. Since we are interested in finding a local criterion, we approximate the cost function by discarding the terms corresponding to lower levels in the hierarchy, which gives

$$C(A, B) \approx 4\big(1 + P(A_1, B_1) + \ldots + P(A_2, B_2)\big). \tag{4.5}$$

Now we will derive an estimate of the probability $P(A_1, B_1)$. For the sake of simplicity, we will assume in the following that AABBs are used as BVs. However, similar arguments should hold for all other kinds of convex BVs.

The event of box A intersecting box B is equivalent to the condition that B's "anchor point" is contained in the Minkowski sum $A \oplus B$. This situation is depicted in Figure 4.8.[2] Because $B_1$ is a child of $B$, we know that the anchor point of $B_1$ must lie somewhere in the Minkowski sum $A \oplus B \oplus \mathbf{d}$, where $\mathbf{d} = \text{anchor}(B_1) - \text{anchor}(B)$. Since $A_1$ is inside $A$ and $B_1$ inside $B$, we know that

[2]In the figure, we have chosen the lower-left corner of $B$ as its anchor point, but this is arbitrary, because the Minkowski sum is invariant under translation.

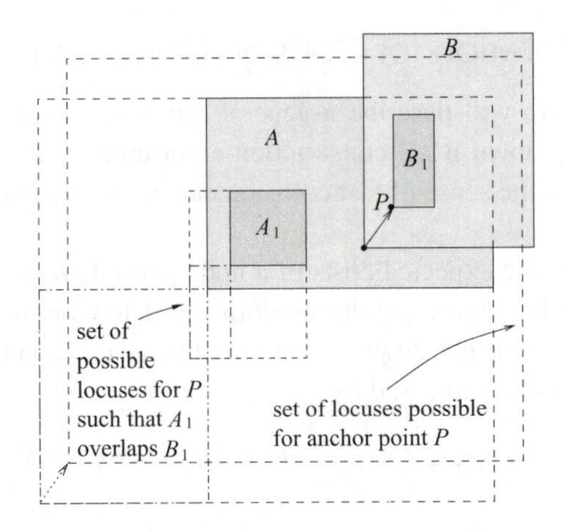


**Figure 4.8.** By estimating the volume of the Minkowski sum of two BVs, we can derive an estimate for the cost of the split of a set of polygons associated with a node. (See Color Plate V.)

$A_1 \oplus B_1 \subset A \oplus B \oplus \mathbf{d}$. So, for arbitrarily convex BVs, the probability of overlap is

$$P(A_1, B_1) = \frac{\mathrm{Vol}(A_1 \oplus B_1)}{\mathrm{Vol}(A \oplus B \oplus \mathbf{d})} = \frac{\mathrm{Vol}(A_1 \oplus B_1)}{\mathrm{Vol}(A \oplus B)}. \tag{4.6}$$

In the case of AABBs, it is safe to assume that the aspect ratio of all BVs is bounded by $\alpha$. Consequently, we can bound the volume of the Minkowski sum by

$$\begin{aligned} \mathrm{Vol}(A) + \mathrm{Vol}(B) + \frac{2}{\alpha}\sqrt{\mathrm{Vol}(A)\,\mathrm{Vol}(B)} & \\ \leq \mathrm{Vol}(A \oplus B) & \\ \leq \mathrm{Vol}(A) + \mathrm{Vol}(B) + 2\alpha\sqrt{\mathrm{Vol}(A)\,\mathrm{Vol}(B)}. & \end{aligned} \tag{4.7}$$

So, we can estimate the volume of the Minkowski sum of two boxes by

$$\mathrm{Vol}(A \oplus B) \approx 2(\mathrm{Vol}(A) + \mathrm{Vol}(B)),$$

yielding

$$P(A_1, B_1) \approx \frac{\mathrm{Vol}(A_1) + \mathrm{Vol}(B_1)}{\mathrm{Vol}(A) + \mathrm{Vol}(B)}. \tag{4.8}$$

Since $\mathrm{Vol}(A) + \mathrm{Vol}(B)$ has already been committed by an earlier step in the recursive construction, Equation (4.5) can be minimized only by minimizing $\mathrm{Vol}(A_1) + \mathrm{Vol}(B_1)$. This is our criterion for constructing restricted box trees.

### 4.1.3. Construction Algorithm

According to the criterion derived in the preceding section, each recursion step in a top-down construction algorithm should try to split the set of polygons so that the cost function (Equation (4.5)) is minimized. Again, finding the optimum is prohibitively expensive, but we can obtain a fairly good solution by kind of a sweep plane approach, as follows (see Figure 4.9).

First, we represent the set of polygons by a set of points, for instance, by replacing each polygon with its barycenter or bounding box center, or by using the set of vertices.

Second, we compute the orientation of the sweep plane(s), for which we just take the largest principal component of the set of points, i. e., the eigenvector of the covariance matrix with the largest eigenvalue. This is the axis that exhibits the largest variance of the set of points. If the three eigenvectors (or two of them) are very similar in magnitude, this means that the set of points does not have a clear direction of largest variance, i. e., it is more or less sphere shaped. In that case, we can just perform the following steps for all three principal axes.

At this point, [Gottschalk et al. 96] uses a very simple heuristic: it divides the set of points into two parts by a splitting plane that is orthogonal to the axis and goes through the barycenter of all points. Alternatively, one could sort all points along the axis and split the set by the median. This would lead to balanced trees, but not necessarily better ones.

If we want to do better, the third step is to consider each polygon and put it in one of the two child BVs (see Figure 4.9, step 3). Each of the two child BVs is initialized by a "seed" polygon from the left and from the right end of the axis, respectively. Then we consider alternatingly one polygon from the left side or from the right side on the axis and put it tentatively in each of the two child BVs. Finally, we put the polygon into the child BV that causes the least increase in volume.[3]

The algorithm and criterion we propose here can also be applied to construct BVHs utilizing other kinds of BVs, such as OBBs, DOPs, and even convex hulls. We suspect that the volume of AABBs would work fairly well as an estimate of the volume of the respective BVs.

This algorithm has proven to be geometrically robust, since there is no error propagation. Therefore, a simple epsilon guard for all comparisons suffices.

[3]Considering the polygons on the left and the right side alternatingly prevents sort of a "creeping greediness," which can happen, for instance, when the sequence of polygons happens to be ordered along the axis such that each polygon would increase one child BV's volume just a little bit, so that the other bounding volume would never get a polygon.

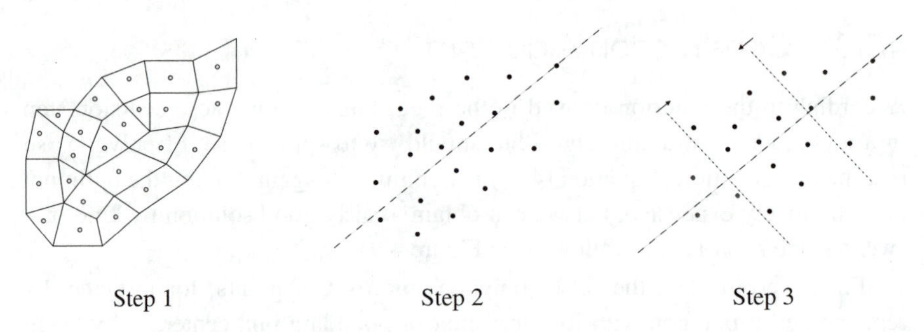


**Figure 4.9.** Algorithm for applying the splitting criterion to BVH construction.

It can be shown that this algorithm is, under certain assumptions,[4] in $O(n)$, where $n$ is the number of polygons [Zachmann 02].

## 4.2. Updating for Morphing Objects

When the geometry underlying a BVH has deformed, the original BVH is no longer valid. Basically, there are two options: rebuilding the BVH (possibly only partially) or refitting. In the latter case, the tree structure itself remains unchanged, and only the extents of the BVs are updated. Refitting is much faster, but the BVs are usually less tight and the overlap between siblings is larger.

An important case of deformation occurs in animation systems, where objects are deformed by *morphing* (or *blending*), i. e., in-between objects are constructed by interpolating between two ore more morph targets. This usually requires the target models to have the same number of vertices and the same topology.

The idea of [Larsson and Akenine-Möller 03] is to construct *one* BVH (for one of the morph targets) and fit this to the other morph targets such that corresponding nodes contain exactly the same vertices. With each node of the BVH, we save all corresponding BVs (one from each morph target; see Figure 4.10). During runtime, we can construct a BVH for the morphed object just by taking the original BVH and interpolating the BVs.

Assume we are given $n$ morph targets, $O^i$, each with vertices $v^i_j$, and $n$ weight vectors $w^i = (w^i_1, \ldots, w^i_m)$. Then each vertex $\bar{v}_j$ of the morphed object is an affine combination

$$\bar{v}_j = \sum_{i=1}^{n} w^i_j v^i_j, \quad \text{with} \quad \sum_{i=1}^{n} w^i_j = 1. \tag{4.9}$$

[4]These assumptions have been valid in all practical cases we have encountered so far.

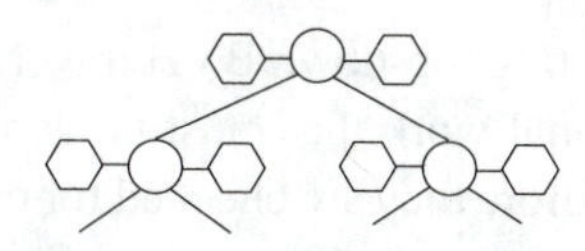

**Figure 4.10.** If the deformation is a predefined morph, then a BVH for in-between objects can be constructed by morphing the BVs. (Courtesy of Blackwell Publishing.)

Let $D^i$ be the $n$ BVs of the corresponding nodes in the BVHs of the $O^i$ (i. e., all $D^i$ contain the same vertices, albeit at different positions). We denote a DOP by $D^i = (S_1^i, \ldots, S_k^i)$, where $S_j^i = (s_j, e_j)$, $s_j \le e_j$, is one interval of the DOP. Let $b^l, l = 1 \ldots k$, denote the set of orientations over which all DOPs are defined.

We can interpolate a new DOP $\bar{D} = (\bar{S}_1, \ldots, \bar{S}_k)$, $\bar{S}_j = (\bar{s}_j, \bar{e}_j)$, from these $n$ DOPs by

$$\bar{s}_j = \sum_{i=1}^{n} w_i s_j^i, \quad \bar{e}_j = \sum_{i=1}^{n} w_i e_j^i. \tag{4.10}$$

This interpolated DOP $\bar{D}$ will enclose all the interpolated vertices beneath its node, i. e.,

$$\forall i : v_j^i \in D^i \Rightarrow \bar{v}_j \in \bar{D}. \tag{4.11}$$

*Proof:* For all $\bar{v}_l \in \bar{D}$, we have

$$\bar{s}_j = \sum_{i=1}^{n} w_i s_j^i \le \sum_{i=1}^{n} w_i \left(v_l^i \cdot b^j\right) = \bar{v}_l \cdot b^j, \tag{4.12}$$

and similarly, we can bound $\bar{v} \cdot b^j$ from above by $\bar{e}_j$. □

This works just the same for AABBs, since AABBs are a special case of DOPs.

It also works for sphere trees. Let $B^i = (c^i, r_i)$ denote a sphere from the sphere tree of morph target $O^i$. Then we can obtain the corresponding bounding sphere $\bar{B} = (\bar{c}, \bar{r})$ by

$$\bar{c} = \sum_{i=1}^{n} w_i c^i, \quad \bar{r} = \sum_{i=1}^{n} w_i r_i. \tag{4.13}$$

*Proof:* For all $\bar{v}_l \in \bar{B}$, we have

$$\begin{aligned} \left\| \bar{v}_l - \bar{c} \right\| &= \left\| \sum_i w_i v_l^i - \sum_i w_i c^i \right\| = \left\| \sum_i w_i (v_l^i - c^i) \right\| \\ &\le \sum_i w_i \left\| v_l^i - c^i \right\| = \sum_i w_i r_i \end{aligned} \tag{4.14}$$

using the triangle inequality. □

Overall, we can use existing top-down BVH traversal algorithms for collision detection. The only additional work that must be done is the interpolation, i. e., morphing of the BVs, just before they are checked for overlap. The exact positions of the morphed vertices enclosed by a bounding volume are not needed.

This deformable collision detection algorithm seems to be faster in practical cases, and its performance depends much less on the polygon count than the more general method presented in [Larsson and Akenine-Möller 01].

The method does have a few drawbacks:

- Since it does not rebuild the BVH, one must somehow construct a BVH that yields good performance for *all* in-between models;

- As explained, it works only for morphing schemes that allow only *one* weight per morph target. One could extend the technique to a more general morph scheme, where each target has a weight vector $w^i = (w^i_1, \ldots, w^i_m)$ ($m$ = number of vertices). Then, however, we would have to morph the DOPs by

$$\bar{s}_j = \sum_i \bar{w}^i s^i_j, \quad \bar{w}^i = \min\{w^i_l \mid v^i_l \in D^i\}. \tag{4.15}$$

  So, we would need to precompute and store those $\bar{w}^i$ for each DOP $\bar{D}$ in the BVH. This may or may not be feasible in practice;

- As we stated, it works for only certain kinds of deformations.

## 4.3. Collision Detection

Fast and exact collision detection of polygonal objects undergoing rigid motions is at the core of many simulation algorithms in computer graphics. In particular, all kinds of highly interactive applications such as virtual prototyping need exact collision detection at interactive speeds for very complex, arbitrary "polygon soups." It is a fundamental problem of dynamic simulation of rigid bodies, simulation of natural interaction with objects, and haptic rendering.

BVHs seem to be a very efficient data structure to tackle the problem of collision detection for rigid bodies. All kinds of different types of BVs have been explored in the past: sphere [Hubbard 95, Palmer and Grimsdale 95], OBBs [Gottschalk et al. 96], DOPs [Klosowski et al. 98, Zachmann 98], AABBs [Zachmann 97, van den Bergen 97, Larsson and Akenine-Möller 01], and convex hulls [Ehmann and Lin 01], to name but a few.

```
traverse(A,B)
if A and B do not overlap then
        return
end if
if A and B are leaves then
        return intersection of primitives
                enclosed by A and B
else
        for all children A[i] and B[j] do
                traverse(A[i],B[j])
        end for
end if
```

**Algorithm 4.2.** General scheme of hierarchical collision detection.

Given two hierarchical BV data structures for two objects $A$ and $B$, almost all hierarchical collision detection algorithms implement the general scheme shown in Algorithm 4.2. This algorithm is essentially a simultaneous traversal of two hierarchies, which induces a so-called *bounding volume test tree* (BVTT) (see Figure 4.12). Each node in this tree denotes a BV overlap test. Leaves in the BVTT denote an intersection test of the enclosed primitives (polygons); whether or not a BV test is done at the leaves depends on how expensive it is compared to the intersection test of primitives.

The characteristics of different hierarchical collision detection algorithms lie in the type of BV used, the overlap test for a pair of nodes, and the algorithm for construction of the BVHs.

During collision detection, the simultaneous traversal will stop at some nodes in the BVTT. Let us call the set of nodes, of which some children are not visited

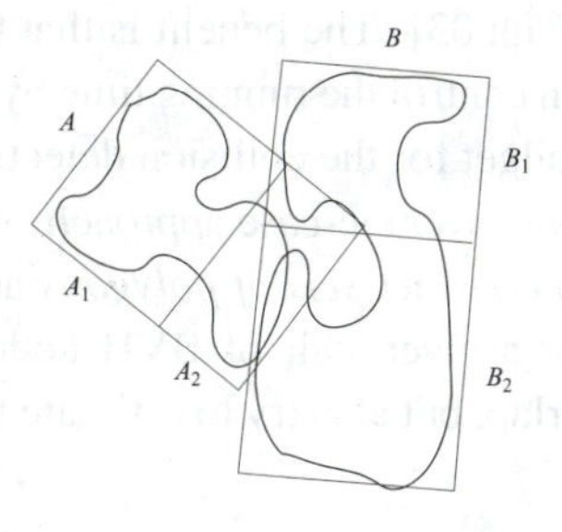


**Figure 4.11.** Hierarchical collision detection can discard many pairs of polygons with one bounding volume check. Here, all pairs of polygons from $A_1$ and $B_2$ can be discarded.

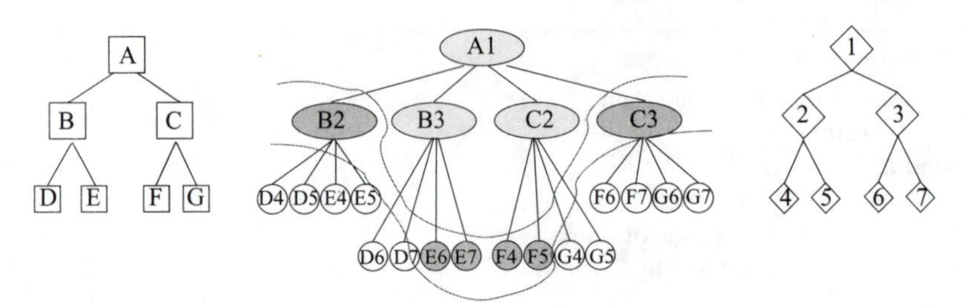


**Figure 4.12.** The bounding volume test tree is induced by the simultaneous traversal of two BVHs. (Courtesy of the editors of the Proceedings of Vision, Modeling, and Visualization 2003.)

(because their bounding volumes do not overlap), the "bottom slice" through the BVTT (see the curved lines in Figure 4.12).

One idea is to save this set for a given pair of objects [Li and Chen 98]. When this pair is to be checked next time, we can start from this set, going either up or down. If the objects have moved only a little relative to each other, the number of nodes that need to be added or removed from the bottom slice should be small. This scheme is called *incremental hierarchical collision detection.*

### 4.3.1. Stochastic Collision Detection

It has often been noted that the *perceived* quality of a virtual environment (and, in fact, most interactive 3D applications) crucially depends on the real-time response to collisions [Uno and Slater 97]. It depends much less on the correctness of the simulation, because humans cannot distinguish between physically correct and physically *plausible* behavior of objects (at least to some degree) [Barzel et al. 96].[5]

Therefore, we can exploit this to develop an "inexact" collision detection method [Klein and Zachmann 03]. The benefit is that the algorithm is time-critical, i. e., the application can control the running time by pre-specifying the desired quality or a certain time budget for the collision detection.

The idea is a fairly simple *average-case approach*: conceptually, the main idea of the new algorithm is to consider *sets of polygons* at inner nodes of the BVH. Then, during traversal, for a given pair of BVH nodes, we not only check the bounding volumes for overlap, but also try to estimate the probability that the two sets of polygons intersect.

[5] Analogously to rendering, a number of human factors determine whether or not the "incorrectness" of a simulation will be noticed. These include the mental load of the viewing person, cluttering of the scene, occlusions, velocity of the objects and the viewpoint, point of attention, etc.

```
traverse(A, B)
while q is not empty do
    A, B ← q.pop
    for all children A_i and B_j do
        p ← P[A_i, B_j]
        if P[A_i, B_j] is large enough then
            return "collision"
        else if P[A_i, B_j] > 0 then
            q.insert(A_i, B_j, P[A_i, B_j])
        end if
    end for
end while
```

**Algorithm 4.3.** Time-critical traversal scheme. The recursive procedure is now replaced by a queue.

Since the exact order in which nodes in the BVTT (see Figure 4.12) are visited is unimportant, we can change the traversal scheme from above and direct the traversal first into those subtrees where a collision is more likely. Thus, the time-critical traversal scheme for two BVHs looks like Algorithm 4.3, where $P[A_i, B_j]$ denotes the probability that there is an intersection between the polygons enclosed by $A_i$ and $B_j$, and $q$ is a priority queue, which is initialized with the root bounding volume pair.

In the following, we will see how to estimate the probability of intersection between two BVs without ever checking any polygons (until we reach the leaves) and without storing any polygons at inner nodes.

Let us first begin with a simple "thought experiment" (*Gedankenexperiment*). Consider a simple cell in space, for instance, a cube (see Figure 4.13, left image). Assume we know that this cell contains a polygon from object $A$ with maximal

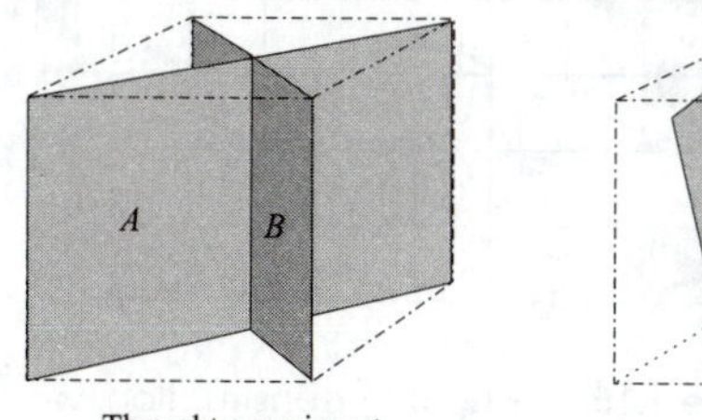


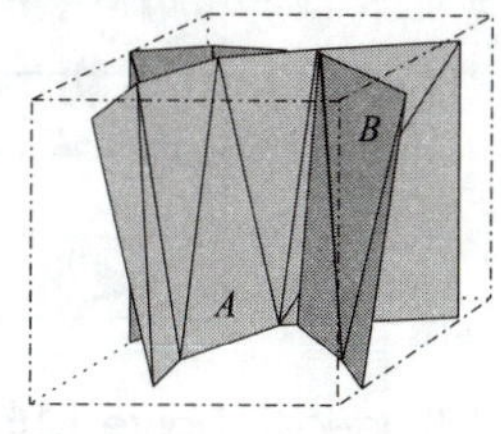


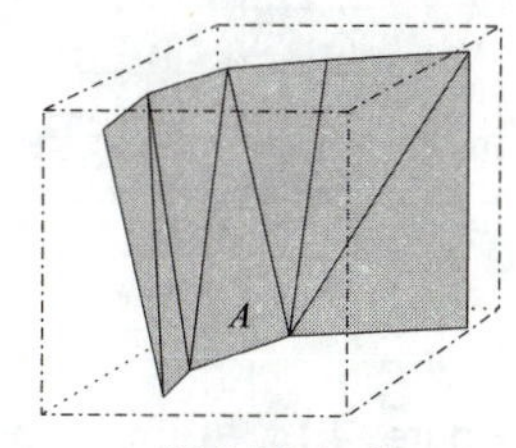


**Figure 4.13.** Left: thought experiment illustrating our approach to estimate the probability of an intersection between the surfaces contained a pair of bounding volumes during traversal. Middle: practical case, defined as *collision cell.* Right: definition of a *well-filled cell.*

surface area such that it just fits completely inside the cell. Assume further that we also have a similar maximal polygon from object $B$ inside the cell. Then we know, without further computation, that there must be an intersection between objects $A$ and $B$.

In practice, of course, we never have such maximal polygons. What does happen, though, is that a part of the *surface* of $A$ is inside the cell that has the same area as the maximal polygon from the thought experiment, and also a similar part of $B$ (see Figure 4.13, middle image). In that case, a collision is at least very likely. We define that case as a *collision cell.*

Conceptually, given a pair of bounding volumes $(A, B)$ (for sake of illustration we use AABBs), we determine the probability of an intersection as follows (see Figure 4.14):

1. Partition $A \cap B$ into a regular grid;
2. Determine the cells that are well-filled by polygons from $A$ (see Figure 4.13, right image);
3. Determine cells well-filled from $B$;
4. Count the number $c(A \cap B)$ of collision cells, i. e., cells that are well-filled by $A$ *and* $B$.

It is well understood that this method would be much too slow. Therefore, we don't actually *count* the number $c$, but rather estimate it as follows.

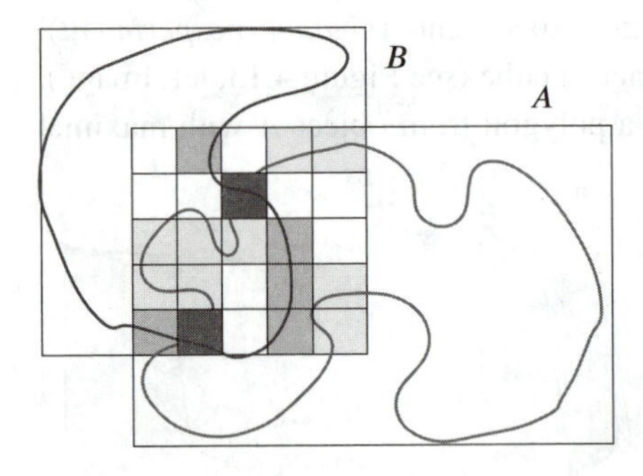


**Figure 4.14.** The (conceptual) idea of the time-critical collision detection method is to count the number of cells that are well-filled by polygons from $A$ *and* from $B$. (Courtesy of Blackwell Publishing.)

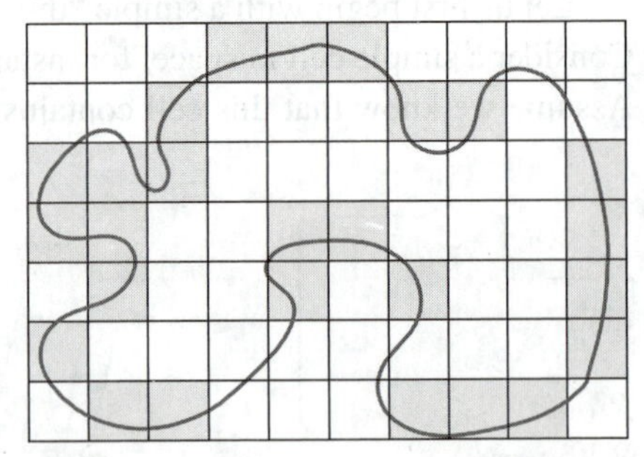

**Figure 4.15.** During BVH construction, we determine the overall number of cells inside bounding volume $A$ that are well-filled.

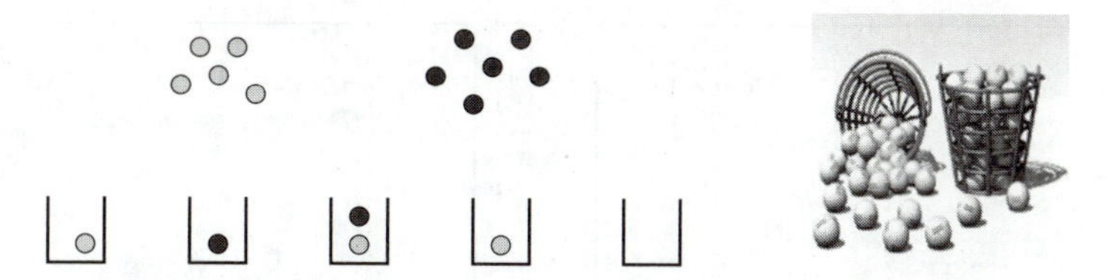

**Figure 4.16.** The balls-into-bins model. (Right image courtesy of Animation Factory.)

During BVH construction (which is preprocessing), we partition each BV by a grid and count the number, $s_A$, of well-filled cells. This number is stored with the node in the BVH. At runtime, we estimate the number of well-filled cells in the volume $A \cap B$ as

$$s'_A = s_A \frac{\mathrm{Vol}(A)}{\mathrm{Vol}(A \cap B)},$$

and, similarly, we estimate $s'_B$.

Now, we reduce the estimation of $P[A, B]$, the probability that there is an intersection, to the estimation of the probability that there are at least $x$ many collision cells, $P[c(A \cap B) \geq x]$, by a combinatorial argument.

This probability can be computed quite easily by the so-called *balls-into-bins model*: given a number of bins, $s$, and a number of light balls, $s_A$, and dark balls, $s_B$, we throw the balls randomly into the bins, with the additional constraint that each bin may contain at most one light and one dark ball (see Figure 4.16).

Using this model, the probability that there are $c \geq x$ collision cells ($x < s_A, s_B$) is

$$P[c(A \cap B) \geq x] = 1 - \sum_{t=0}^{x-1} \frac{\binom{s_A}{t}\binom{s - s_A}{s_B - t}}{\binom{s}{s_B}}. \tag{4.16}$$

Note that we assume that the well-filled cells in each bounding volume are evenly distributed.

*Proof:* Let us assume the $s_A$ light balls have already been thrown into the bins. Without loss of generality, we can assume that the first $s_A$ bins are occupied by one light ball each. The overall number of possibilities to distribute the $s_B$ dark balls into the $s$ bins is $\binom{s}{s_B}$. However, the number of possibilities that exactly $t$ of the $s_B$ dark balls fall into a light bin is $\binom{s_A}{t}\binom{s-s_A}{s_B-t}$, because there are $\binom{s_A}{t}$ many possibilities to pick exactly $t$ light bins and $\binom{s-s_A}{s_B-t}$ many possibilities to pick exactly $s_B - t$ unoccupied bins. Applying the rule "number of favorable

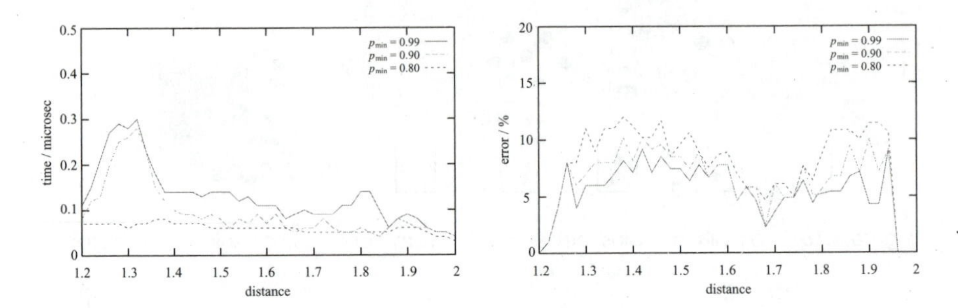


**Figure 4.17.** Running time (left) and error (right) of an example implementation of time-critical collision detection. (Objects: two copies of a car body, 60,000 polygons each.)

divided by number of overall possibilities," it is easy to see that the probability that exactly $t$ bins get a dark and a light ball is

$$\frac{\binom{s_A}{t}\binom{s - s_A}{s_B - t}}{\binom{s}{s_B}}. \qquad \square$$

Obviously, $P$ decreases and the likelihood for an intersection increases as $x$ increases.

The good thing now is that we can precompute a look-up table for $P$. We can partition each BV in the BVH by, say, $8^3$ grid cells, so that $s, s_A, s_B \in [0, 512]$. Exploiting symmetry and the monotonicity of $P$, such a look-up table uses on the order of 10 to 30 MB. For the details, we refer you to [Klein and Zachmann 03].

Note that this approach is a general framework that can be applied to many BVHs with different types of BVs. The BVH must be augmented by a single number: the number of cells in each node that contain "many" polygons.[6]

Figure 4.17 shows an example of an implementation with AABB trees, demonstrating how the running time can be influenced by the probability threshold and how the error changes accordingly.

[6]One could even discard the polygons altogether, if one never wants to perform exact collision detection. In that case, we could just say that there is a collision if the probability computed at a pair of leaves is above a threshold.

# 5

# Distance Fields

*Distance fields* (DFs) can be viewed as a way to represent surfaces (and solids). They are a very powerful data structure, because they contain a huge amount of information (compared to just the surface itself). This is why they usually take fairly long to compute. Consequently, this computation can usually be performed only as a preprocessing step. Thus, they are difficult to adapt for deformable geometry.

**Definition 5.1. (Distance field.)** Let $S$ be a surface in $\mathbb{R}^3$. Then the distance field of a surface $S$ is a scalar function $D_S : \mathbb{R}^3 \rightarrow \mathbb{R}$ such that for all $p \in \mathbb{R}^3$,

$$D_S(p) = \text{sgn}(p) \cdot \min\{d(p,q)|q \in S\}, \tag{5.1}$$

where

$$\text{sgn}(p) = \begin{cases} -1, & \text{if } p \text{ inside}, \\ +1, & \text{if } p \text{ outside}. \end{cases}$$

In other words, $D_S$ tells for any point $p$ the distance to the closest point on the surface $S$.

We can augment the DF further to obtain a *vector distance field* $V_S$ by storing a vector to the closest point $p \in S$ for each point $x$ [Jones and Satherley 01]. So, $D_S(x) = \left| V_S(x) \right|$. This is a vector field. Another vector field that is often used in the context of DFs is the *gradient of the distance field*, or just *gradient field.*

Figure 5.1 shows a DF for a simple polygon in the plane. The distance of a point from the surface is color-coded (the distance for points inside the polygon is not shown). Figure 5.2 shows a vector DF (for the same surface).

Apparently, DFs have been "invented" in many different areas, such as computational physics [Sethian 82, Sethian 99], robotics [Kimmel et al. 98], GIS (see Figure 5.3 for an example), and image processing [Paglieroni 92]. Not surprisingly, DFs come under many other names, such as *distance maps* and *potential*

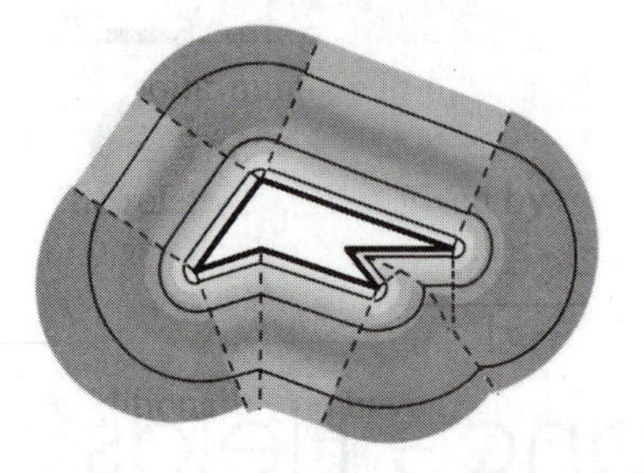

**Figure 5.1.** Example of a distance field in the plane for a simple polygon (thick black lines). The distance field inside the polygon is not shown. The dashed lines show the Voronoi diagram for the same polygon. The thin lines show isosurfaces for various isovalues. (See Color Plate VI.)

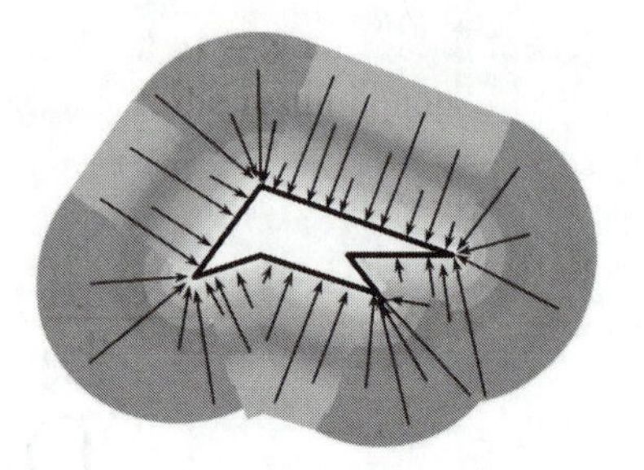

**Figure 5.2.** The vector distance field for the same polygon as shown on the left. Only a few vectors are shown, although (in theory) every point of the field has a vector. (See Color Plate VII.)

*fields*. The process of, or the algorithm for, computing a DF is sometimes called *distance transform*.[1]

DFs have close relationships to isosurfaces and implicit functions. When regarded as an implicit function, the isosurface for the isovalue 0 of a DF is exactly the original surface (see Figure 5.1). However, the converse is not true in general, i.e., the DF of a surface defined by an implicit function is not necessarily identical to the original implicit function.

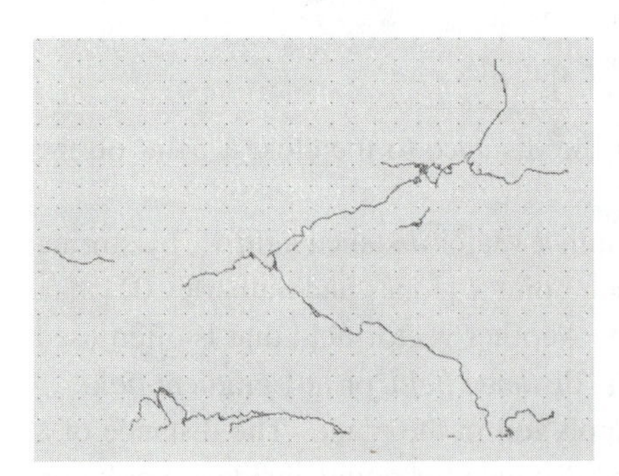

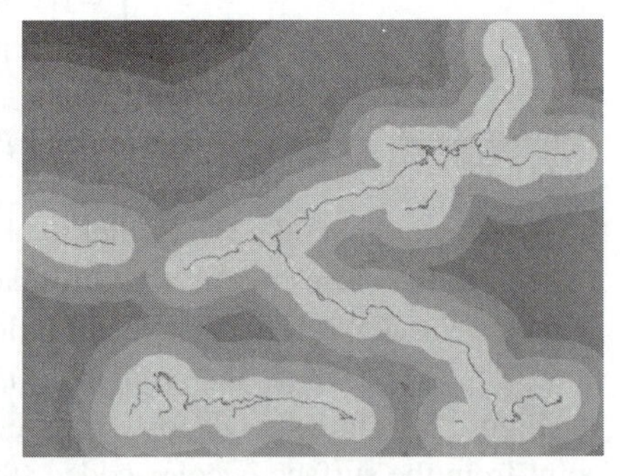

**Figure 5.3.** A network of roads described in the plane by a set of edges (left) and the distance map of these roads. The distance of each point in the plane from a road is color-coded. It could be used, for example, to determine the areas where new houses can be built. (See Color Plate VIII.)

[1]Sometimes, this term bears the connotation of producing *inaccurate* distance fields.

There is another data structure related to DFs, namely Voronoi diagrams (see Chapter 6). Given a vector DF for a set of points, edges, and polygons (not necessarily connected), then all points in space, whose vectors point to the same feature (point, edge, or polygon), are in the same Voronoi cell (see Figures 5.1 and 5.2). (We could also regard the vector of the DF as a kind of ID of the respective Voronoi cell.)

## 5.1. Computation and Representation of DFs

For special surfaces $S$, we may be able to compute $D_S$ analytically. In general, however, we have to discretize $D_S$ spatially, i.e., store the information in a 3D voxel grid, octree, or other space partitioning data structure. Voxel grids and octrees are the most commonly used data structures for storing DFs, and we will describe algorithms for them in the following sections. More sophisticated representations try to store more information at each voxel in order to be able to extract distances quickly from the field [Huang et al. 01].

Since each cell in a discrete DF stores one signed distance to the surface for only one "representative," the distance of other points must be interpolated from these values. One simple method is to store exact distances for the nodes (i.e., the corners of the voxels), and generate all other distances in the interior of voxels by trilinear interpolation. Other interpolation methods could be used as well.

The simplest discrete representation of a DF is the 3D voxel grid. However, for most applications, this is too costly, not only in memory utilization but also in computational efforts. So, usually DFs are represented using octrees (see Figure 5.4), because they are simple to construct, offer fairly good adaptivity, and allow for easy algorithms [Frisken et al. 00, **?**]. This representation is called an *adaptively sampled distance field* (ADF). Actually, the kind of hierarchical space partitioning is a design parameter of the algorithm; for instance, BSPs (see Chapter 3) probably offer much more efficient storage at the price of more complex interpolation.

Constructing ADFs works much like constructing regular octrees for surfaces, except that here, we subdivide a cell if the DF is not well approximated by the interpolation function defined so far by the cell corners.

The following two algorithms for computing a DF produce only flat representations (i.e., grids), but they can be turned into an ADF by coalescing cells in a bottom-up manner if the resulting cell still describes the DF well enough.[2]

[2]Usually, it is faster to compute an ADF top-down, though.

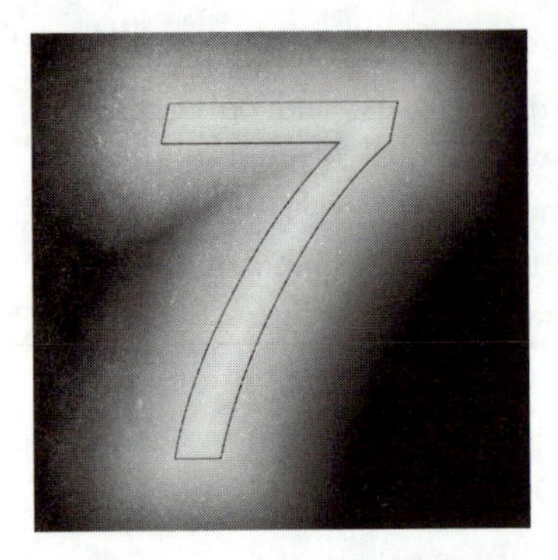

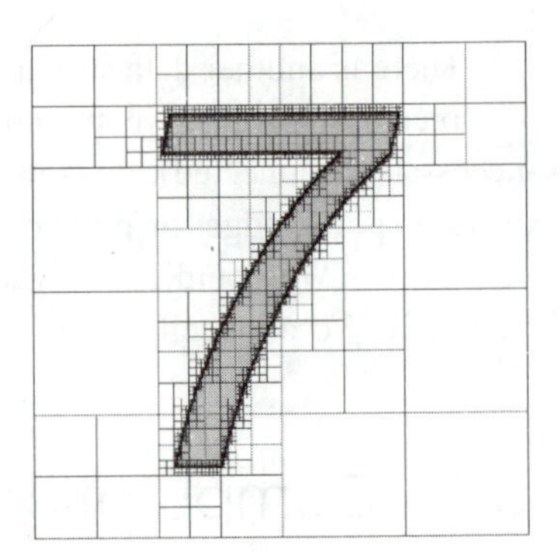

**Figure 5.4.** Representing distance fields (left: 2D field of the number 7, 262,144 distance samples) by octrees (right: quadtree cells are subdivided to their highest resolution along the boundary of the 2D shape, yielding 6681 cells) has both memory and computational advantages. (Courtey of Sarah Frisken, Tufts University, and Ronald Perry, Mitsubishi Electric Research Laboratories.)

### 5.1.1. Propagation Method

The propagation method bears some resemblance to region growing and flood filling algorithms, and is also called the *chamfer method*. It can produce only approximations of the exact DF.

The idea is to start with a binary DF, where all voxels intersecting the surface are assigned a distance of 0, and all others are assigned a distance of $\infty$. Then we somehow "propagate" these known values to voxels in the neighborhood, taking advantage of the fact that we already know the distances between neighboring voxels.[3] Note that these distances depend on only the local neighborhood constellation, no matter where the "propagation front" currently is.

More formally speaking, since we want to compute the DF for only a discrete set of points in space, we can reformulate Equation (5.1) as follows:

$$\tilde{D}_S(x, y, z) = \min_{i,j,k \in \mathbb{Z}^3} \{D(x+i, y+j, z+k) \\ + d_M(i, j, k)\}' \tag{5.2}$$

where $d_M$ is the distance of a node $(i, j, k)$ from the center node $(0, 0, 0)$. Actually, this is already a slight approximation of the exact DF.

We can now further approximate this by not considering the infinite "neighborhood" $(i, j, k) \in \mathbb{Z}^3$, but only a local one $\mathbb{I} \subset \mathbb{Z}$. Since a practical computation of $\tilde{D}_S$ would compute each node in some kind of scan order, we choose $\mathbb{I}$ such that it contains only nodes that have already been computed by the respective

[3]This idea has been proposed as early as 1984 [Borgefors 84], and maybe even earlier.

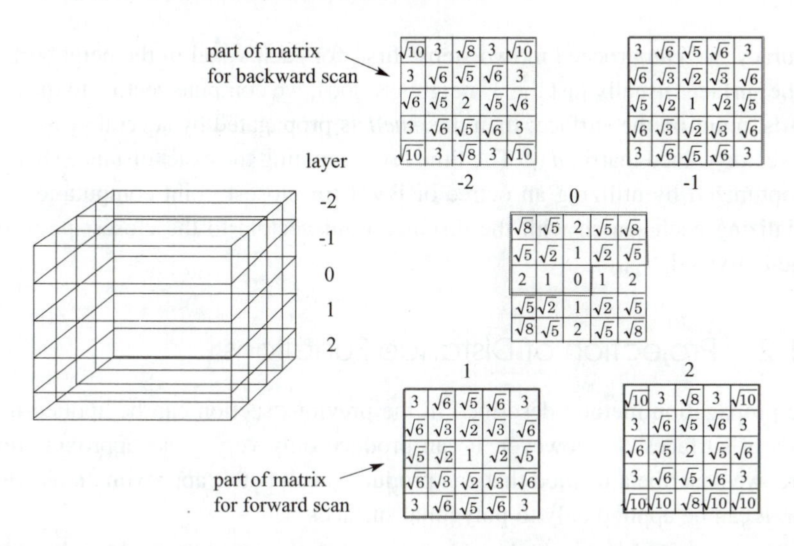


**Figure 5.5.** By convoluting a "distance matrix" with an initially binary distance field (0 or $\infty$), one gets an approximate distance field. Here, a $5 \times 5 \times 5$ example of such a matrix is shown.

scan. Thus, $d_M$ can be precomputed and stored conveniently in a 3D matrix (see Figure 5.5).

This process looks very similar to conventional convolution, except that here we perform a minimum operation over a number of sums; the conventional convolution performs a sum over a number of products.

Obviously, a single scan will not assign meaningful distance values to all nodes, no matter in what order the scan performs. Therefore, we need at least two scans: for instance, first, a 3D scanline order from top to bottom ("forward"), and second, backward from bottom to top. In each pass, only one half of the 3D matrix $d_M$ needs to be considered (see Figure 5.5). Of course, there could still be nodes with distance $\infty$ in the respective part of the chamfer matrix.

The time complexity of this method is in $O(m)$, $m$ = number of voxels. However, the result is only an approximation of the DF; in fact, the accuracy of the distance values can be pretty low.[4]

In order to further improve accuracy, one can make a few more passes, not only from top to bottom and back, but also from left to right, etc. Another option is to increase the size of $d_M$.

Analogously, an approximate vector DF can be computed [Jones and Satherly 01]. In that case, the matrix $d_M$ contains vectors instead of scalars. To improve the

[4]Furthermore, the distance values are not even an upper bound on the real distance.

accuracy, we can proceed in two steps: first, for each voxel in the neighborhood of the surface (usually just the $3^3$ neighborhood), we compute vectors to the *exact* closest point on the surface; then this *shell* is propagated by several passes with the vector-valued matrix $d_M$. The time for computing the exact distance shell can be optimized by utilizing an octree or BVH for closest point computation, and initializing each search with the distance (and vector) to the closest point of a nieghbor voxel.

### 5.1.2. Projection of Distance Functions

The propagation method described in the previous section can be applied to all kinds of surfaces $S$. However, it can produce only very gross approximations. Here, we describe a method that can produce much better approximations. However, it can be applied only to polygonal surfaces $S$.

The key technique applied in this approach is to embed the problem into more dimensions than are inherent in the input/output itself. This is a very general technique that often helps to get a better and simpler view on a complex problem.

The method described here was presented in [Hoff III et al. 99] and [Haeberli 90], and, from a different point of view, in [Sethian 96].

Let us consider the DF of a single point in 2D. If we consider this a surface in 3D, then we get a cone with the apex in that point. In other words, the DF of a point in a plane is just the orthogonal projection of suitable cone onto the plane $z = 0$.

Now, if there are $n$ point sites in the plane, we get $n$ cones, so each point in the plane will be "hit" by $n$ distance values from the projection (see Figure 5.6(left)). Obviously, only the smallest one wins and gets "stored" with that point—and this is exactly what the Z-buffer of graphics hardware was designed for.

These cones are called the *distance function* for a point. The distance function for other features (line segments, polygons, and curves) is a little more complicated (see Figure 5.6(middle)).

So, computing a discrete DF for a set of sites in the plane can be done by rendering all the associated distance functions (represented as polygonal meshes) and reading out the Z-buffer (and possibly the frame buffer if we also want the site IDs, i.e., discretized Voronoi regions). If we want to compute a 3D DF, then we proceed slice by slice. One noteworthy catch, though, is that the distance functions of the sites *change* from slice to slice. For instance, the distance function for a point that lies *not* in the current slice is a hyperboloid, with the point site being coincident with the vertex of the hyperboloid (see Figure 5.6(right)).

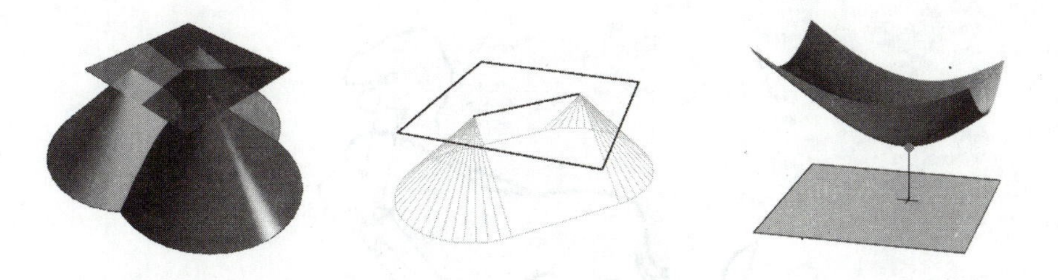

**Figure 5.6.** (left) The distance function of a point site in the plane is a cone. (middle) More complex sites have a bit more complex distance functions. (right) The distance function of sites in 3D is different for the different slices of the volume (Hoff III et al. 99). (See Color Plate IX.) (Courtesy of K. Hoff, T. Culver, J. Keyser, M. Lin, and D. Manocha, Copyright ACM.)

# 5.2. Applications of DFs

Due to their high information density, DFs have a huge number of applications. Among them are motion planning in robotics [Kimmel et al. 98, Latombe 91], collision detection, shape matching [Novotni and Klein 01], morphing [Cohen-Or et al. 98b], volumetric modeling [Frisken et al. 00, Bremer et al. 02, Youngblut et al. 02], navigation in virtual environments [Wan et al. 01], reconstruction [Klein et al. 99], offset surface construction [Payne and Toga 92], and dynamic levels of detail, to name but a few. In the following, we will highlight two easy applications of DFs.

## 5.2.1. Morphing

One interesting application of DFs is morphing, i.e., the problem of finding a "smooth" transition from a shape $S \subset \mathbb{R}^d$ to a shape $T \subset \mathbb{R}^d$, i.e., we are looking for a shape transformation $M(t, \mathbb{R}^d), t \in [0, 1]$, such that $M(0, S) = S$, $M(1, S) = T$.

Sometimes, there is a bit of confusion about terms. Sometimes, morphing refers to any smooth transition. Other times, it refers only to those transitions that are bijective, continuous, and have a continuous inverse. In the latter case, the term morphing is equivalent to the notion of *homeomorphism* in the area of topology, and the term *metamorphosis* is used to refer to the broader definition. The difference between a homeomorphism and a metamorphosis is that the former does not allow you to cut or glue the shape during the transition (this would change the topology).

In the following, we will describe a simple method for metamorphosis of two shapes given as volumetric models; it was presented in [**?**]. The nice thing about

**Figure 5.7.** By specifying correspondences, the user can identify feature points that should transition into each other during the morph (?). (Courtesy of Daniel Cohen-Or.)

a volumetric representation is that it can naturally handle genus changes (i.e., the number of "holes" in $S$ and $T$ is different).

In the simplest form, we can just interpolate linearly between the two DFs of $S$ and $T$, yielding a new DF

$$M(t, S) = J(t, S, T) = tD_S + (1-t)D_T. \tag{5.3}$$

In order to obtain a polygonal representation, we can compute the isosurface for the value 0.

This works quite well, if $S$ and $T$ are "aligned" with each other (in an intuitive sense). However, when morphing two sticks that are perpendicular to each other, there will be in-between models that almost vanish.

Therefore, in addition to the DF interpolation, we usually want to split the morphing into two steps (for each $t$): first, $S$ is *warped* by some function $W(t, x)$ into $S'$, then $D_{S'}$ and $D_T$ are interpolated, i.e., $M(t, S) = tD_{W(t,S)} + (1-t)D_T$.

The idea of the warp is that it encodes the expertise of the user who usually wants a number of prominent feature points from $S$ to transition onto other feature points on $T$. For instance, she wants the corresponding tips of nose, fingers, and feet of two different characters to transition into each other.

Therefore, we must specify a small number of *correspondences* $(p_{0,i}, p_{1,i})$, where $p_{0,i} \in S$, and $p_{1,i} \in T$ (see Figure 5.7). Then we determine a warp function $W(t, x)$ such that $W(1, p_{0,i}) = p_{1,i}$.

Since the warp function should distort $S$ as little as possible (just "align" them), we can assemble it from a rotation, then a translation, and finally an *elastic* part that is not a linear transformation:

$$W(t, \mathbf{x}) = \big((1-t)I + tE\big)\big(R(\mathbf{a}, t\theta)\mathbf{x} + t\mathbf{c}\big), \tag{5.4}$$

where $E$ is the elastic transformation, $R(\mathbf{a}, \theta)$ is a rotation about axis $\mathbf{a}$ and angle $\theta$, and $\mathbf{c}$ is the translation.

The rotation and translation of the warp function can be determined by least-squares fitting, while the elastic transformation can be formulated as a scattered data interpolation problem. One method that has proven quite robust is the approach using radial basis functions. The interested reader will find the details in [?].

### 5.2.2. Modeling

A frequently used modeling paradigm in CAD is *volumetric modeling*, which means that objects are, conceptually, defined by specifying for each point whether or not it is a member of the object. The advantage is that it is fairly simple to define Boolean operations on such a representation, such as union, subtraction, and intersection. Examples of volumetric modeling is *constructive solid geometry* (CSG) and *voxel-based modeling*. The advantage of using DFs for volumetric modeling is that they are much easier to implement than CSG, but provide much better quality than simple voxel-based approaches.

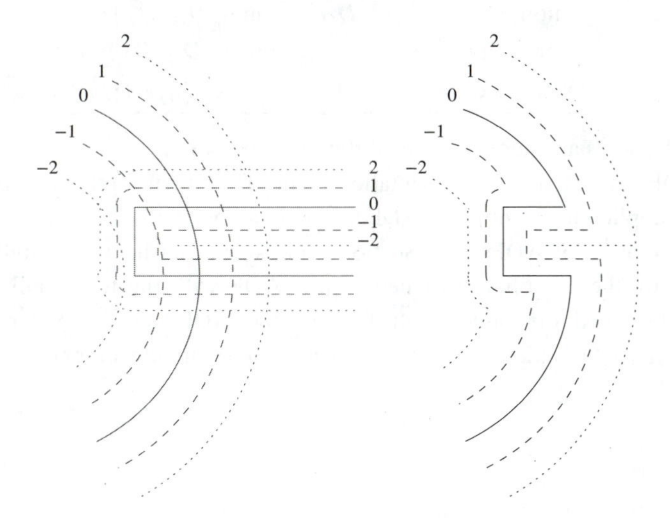


**Figure 5.8.** Example of difference of two DFs. On the left, the two DFs for tool and object are shown superimposed; on the right, the result is shown.

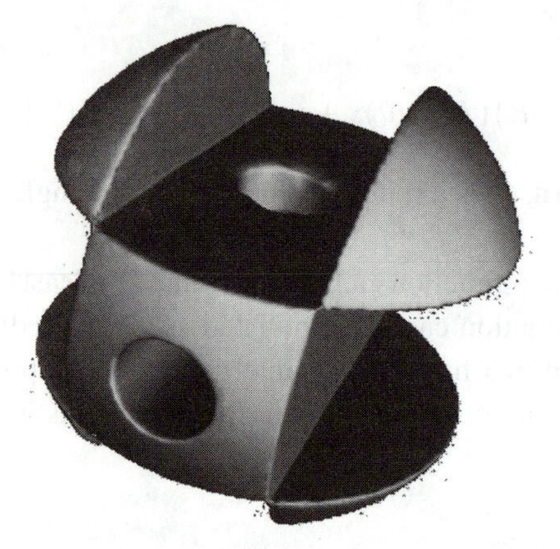

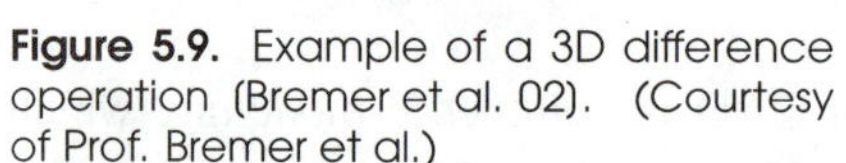

**Figure 5.9.** Example of a 3D difference operation (Bremer et al. 02). (Courtesy of Prof. Bremer et al.)

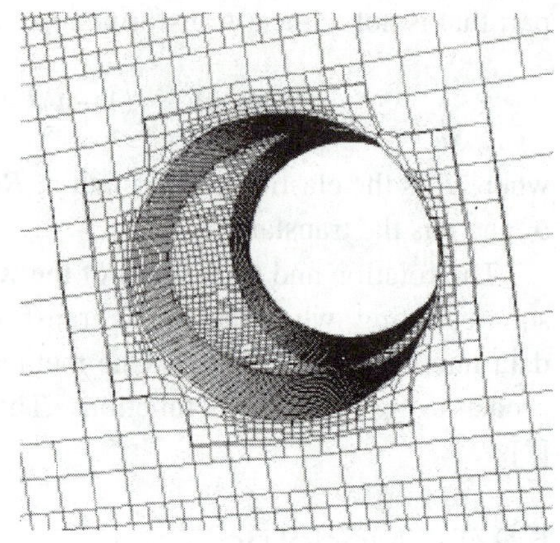

**Figure 5.10.** Close-up view showing the adaptivity (Bremer et al. 02). (Courtesy of Prof. Bremer et al.)

Computing the Boolean operation on two DFs basically amounts to evaluating the appropriate operator on the two fields. The following table gives these operators for a field $D_O$ (for the object) and a field $D_T$ (for the tool):

| Boolean operation | Operator on DF |
|---|---|
| Union | $D_{O \cup T} = \min(D_O, D_T)$ |
| Intersection | $D_{O \cap T} = \max(D_O, D_T)$ |
| Difference | $D_{O-T} = \max(D_O, -D_T)$ |

Figure 5.8 shows an example of the difference operation.

If ADFs are used for representation of the DFs, then a recursive procedure has to be applied to the object's ADF to compute the resulting ADF. It is similar to the generation of ADFs (see Section 5.1) except that here the subdivision is governed by the compliance of new object ADF with the tool ADF: a cell is further subdivided if it contains cells from the tool ADF that are smaller and if the trilinear interpolation inside the cell does not approximate well enough the ADF of the tool.

# 6 Voronoi Diagrams

For a given set of sites inside an area, the *Voronoi diagram* is a partition of the area into regions of the same neighborship. The Voronoi diagram is used for solving numerous problems in many fields of science. In the 2D Euclidean plane the Voronoi diagram is represented by a planar graph that divides the plane into cells.

We will concentrate on applications to geometric problems in 2D and 3D. An overview of the Voronoi diagram and its graph theoretic dual, the Delaunay triangulation or Delaunay tesselation, is presented in the surveys [Aurenhammer 91], [Bernal 92], [Fortune 92b], [Aurenhammer and Klein 00], and [Okabe et al. 92]. Additionally, Chapters 5 and 6 of [Preparata and Shamos 90] and Chapter 13 of [Edelsbrunner 87] could be considered.

We start in Section 6.1 with the simple case of the Voronoi diagram and the Delaunay triangulation of $n$ points in the plane, under the Euclidean distance. Additionally, we will mention some of the elementary structural properties that follow from the definitions.

In Section 6.2, different algorithmic schemes for computing the structures are mentioned. We present a simple *incremental construction* approach which can be easily generalized to 3D; see Section 6.3.1.

We present generalizations of the Voronoi diagram and the Delaunay triangulation in Section 6.3. In Section 6.3.1, transformations to three dimensions are given, and in Section 6.3.2 the concept of *constrained Voronoi diagrams* is introduced. A collection of other interesting generalizations is presented in Section 6.3.3.

In Section 6.4, applications of the Voronoi diagram and the Delaunay triangulation in 2D and 3D are shown. First, in Section 6.4.1 we discuss the famous post office problem and present data structures for the nearest neighbor search based on the Voronoi diagram. Finally, in Section 6.4.2, a collection of applications is shown.

The presented pseudocode algorithms make use of straightforward and self-explanatory operations, for example, for list and array manipulations. They should be clear from the context.

# 6.1. Definitions and Properties

## 6.1.1. Voronoi Diagram in 2D

We start with the discussion of the simple Voronoi diagram in the plane. Let $S$ be a set of $n \geq 3$ points $p_1, p_2, \ldots, p_n$ in the plane with $p_i = (p_{i_x}, p_{i_y})$. In the following, we assume that the points are in *general position*, i. e., no four of them lie on the same circle and no three of them on the same line; see Section 9.5 for handling degenerate situations in geometric computing.

The Voronoi diagram in 2D is represented by a graph. Each face of the graph is dedicated to a unique point $p_i$. Note that the points are sometimes denoted as sites. The face of $p_i$ represents all points in 2D that are closer to $p_i$ than to any other $p_j$. The representation of the graph can be given by a double-connected edge list (DCEL); see Section 9.1. An example of a Voronoi diagram in 2D is shown in Figure 6.1.

Following the lines of [Okabe et al. 92] or [Aurenhammer and Klein 00], we define the Voronoi diagram more formally so that the concept can be easily generalized. For two points $p_i = (p_{i_x}, p_{i_y})$ and $p_j = (p_{j_x}, p_{j_y})$, let $d(p_i, p_j)$ denote their Euclidean distance. By $\overline{B}$, we denote the closure of a set $B$. The closure

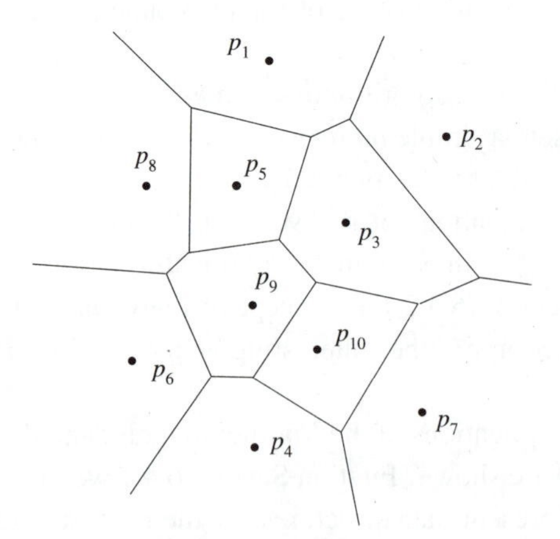


**Figure 6.1.** The Voronoi diagram of a set of sites in the Euclidean plane.

of $B$ also contains the points on the boundary of a given area; for example, the closure of an open interval $(a, b) \in \mathbb{R}$ is given by $[a, b]$.

**Definition 6.1. (Voronoi diagram.)** For $p_i, p_j \in S$ let

$$\text{Bis}(p_i, p_j) = \{x \mid d(p_i, x) = d(p_j, x)\}$$

denote the *bisector* of $p_i$ and $p_j$. $\text{Bis}(p_i, p_j)$ represents the perpendicular line through the midpoint of the line segment $p_i p_j$. The bisector separates the plane into two open half-planes

$$\text{H}(p_i, p_j) = \{x \mid d(p_i, x) < d(p_j, x)\}$$

and

$$\text{H}(p_j, p_i) = \{x \mid d(p_j, x) < d(p_i, x)\},$$

where $\text{H}(p_i, p_j)$ contains $p_i$ and $\text{H}(p_j, p_i)$ contains $p_j$.

The *Voronoi region* of $p_i$ with respect to $S$ is defined by the intersection of $n - 1$ half-planes as follows:

$$\text{VoR}(p_i, S) = \bigcap_{p_j \in S, p_j \neq p_i} H(p_i, p_j).$$

The *Voronoi diagram* $\text{VD}(S)$ of $S$ itself is defined by

$$\text{VD}(S) := \bigcup_{p_i, p_j \in S, p_i \neq p_j} \overline{\text{VoR}(p_i, S)} \cap \overline{\text{VoR}(p_j, S)}.$$

By definition, the Voronoi diagram represents a graph. An illustration of the Voronoi diagram is given in Figure 6.1. It shows how the plane is decomposed by $\text{VD}(S)$ into open *Voronoi regions*. Let us discuss the graph structure more precisely. The boundary shared by two Voronoi regions belongs to the diagram and is called a *Voronoi edge*. For the Voronoi edge $e$ bordering the regions of $p_i$ and $p_j$, we easily conclude $e \subset \text{Bis}(p_i, p_j)$, i.e., every Voronoi edge is part of a corresponding bisector. The endpoints of a Voronoi edge are called *Voronoi vertices*. A Voronoi vertex must belong to the common boundary of three Voronoi regions.

For every point $p$ in the plane, we can uniquely determine its role within a Voronoi diagram of a set of sites $S$. We expand the circle $\text{Circle}(p, r)$ with center $p$ and radius $r$ by increasing $r$ continuously. One of the following events will uniquely determine the role of point $p$ in the Voronoi diagram:

- If $\text{Circle}(w, r)$ first hits exactly one of the $n$ sites, say $p_i$, then $w \in \text{VoR}(p_i, S)$;

- If Circle$(w, r)$ first hits exactly two sites $p_i$ and $p_j$ simultaneously, $w$ belongs to the Voronoi edge of $p_i$ and $p_j$;

- If Circle$(w, r)$ first hits exactly three sites $p_i$, $p_j$, and $p_k$ simultaneously, $w$ is the Voronoi vertex of $p_i$, $p_j$, and $p_k$.

The *expanding circle* characterization can be easily proven by the definition of the Voronoi diagram. Furthermore, we will enumerate the following three elementary properties of a Voronoi diagram:

- Each Voronoi region VoR$(p_i, S)$ is the intersection of at most $n - 1$ open half-planes containing the site $p$. Every VoR$(p_i, S)$ is open and convex. Different Voronoi regions are disjoint;

- A point $p_i$ of $S$ lies on the convex hull of $S$ iff its Voronoi region VoR$(p_i, S)$ is unbounded;

- The Voronoi diagram VD$(S)$ has $O(n)$ many edges and vertices. The average number of edges in the boundary of a Voronoi region is less than 6.

Apart from the last fact, which needs an application of the Eulerian formula for planar graphs, see [Gibbons 85], the structural properties can be easily deduced from the definition of the Voronoi diagram and the convex hull. The Voronoi diagram is a simple linear structure and provides for a partition of the plane into cells of the same neighborship. We omit the proofs and refer to the surveys mentioned in the beginning; for example, see [Aurenhammer and Klein 00].

### 6.1.2. Delaunay Triangulation in 2D

We consider the dual graph of the Voronoi diagram, the so-called *Delaunay triangulation.* The *dual graph* dual$(G)$ of a geometric graph $G$ is defined as follows. Every face $F$ of $G$ represents a vertex in dual$(G)$. Two faces that share a common edge in $G$ build an edge in dual$(G)$. If $G'$ is the dual graph of $G$, $G$ is called the *primal graph* of $G'$, primal$(G') = G$ for short. Figure 6.2 shows an example of a graph and its dual. If the graph $G$ is represented by a DCEL (see Section 9.1.1 and Section 9.1.2), one can easily build a DCEL for the graph dual$(G)$ in time $O(|G|)$ using the primal DCEL. Note that the dual of the Voronoi diagram is given logically by a set of vertices and edges. In the following, we will consider a *geometric interpretation* that uses the sites as vertices.

In general, a *triangulation* of a set $S$ of points in the plane can be defined to be a planar graph with vertices from $S$ and with a maximal number of edges so that all closed faces are triangles. The triangulation is convex if the complement of the outer face is convex. The edges adjacent to the outer face of the graph represent the convex hull of the point set. Since a triangulation $T$ is a graph, it can be represented by a DCEL. The triangulation of a point set $S$ has no more than $O(|S|)$ triangles, which can be easily shown by induction, starting with a triangle. It is easy to see that a point set might have many triangulations. By simple induction starting with a single triangle, one can show that for a fixed point set $S$, every triangulation of $S$ has the same number of triangles and edges.

**Definition 6.2. (Delaunay triangulation.)** For a set of sites $S$ and the Voronoi diagram $\mathrm{VD}(S)$, let every face be represented by its site. The *Delaunay triangulation* $\mathrm{DT}(S)$ is the dual graph of the Voronoi diagram. The edges of $\mathrm{DT}(S)$ are called *Delaunay edges*.

We will show that the Delaunay triangulation $\mathrm{DT}(S)$ is in fact a triangulation if the set of sites (vertices) is in general position. An example is shown in Figure 6.2. We present two equivalent definitions of the Delaunay triangulation. They can be applied for the computation of the diagram and give rise also to generalizations. They are applicable for non-general situations, also.

1. Two points $p_i, p_j$ of $S$ build a Delaunay edge iff a circle $C$ exists that passes through $p_i$ and $p_j$ and does not contain any other site of $S$ in its interior or boundary.

2. Three points $p_i$, $p_j$, and $p_k$ of $S$ build a Delaunay triangle iff their circumcircle does not contain a point of $S$ in its interior or boundary.

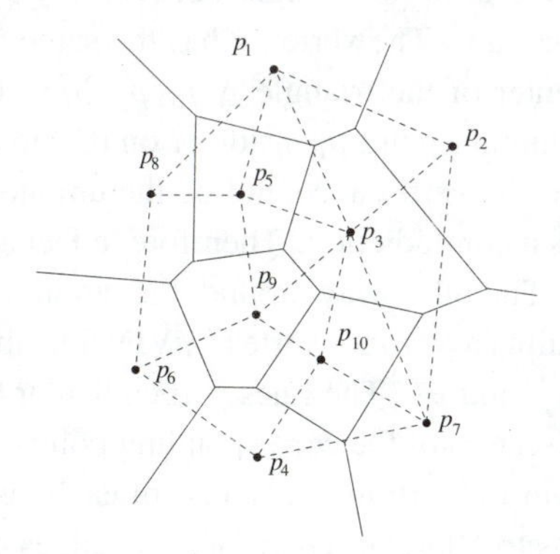


**Figure 6.2.** The Voronoi diagram and the Delaunay triangulation of a set of sites in the Euclidean plane.

We will later concentrate on the last characterization of a Delaunay triangulation. It can be easily generalized and is highly recommended for construction schemes since we can make use of a simple incircle test and apply the paradigm of exact geometric computation; see also Chapter 9. We have to prove the following statements.

**Lemma 6.3.** *Let S be a set of sites. Three points $p_i$, $p_j$, and $p_k$ of S build a triangle in* DT(*S*) *iff the circle passing through the endpoints of the triangle does not contain another point of S in its interior or boundary.*

*Two points $p_i, p_j$ of S build a Delaunay edge in* DT($S$) *iff a circle C exists that passes through $p_i$ and $p_j$ and does not contain any other site of S in its interior or boundary.*

*A Delaunay triangulation is a convex triangulation iff no four points lie on a common circle; see Figure 6.2.*

*Proof:* The second part of the lemma can be proven as follows. If such a circle exists, the circumcenter $c$ belongs to the boundary of the regions of $p_i$ and $p_j$, and there is no other site in $S$ that is closer to $c$. Therefore, a part of the bisector (including $c$) belongs to VD($S$). Conversely, if $p_i$ and $p_j$ build a common edge that is part of the bisector of $p_i$ and $p_j$, there is a point $c$ on the bisector that is closer to $p_i$ and $p_j$ than to any other site. Then $c$ is the circumcenter of the corresponding circle.

The first part of the lemma is proven as follows. If three Delaunay edges build a triangle for the points $p_i$, $p_j$, and $p_k$, the corresponding Voronoi edges share a common vertex $v$ and exactly three Voronoi edges emanate from the vertex $v$; otherwise, there is not a triangle between $p_i$, $p_j$, and $p_k$ in the dual graph of the Voronoi diagram. The vertex $v$ has the same distance to $p_i$, $p_j$, and $p_k$ and is the circumcenter of the triangle $\Delta(p_i, p_j, p_k)$. If the circumcenter of $\Delta(p_i, p_j, p_k)$ contains another point $p_l$ inside or on the boundary, then $v$ will belong to VoR($p_l, S$), and the vertex $v$ cannot be the unique Voronoi vertex of $p_i$, $p_j$, and $p_k$, which gives a contradiction. Therefore, a Delaunay triangulation fulfills the given property. The other way around, the circumcenter $v$ of the triangle $\Delta(p_i, p_j, p_k)$ exactly builds a Voronoi vertex in VD($S$), since no other point of $S$ is closer to $v$ than $p_i$, $p_j$, and $p_k$. The lines perpendicular to the edges of the triangle represent three bisectors of the corresponding points, and there are no other bisectors emanating from $v$. Starting at $v$, a part of each bisector has to belong to VD($S$), since $v$ belongs to VD($S$). Therefore, the edges of $\Delta(p_i, p_j, p_k)$ correspond to primal edges in VD($S$). Altogether, the triangle is a triangle in DT($S$), the dual of VD($S$).

Now, the last part of the lemma follows immediately. Every vertex of the Voronoi diagram represents exactly one triangle in the dual graph, which consists of triangles only. □

The Delaunay triangulation has some interesting properties that are helpful in computer graphics applications.

For example, triangulations are often used for surface reconstructions. The chosen triangulation should not have small angles, which turn out to be more intractable. It can be shown that among all triangulations of $S$, the Delaunay triangulation has the best angle sequence. More precisely, for a triangulation $T(S)$, we can insert all internal angles of the triangles in a vector $T(S)_a$ by increasing order. It can be shown that $T(S)_a < \mathrm{DT}(S)_a$ holds for all triangulations $T(S) \neq \mathrm{DT}(S)$. The minimal internal angle in $\mathrm{DT}(S)$ is bigger than the minimal internal angle of every other triangulation. In other words, the Delaunay triangulation maximizes the minimal internal angle.

On the other hand, the Delaunay triangulation has some good graph properties. The Delaunay triangulation results in a graph of small graph-theoretical dilation; that is, for two vertices $p_i$ and $p_j$ in $S$, the ratio of the shortest graph distance over the shortest Euclidean distance of $p_i$ and $p_j$ is bounded by

$$\frac{2\pi}{3\cos\left(\frac{\pi}{6}\right)},$$

which was shown in [Keil and Gutwin 89]. Therefore, the Delaunay triangulation results in a network of known quality. Note that the graph or triangulation with the lowest dilation might be different from $\mathrm{DT}(S)$, which can be easily shown.

It can be shown also that the graph $\mathrm{DT}(S)$ contains the *minimum spanning tree* of the given point set $S$. The minimum spanning tree is the shortest tree with respect to edge length that connects all sites $p_i$ (see Section 7.1.3).

## 6.2. Computation

The construction of the Voronoi diagram has time complexity $\Theta(n \log n)$. The lower bound $\Omega(n \log n)$ can be achieved by one of the following reductions:

- a reduction to the convex hull problem is given in [Shamos 78];
- a reduction to the $\epsilon$-closeness problem is given in [Djidjev and Lingas 91] and [Zhu and Mirzaian 91].

Let us briefly discuss the simple convex hull reduction. For a sequence of values $x_1, x_2, \ldots, x_n$ in $\mathbb{R}$, we can choose $n$ points $p_i = (x_i, x_i^2)$ on the parabola $y = x^2$ in 2D. In the Voronoi diagram of $p_1, p_2, \ldots, p_n$, the order of the unbounded regions will represent the order of the values $x_i$. We can find the $x$ order of the points $p_1, p_2, \ldots, p_n$ by traversing the DCEL of $\text{VD}(\{p_1, p_2, \ldots, p_n\})$ adequately. The DCEL has linear size in $n$. If we can compute the VD($S$) faster than in $\Omega(n \log n)$, we can sort a sequence of values faster than in $\Omega(n \log n)$, which gives a contradiction.

The well-known computation paradigms *incremental construction*, *divide-and-conquer*, and *sweep* are convenient for the construction of the Voronoi diagram. They can also be generalized to other metrics and sites other than points, such as line segments and polygonal chains. The output of the algorithms is stored in a graph of linear size as mentioned earlier. The given approaches run in deterministic time $O(n \log n)$. The incremental construction and divide-and-conquer approaches for the Voronoi diagram are explained in great detail in [Okabe et al. 92]. A detailed description of the sweepline algorithm can be found in [de Berg et al. 00]. See also [Aurenhammer and Klein 00].

We will concentrate on a simple randomized *incremental construction* technique, which runs in $O(n \log n)$ expected time and computes the Delaunay triangulation; see [Guibas et al. 92b]. We make use of an edge-flipping technique introduced in [Lawson 77] and used in [Guibas and Stolfi 85]. Fortunately, the incremental construction technique for the dual of the Voronoi Diagram can easily be generalized to the 3D case, as we will see in Section 6.3.1. But there are some differences. An arbitrary triangulation in 2D can be transformed into a Delaunay triangulation by a sequence of edge flips. It was shown in [Joe 91a] that this is not true in 3D. Fortunately, there are flipping sequences that are successful; see [Rajan 91].

## 6.2.1. Simple Randomized Incremental Construction for Delaunay Triangulations

Let us assume that the Delaunay triangulation for $\{p_1, \ldots, p_{i-1}, p_i\}$ was already computed. Let $\text{DT}_i := \text{DT}(\{p_1, \ldots, p_{i-1}, p_i\})$. We have to insert the site $p_i$ into $\text{DT}_{i-1}$ in order to obtain $\text{DT}_i$.

We will perform *edge flips* in $\text{DT}_{i-1}$ until $\text{DT}_i$ is finally constructed. A triangle $T$ of $\text{DT}_{i-1}$ whose circumcircle contains the new site $p_i$ is *in conflict* with $p_i$. By Lemma 6.3, every triangle of $\text{DT}_{i-1}$, which is in conflict with $p_i$, will no longer be a *Delaunay triangle* in $\text{DT}_i$. We consider two situations for the location of $p_i$ with respect to $\text{DT}_{i-1}$.

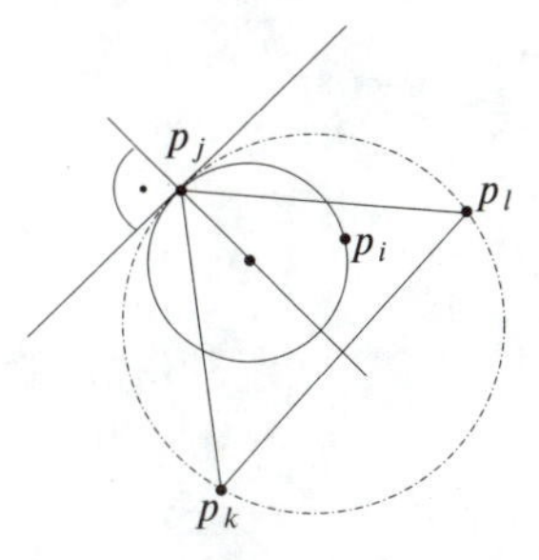


**Figure 6.3.** Adequately shrinking the circumcircle Circle$(p_j, p_k, p_l)$ and holding the contact with $p_j$ shows that there is a circle inside Circle$(p_j, p_k, p_l)$ that has only $p_j$ and $p_i$ on its boundary.

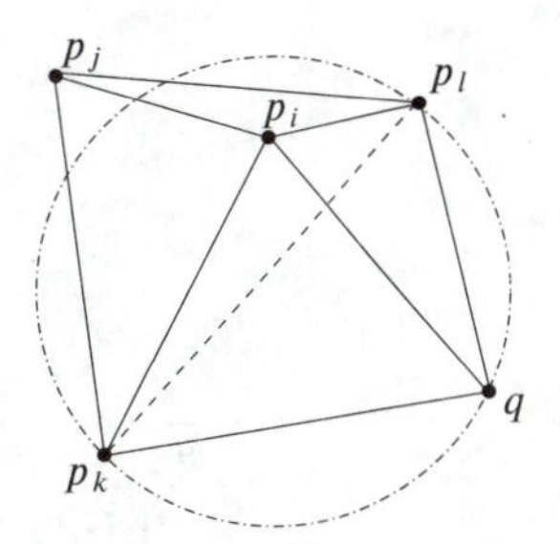


**Figure 6.4.** An edge flip between $p_i q$ and $p_l p_k$.

Case 1: $p_i$ lies inside a triangle $T = \Delta(p_j, p_k, p_l)$. This means that $p_i$ is in conflict with $T$. Surprisingly, the edges $p_i p_j$, $p_i p_k$, and $p_i p_l$ will be Delaunay edges in $\text{DT}_i$. This can be proven as follows. The edge $p_i p_j$ will be a Delaunay edge, because there exists a circle contained in Circle$(p_j, p_k, p_l)$ that contains only $p_i$ and $p_j$ on its boundary. For the construction of such a circle, we shrink the circumcircle Circle$(p_j, p_k, p_l)$ holding the contact with $p_j$ adequately. Finally, there is a circle inside Circle$(p_j, p_k, p_l)$ that has only $p_j$ and $p_i$ on its boundary; see Figure 6.3. The same argument holds for the pairs $p_i$ and $p_k$ and $p_i$ and $p_l$. Therefore, we can insert the Delaunay edges $p_i p_j$, $p_i p_k$, and $p_i p_l$, first.

Next, we concentrate on the triangles adjacent to the edges of $\Delta(p_j, p_k, p_l)$ and opposite $p_i$; for example, let $p_l p_k$ be an edge of the triangle $T = \Delta(q, p_k, p_l)$; see Figure 6.4. If $p_i$ is in conflict with $T$, we *flip* $p_l p_k$ by $p_i q$. That is, $p_l p_k$ is replaced by $p_i q$. We can prove that this edge flip is correct. Since $q$ is opposite $p_i$ with respect to $p_l p_k$ and $p_i$ is in conflict with $T$, $q$ is also in conflict with $\Delta(p_i, p_l, p_k)$. This is true because every circle passing through $p_k$ and $p_l$ either contains $p_i$ or $q$ (or both); see Figure 6.5. Therefore, $p_l p_k$ will never be a Delaunay edge again.

To see that the edge $p_i q$ is a Delaunay edge, we apply the argument of the shrinking circle. We can simply shrink Circle$(q, p_k, p_l)$ holding the contact with $q$ only. Circle$(q, p_k, p_l)$ has only $p_i$ inside because $\Delta(q, p_k, p_l)$ was a Delaunay triangle before $p_i$ was inserted. Finally, there will be a circle passing through $p_i$ and $q$, which has no other point of $S$ inside or on the boundary.

Altogether, we proceed as follows. We successively extend the *star of* $p_i$, $\star(p_i)$, the set of new Delaunay edges emanating from $p_i$. As an invariant, the

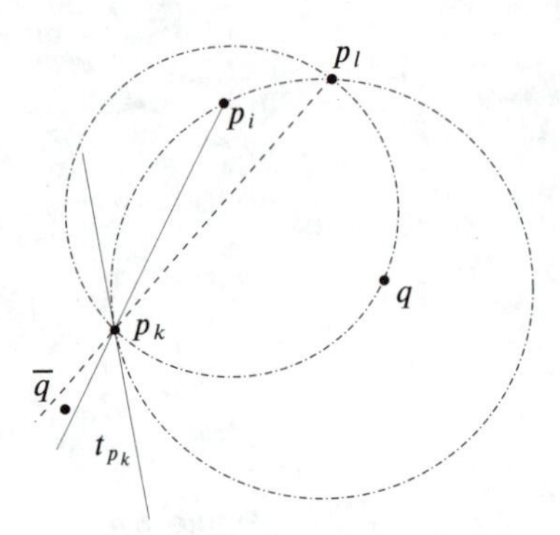


**Figure 6.5.** Every circle passing through $p_k$ and $p_l$ will contain either $p_i$ or $q$. If $\overline{q}, p_i, p_l$, and $p_k$ are in non-convex position, then $\text{Circle}(p_i, p_l, p_k)$ cannot contain $\overline{q}$ because the tangent $t_{p_k}$ *blocks* $\overline{q}$.

outer points of $\star(p_i)$ build a polygon $P(\star(p_i))$; see Figure 6.6. In the beginning, $P(\star(p_i))$ equals $\Delta(p_j, p_k, p_l)$. For every edge $e$ of $P(\star(p_i))$, we test the triangles $T$ of $\text{DT}_{i-1}$ with $e \in T$ and opposite $p_i$. Opposite means that there is a point $q \neq p_i$ such that $q$ and $e$ build $T$. If $q$ and $p_i$ and the endpoints of $e$ are in convex position, $T$ is called *flippable*. If $T$ is not flippable, $T$ cannot be in conflict with $p_i$. This is shown as follows. If $q$ and $p_i$ and the endpoints $p_l$ and $p_k$ of $e$ are in non-convex position, the circle $\text{Circle}(p_i, p_l, p_k)$ cannot contain $q$. Either the chain $p_i, p_l, q$ or the chain $p_i, p_k, q$ is concave. Then $q$ lies beyond the prolongation of a secant $p_ip_k$ or $p_ip_l$ at $\text{Circle}(p_i, p_k, p_l)$. A tangent of $\text{Circle}(p_i, p_l, p_k)$ at $p_l$ or $p_k$ separates $q$ from $\text{Circle}(p_i, p_l, p_k)$; see Figure 6.5 for the point $\overline{q}$ instead of $q$.

If $T$ is flippable, we test whether $T$ is in conflict with $p_i$. If $T$ is in conflict with $p_i$, we perform an edge flip of $e$ with the edge $p_iq$. $\star(p_i)$ and $P(\star(p_i))$ are updated adequately. The process stops if the adjacent outer triangles of $P(\star(p_i))$ are no longer in conflict with $p_i$; see Figure 6.6.

Case 2: $p_i$ lies ouside the convex hull of $\text{DT}_{i-1}$. In this case, all points $q$ on the boundary of $\text{DT}_{i-1}$ *visible* from $p_i$ build an edge $pq$ in $\text{DT}_i$; see Figure 6.7. This can be easily seen as follows. The part of the boundary of $\text{DT}_{i-1}$ visible from $p_i$ builds a convex chain. Visible means that the segment $p_iq$ is not crossed by segments from $\text{DT}_{i-1}$. For every $q$, there is a tangent $t_q$ such that $\text{DT}_{i-1}$ lies fully on one side and $p_i$ lies on the other side. First, we make use of a prolongation of one of the adjacent outer edges of $\text{DT}_{i-1}$ at $q$ and obtain $t_q$. Then we can expand a circle at $q$ that has tangent $t_q$ and meets $p_i$; see Figure 6.7. With Lemma 6.3, we conclude that $qp_i$ is a Delaunay edge. If $\star(p_i)$ is computed for all boundary

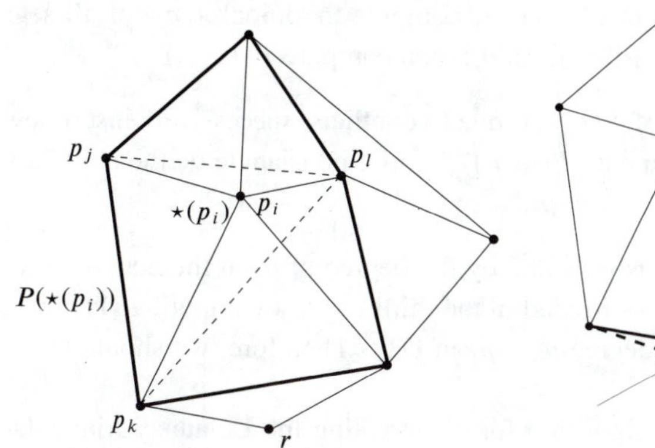


**Figure 6.6.** Successively check the outer triangles of $P(\star(p_i))$ for conflicts with $p_i$. The point $r$ is the last point that will evoke an edge flip.

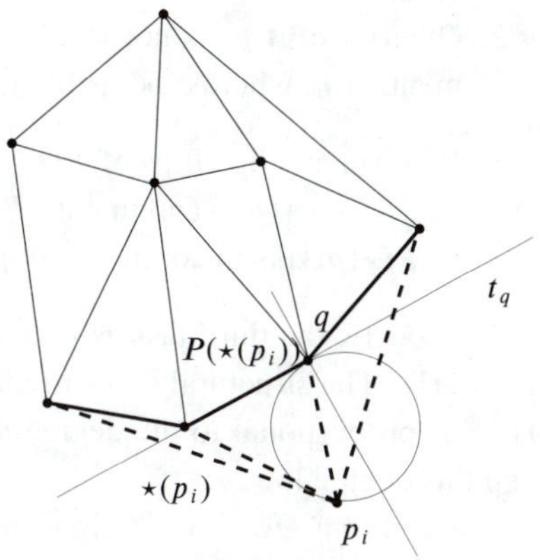

**Figure 6.7.** If $p_i$ is outside the convex hull of $\mathrm{DT}_{i-1}$, the starting $\star(p_i)$ is given by the visible points on the hull. $P\star(p_i))$ is a convex chain opposite $p_i$.

sites $q$, we proceed as shown in Case 1, doing some edge flips until the procedure finally stops.

For both cases, we have the following result.

**Lemma 6.4.** *If all segments of $P(\star(p_i))$ are Delaunay edges, the construction of $\mathrm{DT}_i$ is completed.*

*Proof:* The circumcircles of all triangles around $P(\star(p_i))$ do not contain $p_i$, and all segments of $P(\star(p_i))$ are Delaunay edges. Let $T = \Delta(p,q,r)$ be a triangle outside $P(\star(p_i))$ that is in conflict with $p_i$; i.e., the circumcircle Circle$(p,q,r)$ contains $p_i$. The triangle $T$ was a Delaunay triangle of $\mathrm{DT}_{i-1}$; Circle$(p,q,r)$ contains only $p_i$. By the same shrinking circle arguments as in Case 1, we know that $p_ip$, $p_iq$, and $p_ir$ must be Delaunay edges of $\mathrm{DT}_i$. Since $T$ is outside $P(\star(p_i))$ and all edges of $P(\star(p_i))$ are Delaunay edges, the edges $p_ip$, $p_iq$, and $p_ir$ will cross $P(\star(p_i))$, which gives a contradiction to the definition of a Delaunay triangulation. $T$ cannot be in conflict with $p_i$. □

The edge flips can be carried out efficiently if $\star(p_i)$ is computed efficiently. Altogether, three tasks must be considered.

The following is the edge flip algorithm sketch.

1. Find the triangle $T = T(p_j, p_k, p_l)$ of $\mathrm{DT}_{i-1}$ with $p_i \in T$ and compute the initial $\star(p_i)$ by $p_ip_j$, $p_ip_k$, and $p_ip_l$; then compute $P(\star(p_i))$.

2. Otherwise, if $p_i$ is not inside $\mathrm{DT}_{i-1}$, compute the initial $\star(p_i)$ of all segments $p_i q$, where $q$ is visible from $p_i$; then compute $P(\star(p_i))$.

3. Perform all edge flips of $P(\star(p_i))$ due to conflicts, successively insert new edges into $\star(p_i)$, and update $P(\star(p_i))$. Stop if no triangle on the boundary of $P(\star(p_i))$ is in conflict with $p_i$.

Obviously, the third task is bounded by the degree of $p_i$ in the new triangulation $\mathrm{DT}_i$. The structural work needed in the third task for computing $\mathrm{DT}_i$ from $\mathrm{DT}_{i-1}$ is proportional to the degree $d$ of $p_i$ in $\mathrm{DT}_i$. Therefore, we should try to keep this degree low.

We yield an $O(n^2)$ time algorithm for constructing the Delaunay triangulation of $n$ points. We can determine the triangle of $\mathrm{DT}_{i-1}$ containing $p_i$ within linear time by inspecting all existing candidates. If $p_i$ is not inside $\mathrm{DT}_{i-1}$, we can compute the starting $\star(p_i)$ in $O(k + \log n)$, computing the outer tangents first and moving along $k$ edges for $k$ segments of the initial $\star(p_i)$. The degree of $p_i$ is trivially bounded by $O(n)$ because any triangulation has no more than $O(n)$ edges.

Algorithm 6.1 successively inserts new sites, and Algorithm 6.2 handles the edge flips.

Altogether, we get the following result.

**Theorem 6.5.** *The Delaunay triangulation of a set of n points in the plane can be constructed incrementally in time $O(n^2)$, using linear space.*

We can even do better. The main thing is that we insert the points randomly, thus avoiding worst-case scenarios for the degree of $p_i$ in $\mathrm{DT}_i$. There can be single vertices in $\mathrm{DT}_i$ that do have a high degree, but their *average* degree is bounded by 6. It was shown in [Guibas et al. 92b] that the expected overall number of triangles that appear during a randomized incremental insertion is bounded by $O(n)$.

---

Delaunay($S$) ($S$ represents a set of sites in 2D)

---

```
T := new DCEL
while S ≠ ∅ do
        p := S.First
        S.DeleteFirst
        T.InsertSite(p)
end while
```

---

**Algorithm 6.1.** The incremental construction of the Voronoi diagram.

$T$. InsertSite($p$) ($T$ represents the DCEL of the current Delaunay triangulation; $p$ is a new site)

```
t := T. FindTriangle(p)
if t = Nil then
        ⋆(p) := T. OuterTriangle(p)
else
        ⋆(p) := T. Edges(t, p)
end if
T. DCELInsert(⋆(p), T)
P(⋆(p)) := T. Polygon(⋆(p))
while P(⋆(p)) ≠ do
        e := P(⋆(p)). First
        P(⋆(p)). DeleteFirst
        q := p. Flippable(e)
        if q ≠ Nil then
                (r, s) := e. EndPoints
                if InCircleTest(r, s, p, q) then
                        T. Flip(e, p, q)
                        P(⋆(p)). Extend(q)
                end if
        end if
end while
```

**Algorithm 6.2.** Inserting a new site in a Delaunay triangulation.

Additionally, with a special implementation using a *directed acyclic graph* (DAG), denoted also as *Delaunay tree* due to [Boissonnat and Teillaud 93], we can detect the triangles of $\mathrm{DT}_{i-1}$ that are in conflict with $p_i$ in $O(\log i)$ expected time. Due to the expected number of overall triangles, the DAG has the expected size $O(i)$ after $i$ insertions.

We briefly explain the DAG idea. The DAG contains one entry for every Delaunay triangle ever constructed. For convenience, we start with an outer triangle $\Delta(p_{-1}, p_{-2}, p_{-3})$ that contains all points in $S$. The root of the DAG represents this triangle. All points of $S$ are in conflict with the root. We exemplify the construction of the DAG. If a point $p$ is inserted, we traverse the DAG and will find the triangle $T$ with $p \in T$.

The following is the DAG construction sketch.

1. The first point $p_{i_1}$ is inserted, and $p_{i_1}$ is in conflict with the root. Three new triangles $\Delta_1$, $\Delta_2$, and $\Delta_3$ are inserted in the Delaunay triangulation and in the DAG; see Figure 6.8.

2. The point $p_{i_2}$ is inserted. Starting from the root, we check that $p_{i_2} \in \Delta_1$. Now, we start the flipping process recursively and begin with the initial $\star(p_{i_1})$. This means that three new triangles $\Delta_4$, $\Delta_5$, and $\Delta_6$ are created. They become children of $\Delta_1$. Since for $P(\star(p_{i_1}))$ there are no more conflicts, we are finished; see Figure 6.8.

3. The point $p_{i_3}$ is inserted. Starting from the root, we check that $p_{i_2} \in \Delta_2$. Again, the flipping process starts with the initial $\star(p_{i_2})$, and three additional triangles $\Delta_7$, $\Delta_8$, and $\Delta_9$ become children of $\Delta_2$. There is a conflict for $P(\star(p_{i_2}))$, which changes $\Delta_6$ and $\Delta_9$ to $\Delta_{10}$ and $\Delta_{11}$. Therefore, we have

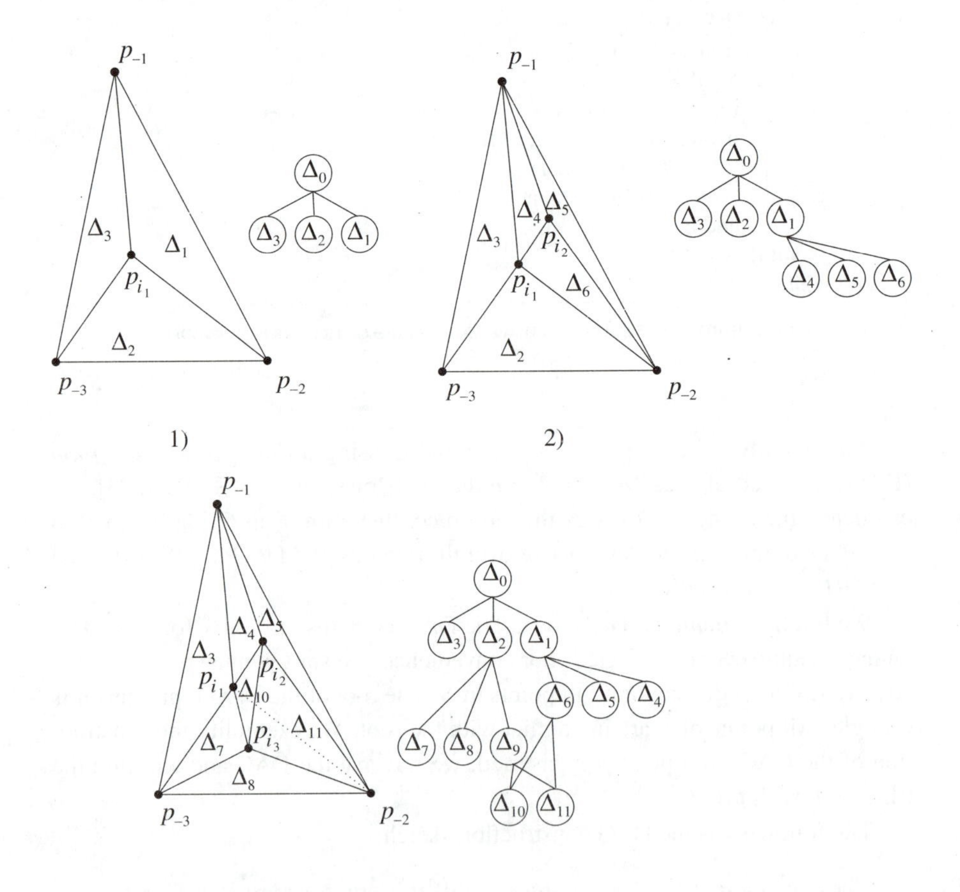


**Figure 6.8.** The Delaunay DAG stores the history of all triangles ever constructed.

DAG . FindTriangle($p$) (DAG represents a starting node of the Delaunay DAG)

```
if p ∉ DAG . Δ then
        D := Nil
else if DAG is a leaf then
        D := DAG . Δ
else
        d := DAG . BranchToChild(p)
        D := d . FindTriangle(p)
end if
return D
```

**Algorithm 6.3.** Searching for the triangle in the Delaunay diagram that contains $p$.

two new leaves and pointers to the leaves from the nodes of $\Delta_6$ and $\Delta_9$; see Figure 6.8.

Obviously, the leaves of the DAG represent the current Delaunay triangulation. Additionally, we would like to store the leaves in a DCEL, denoted by $T$(DAG) in the following. $T$(DAG) is required because we need to find adjacent triangles of an edge for the reconstruction of the DAG; i.e., for a flip operation.

For incorporating the DAG and its DCEL, we have to implement DAG . FindTriangle($p$), DAG . Edges($t, p$), and DAG . Flip($e, p, q$) in Algorithm 6.2 adequately for the DAG instead of $T$. Additionally, we have to initialize the DAG in Algorithm 6.1. The procedure DAG . FindTriangle($p$) uses

DAG . Flip($e, p, q$) (DAG represents a starting node of the Delaunay DAG, $T$(DAG) represents the current triangulation in a DCEL)

```
(t_1, t_2) := T(DAG). AdjacentTriangles(e)
D_{t_1} := DAG . Node(t_1)
D_{t_2} := DAG . Node(t_2)
D_1 := new node
D_2 := new node
r := D_{t_1} . OppositeVertex(e)
s := D_{t_2} . OppositeVertex(e)
D_1 . Triangle(p, q, r)
D_2 . Triangle(p, q, s)
D_{t_1} . ChildrenInsert(D_1, D_2)
D_{t_2} . ChildrenInsert(D_1, D_2)
T(DAG). FlipDCEL(D_1, D_2, e)
```

**Algorithm 6.4.** The DAG and its DCEL are updated by flip operations, cross-references between triangles in DAG and $T$(DAG) are created by FlipDCEL.

DAG . Edges($t, p$) (DAG represents a starting node of the Delaunay DAG, $T$(DAG) represents the current triangulation in a DCEL, and $p$ lies in triangle $t$)

$(v_1, v_2, v_3)$ := $T$(DAG). GetVertices($t$)
Update($\{p, v_1, v_2, v_3\}, T$(DAG), DAG)
**return** $T$(DAG). Boundary($t$)

**Algorithm 6.5.** Update of the DAG and its DCEL by inserting $p$ into triangle $t$.

the DAG for finding the starting triangle, recursively branching to the child that contains $p$. The procedure DAG . Edges($t, p$) inserts three new triangles into the DAG and also into $T$(DAG); see Algorithm 6.5. If two new triangles for the DAG are created, let $T$(DAG). FlipDCEL($D_1, D_2, e$) express the flip operation on the DCEL, where $D_1$ and $D_2$ denote the new triangles and nodes of the DAG, and $e$ is the edge that must be eliminated. Thus, we can always create some cross references between the triangles in the DAG and its DCEL in $T$(DAG); see Algorithm 6.4. For a triangle in $T$(DAG), we require its corresponding node in the DAG. In Delaunay($S$), the DAG is initialized by DAG := $S$. Init, which means that the starting node with a triangle of vertices $p_{-1}$, $p_{-2}$, and $p_{-3}$ is computed. Note that there are no outer triangles and OuterTriangle is not implemented for the DAG.

In general, if we insert $p_{i_j}$ into the DAG of $\mathrm{DT}_{j-1}$, we traverse a set of triangles from the root to the triangle $T$ with $p_{i_j} \in T$. All these triangles are in conflict with $p_{i_j}$. Since the expected number of edge flips (or new triangles) from $\mathrm{DT}_{i-1}$ to $\mathrm{DT}_i$ is a constant and $p_{i_j}$ is chosen randomly, the expected number of triangles in the DAG of $\mathrm{DT}_{j-1}$ in conflict with $p_{i_j}$ lies only in $O(\log j)$.

Altogether we get the following result.

**Theorem 6.6.** *The Delaunay triangulation of a set of n points in the plane can be constructed incrementally in expected time $O(n \log n)$, using expected linear space. The average is taken over the different orders of inserting the n sites.*

## 6.3. Generalization of the Voronoi Diagram

### 6.3.1. Voronoi Diagram and Delaunay Triangulation in 3D

The Voronoi diagram in 3D for a set of points $S$ subdivides the 3D space into cells of the same neighborship and can be represented by a graph in 3D. For convenience, we assume that the points are in general position; for example, no five points lie on a common sphere. The corresponding data structure is given by a DCEL in 3D. We will see that the incremental construction approach is a simple and effective construction scheme in the 3D case.

**Formal description.** Formally, we extend the notations of the 2D case. The set $S$ denotes the set $\{p_1, p_2, \ldots, p_n\}$ of $n$ points in 3D. The 3D bisector $\mathrm{Bis}(p_i, p_j)$ of two points $p_i, p_j \in S$ is defined to be a plane through the midpoint of the line segment $p_ip_j$ that is perpendicular to the segment $p_ip_j$. A Voronoi region $\mathrm{VoR}(p_i, S)$ is given by the intersection of the half-spaces bounded by bisectors $\mathrm{Bis}(p_i, p_j)$ for all $j \neq i$. As an intersection of half-spaces, the boundary of $\mathrm{VoR}(p_i, S)$ consists of faces, edges, and vertices, and represents a 3D convex polyhedron.

**Complexity.** The complexity is as follows.

1. Obviously, the number of faces of $\mathrm{VoR}(p_i, S)$ is at most $n - 1$. Each half-space defined by $\mathrm{Bis}(p_i, p_j)$ will contribute to at most one face.

2. Applying the Eulerian formula, the number of edges and vertices of $\mathrm{VoR}(p_i, S)$ is in $O(n)$.

3. Counting the complexity of all regions, the total number of components of the diagram $\mathrm{VD}(S)$ in 3D is in $O(n^2)$.

4. Unfortunately, there are configurations of $S$ with $\Omega(n^2)$ components; see [Dewdney and Vranch 77].

5. Fortunately, in practical situations, the complexity of $\mathrm{VD}(S)$ in higher dimensions is linear. For example, if the points of $S$ are drawn uniformly at random in a unit ball, the expected size of the diagram is in $O(n)$; see [Dwyer 91].

Altogether, we can assume that, in practical situations, $\mathrm{VD}(S)$ is a geometric graph in 3D of linear size.

**Delaunay triangulation $\mathrm{DT}(S)$ in 3D.** In analogy to the 2D case, the Delaunay triangulation $\mathrm{DT}(S)$ in 3D is defined to be the geometric dual of $\mathrm{VD}(S)$ with vertices from $S$.

1. For two points $p_i$ and $p_j$ that share a common face in the $\mathrm{VD}(S)$, there is an edge $p_ip_j$ in $\mathrm{DT}(S)$.

2. Every edge of the VD($S$) subdivides three faces. Thus, for every edge in the VD($S$), there is a triangle in DT($S$).

3. At a vertex $v$ of VD($S$), four points of $S$ have the same distance to $v$ and the four points share four edges with each other. Therefore, there is a tetrahedron for the vertex $v$.

Note that we have assumed general position. Altogether, DT($S$) is a graph in 3D consisting of tetrahedra.

Equivalently, DT($S$) may be defined extending the empty circle property. If the circumsphere of four points of $S$ does not contain another point of $S$ inside or on the boundary, the corresponding tetrahedron of the four points is a Delaunay tetrahedron. The circumcenters of these empty spheres are just the vertices of VD($S$). In analogy to the triangulation in 2D, the Delaunay triangulation DT($S$) is a partition of the convex hull of $S$ into tetrahedra, provided that $S$ is in general position.

There are robust implementations of the empty sphere property (see Chapter 9), and the general position condition can be easily fulfilled for the test in 3D (see Section 9.5). Therefore, it is highly recommended to extend the incremental construction scheme of the previous section to 3D.

**Simple incremental construction in 3D.** We consider an incremental construction of the dual DT($S$). Some algorithms for the incremental construction of VD($S$) in 3D also exist; see [Watson 81], [Field 86], [Tanemura et al. 83], or [Inagaki et al. 92].

In the dual environment, analogously to the 2D case, we have to perform edge flips for removing all tetrahedra of $\mathrm{DT}_{i-1}$ that are in conflict with $p_i$. That is, the circumsphere of the tetrahedra $\Delta(p_j, p_k, p_l, p_m)$ of four points $p_j$, $p_k$, $p_l$, and $p_m$ contains $p_i$. In analogy to the 2D case, this can happen only if $p_j$, $p_k$, $p_l$, $p_m$, and $p_i$ are in convex position.

Similar to the 2D case, we make use of a 3D tetrahedral DAG and find the tetrahedron, say $\Delta(p_j, p_k, p_l, p_m)$, of $\mathrm{DT}_{i-1}$ that contains $p_i$. With similar arguments, we first insert the edges $p_ip_j$, $p_ip_k$, $p_ip_l$, and $p_ip_m$, thus obtaining three new tetrahedra; see Figure 6.9.

Now we obtain $P(\star(p_i))$, the set of outer triangles of $\Delta(p_j, p_k, p_l, p_m)$ opposite $p_i$. In the 2D case, $P(\star(p_i))$ was a set of edges opposite $p_i$. Similar to the 2D case, we go on with $p_i$ and with $P(\star(p_i))$. In the 2D case, every edge of $P(\star(p_i))$ may subdivide two triangles. Analogously, every triangle of $P(\star(p_i))$ may subdivide two tetrahedra as follows. Every triangle of $P(\star(p_i))$ can be a *ground face* $G$ of a tetrahedron with a point $q$ outside $P(\star(p_i))$ and opposite $p_i$,

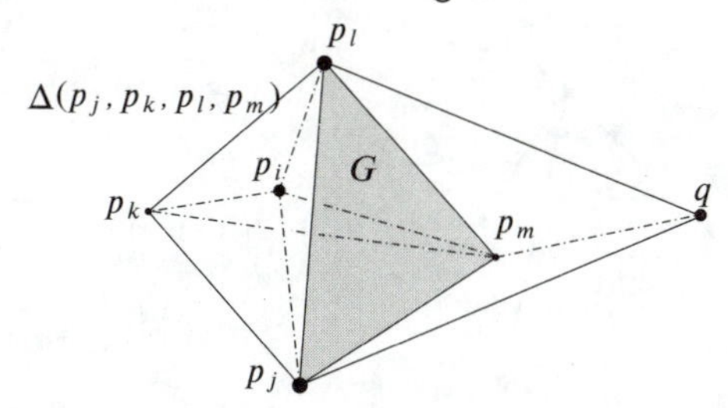


**Figure 6.9.** The tetrahedron $\Delta(p_j, p_k, p_l, p_m)$ contains $p_i$ and four edges $p_ip_j$, $p_ip_k$, $p_ip_l$, and $p_ip_m$ are inserted. For every face of $P(\star(p_i))$, we check the outer tertrahedra; in this case, $\Delta(p_j, p_l, p_m, q)$ with the opposite $p_i$.

and $G$ builds a tetrahedron with $p_i$. Together with $p_i$, we have to check whether the four points of the outer tetrahedron and $p_i$ are in conflict. That is, does the circumcenter of $G$ and $q$ contain $p_i$? This cannot happen if the five points are not in convex position, since $p_i$ lies outside the tetrahedron of $G$ and $q$ in this case. This fact can be proven similar to the 2D case; see Figure 6.5. If the five points are in convex position, the triangle $G$ is called *flippable*.

If $G$ is flippable, we apply the circumsphere test. If $G$ is no longer Delaunay, i.e., the circumsphere test returns true, then we can either remove an edge or we can insert an edge. Note that in the 2D case, we replace one edge by another, so here we have one of the main differences between 2D and 3D. The nature of the 3D *edge flip* depends on the situation; see Figure 6.10. In the first case, we replace two tetrahedra by three tetrahedra in the tetrahedral decomposition. In the second case, we replace three tetrahedra by two tetrahedra. We refer to the first case as a $2 \rightarrow 3$ edge flip; the second case is denoted by a $3 \rightarrow 2$ edge flip. We change the number of tetrahedra of the five points by $+1$ or $-1$. Obviously, other edge flippings are impossible!

As shown in Figure 6.10, it might happen that $T$ is in conflict with $p_i$ and the corresponding vertex $q$ opposite $p_i$ belongs to the ground face $G$ of another triangle of $P(\star(p_i))$. This is another significant difference between 2D and 3D. In this case, there is an outer tetrahedra with two triangles stemming from $P(\star(p_i))$ that has to vanish. Note that $P(\star(p_i))$ can be non-convex; see Figure 6.10.

If an edge flip is performed, new ground faces (triangles) in $P(\star(p_i))$ might become available. A triangle $T$ in $P(\star(p_i))$ is uniquely determined by the fact that $p_i$ is the apex of a tetrahedron in the current triangulation and $T$ lies opposite $p_i$; see Figure 6.9.

In principle, the presented algorithm can be implemented in analogy to the 2D case; we store the history of a tetrahedral subdivisions in a DAG. Now, a DAG node has four children for the triangles of the corresponding tetrahedron, a

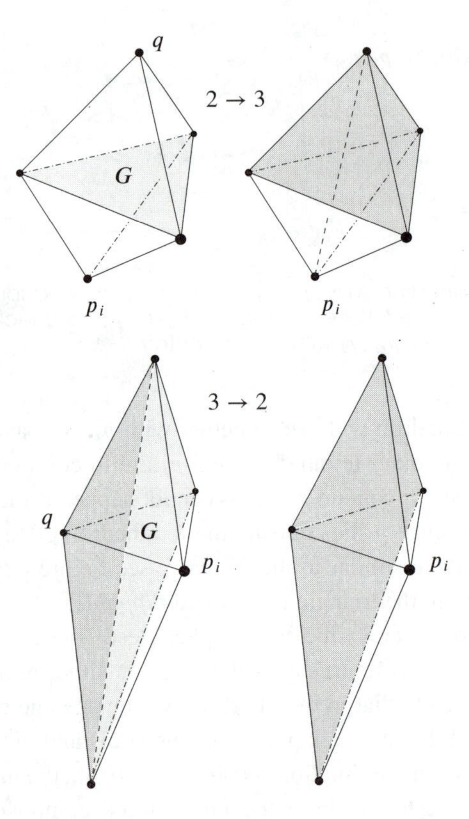


**Figure 6.10.** The first figure shows the $2 \to 3$ edge flip, whereas the second figure shows a $3 \to 2$ edge flip. The ground face $G$ is flippable because the endpoints of $G$, $p_i$, and $q$ are in convex position. New triangles (shaded) of $P(\star(p_i))$ opposite $p_i$ are found after the flipping.

triangle $t$ uniquely separates two tetrahedra $T_1$ and $T_2$, and so on. If there are no more flippable tetrahedra beyond the triangles of the current $P(\star(p_i))$ opposite $p_i$, the process stops and $\mathrm{DT}_i$ is constructed.

Unfortunately, not every order of edge flips in a single insertion step might finally succeed. Fortunately, [Rajan 91] showed that the flipping process will always result in a current Delaunay triangulation if the flips are done in an appropriate order.[1] This result was used for an $O(n^2)$ implementation by [Joe 91b].

It was shown in [Guibas et al. 92b] that the expected number of tetrahedra occurring during an incremental construction of the Delaunay triangulation in 3D is bounded by $O(n^2)$. A drawback is that the number of final tetrahedra might be

[1]The new site $p_i$ might be in conflict with many tetrahedra along the boundary of $P(\star(p_i))$.

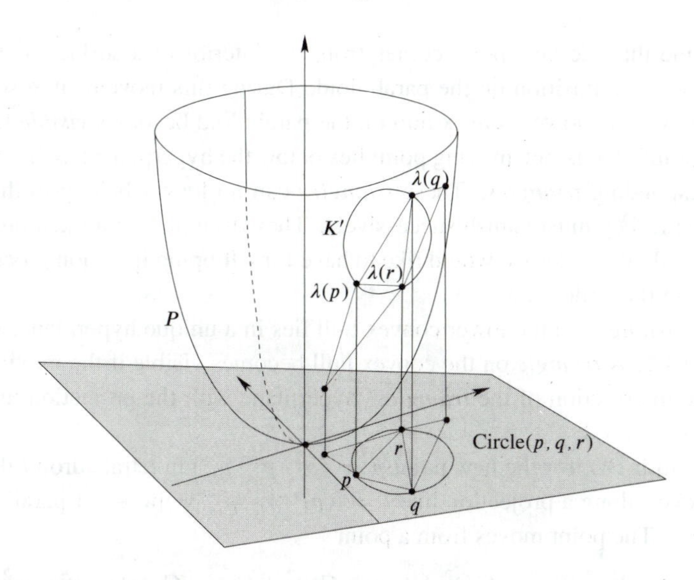


**Figure 6.11.** The projection of points from 2D onto the paraboloid in 3D. The lower convex hull on the paraboloid can be projected back to 2D and represents the Delaunay triangulation in 2D. In general, the Delaunay triangulation in $\mathbb{R}^d$ is equivalent to the convex hull in $\mathbb{R}^{d+1}$.

in $O(n)$, whereas the expected number of edge flips remains to be in $O(n^2)$. In [Edelsbrunner and Shah 96], the above results were collected and the randomized incremental construction was extended to regular triangulations.

It remains to show how to compute the flipping order after $P(\star(p_i))$ was initialized. Among all flippable tetrahedra opposite $p_i$ having a ground face in $P(\star(p_i))$, we compute an order of the candidates by the following elegant idea. The order can be easily computed and incorporated into the DAG algorithm.

The convex hull in $\mathbb{R}^{d+1}$ and the Delaunay triangulation in $\mathbb{R}^d$ defined by empty circle predicates are identical with respect to a lifting transformation; see [Brown 79] or [Edelsbrunner 87]. As will be shown in Section 9.4.2, we can project points $p = (p_1, p_2, \ldots, p_d)$ in $\mathbb{R}^d$ onto the paraboloid $P$ in $\mathbb{R}^{d+1}$ by

$$\lambda(p) = (p_1, p_2, \ldots, p_d, p_1^2 + p_2^2 + \ldots + p_d^2) \in P;$$

see Figure 6.11 and Figure 9.19 for $d = 2$. The lower convex hull of the transformed points on the paraboloid in $\mathbb{R}^{d+1}$ can be projected back into $\mathbb{R}^d$, and we get the Delaunay triangulation by the empty circle property; see Theorem 9.30 in Section 9.4.2.

Let us assume that a new point $p_i$ in $\mathbb{R}^d$ is inserted, and let us consider the situation on the convex hull. We use a *kinetic* interpretation. On the hull, we can assume that the new point comes from the interior of a surface and moves down to its final position on the paraboloid. During this movement, a sequence of *triangles* of the lower convex hull on the paraboloid becomes *visible* from the moving point; that is, the moving point lies below the hyperplane passing through the corresponding *triangles*. These *triangles* can no longer belong to the lower convex hull; they must vanish successively. They disappear in a fixed order, and this is exactly the order we would like to have for a flipping insertion process. Let us compute this order now.

Every *triangle* on the lower convex hull lies in a unique hyperplane; see also Section 9.4.2. A *triangle* on the convex hull becomes visible if the moving point meets the intersection of the *triangles'* hyperplane with the projection line of $p_i$ in dimension $d + 1$.

In detail in 3D, for the new point $p_i = (p_{i_x}, p_{i_y}, p_{i_z})$ in tetrahedron $t$ the point in 4D moves along a projection line $l = \{(p_{i_x}, p_{i_y}, p_{i_z}, w) | w \in \mathbb{R}\}$ parallel to the fourth axis. The point moves from a point

$$(p_{i_x}, p_{i_y}, p_{i_z}, R_t^2 - ((p_{i_x} - C_{t_x})^2 + (p_{i_y} - C_{t_y})^2 + (p_{i_z} - C_{t_z})^2) + p_{i_x}^2 + p_{i_y}^2 + p_{i_z}^2)$$

in a 3D hyperplane, stemming from the projection of the vertices $t$, to a point

$$(p_{i_x}, p_{i_y}, p_{i_z}, p_{i_x}^2 + p_{i_y}^2 + p_{i_z}^2)$$

on the paraboloid. Here, $C_t = (C_{t_x}, C_{t_y}, C_{t_z})$ is the circumcenter of the tetrahedron $t$, and $R_t$ denotes the radius of the circumsphere of $t$ in 3D. Let $W$ denote the fourth coordinate. For the neighboring tetrahedra of $P(\star(p_i))$ in 3D, we consider the corresponding projected tetrahedra on the lower convex hull on the paraboloid in 4D. The hyperplane of a projected tetrahedron $t$ has an intersection $(p_{i_x}, p_{i_y}, p_{i_z}, w_t)$ with $l$.

We compute all intersections $w_t$ with the line $l$ for every neighboring tetrahedron in $P(\star(p_i))$ and sort them in decreasing order. This gives the flipping order. If $w_t < p_{i_x}^2 + p_{i_y}^2 + p_{i_z}^2$, we can neglect the tetrahedron $t$. The details of the computations can be found in [Rajan 91] and for a tetrahedron $t$ on the boundary of $P(\star(p_i))$ with circumcenter $C_t = (C_{t_x}, C_{t_y}, C_{t_z})$ and circumradius $R_t$, we can show

$$w_t = R_t^2 - ((p_{i_x} - C_{t_x})^2 + (p_{i_y} - C_{t_y})^2 + (p_{i_z} - C_{t_z})^2) + p_{i_x}^2 + p_{i_y}^2 + p_{i_z}^2 \, .$$

We can compute the values $w_t$ for every tetrahedron around $P(\star(p_i))$; sorting the values gives an appropriate flipping order. Altogether, we get the following result.

**Theorem 6.7.** *The Delaunay triangulation of a set of $n$ points in* 3D *can be constructed incrementally in expected time $O(n^2)$, using expected quadratic space. The average is taken over the different orders of inserting the $n$ sites.*

## 6.3.2. Constrained Voronoi Diagrams

For many applications, it is not sufficient to consider neighborship relations without any constraints. For example, a small distance between two cities may be unimportant if there is an obstacle (natural or artificial) that blocks their sight to each other. To overcome such problems, the concept of constrained Voronoi diagrams was introduced by [Lee and Lin 86].

We discuss the case of a Voronoi diagram of points in the plane with additional line segment obstacles. Formally, let $S$ be a set of $n$ points in the plane, and let $L$ be a set of non-crossing line segments spanned by $S$. By induction, it is easy to show that $L$ contains at most $3n - 6$ line segments. The segments in $L$ may be considered as obstacles.

The *constrained distance* between two points $x$ and $y$ takes visibility information between the points with respect to the segments of $L$ into account. The constrained distance is defined by

$$b(x, y) = \begin{cases} d(x, y), & \text{if } \overline{xy} \cap L = \emptyset, \\ \infty, & \text{otherwise,} \end{cases}$$

where $d(x, y)$ denotes the Euclidean distance between $x$ and $y$. The distance function gives rise to the *constrained Voronoi diagram* of $S$ and $L$, ConstrVD($S$, $L$) for short. Regions of sites that are close but not visible from each other are separated by segments in $L$; an example is shown in Figure 6.12.

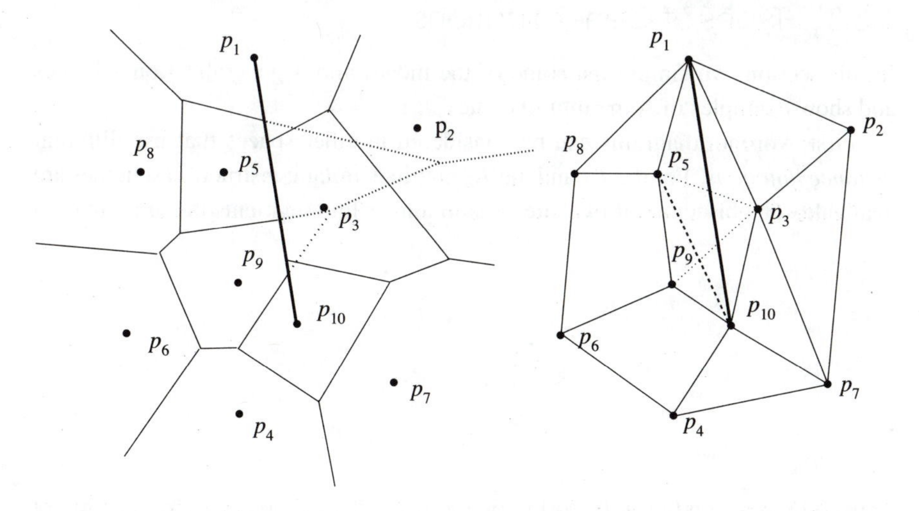


**Figure 6.12.** The constrained Voronoi diagram ConstrVD($S$, $L$) and its dual.

The exact dual of VD($S, L$) may no longer be a full triangulation of $S$ (even if $S$ is included). For example, in Figure 6.12, in the dual of VD($S, L$) including $L$, the face given by $p_1$, $p_{10}$, $p_9$, and $p_5$ is not a triangle. The edges $p_5p_3$ and $p_9p_3$ of DT($V$) are not part of the dual of ConstrVD($S, L$).

Fortunately, we can modify ConstrVD($S, L$) in order to dualize it into a *"near to Delaunay"* triangulation $DT(S, L)$, which includes $L$.

For every line segment $l$, we proceed as follows. All sites belonging to clipped regions lying to the left of a segment $l$ build a Voronoi diagram to the right of $l$ and vice versa; see Figure 6.12 for an example. The neighborship within these new diagrams leads to additional edges, which are inserted into the dual of ConstrVD($S, L$). For example, in Figure 6.12, in the diagram of $p_1$, $p_5$, $p_9$, and $p_{10}$ extended to the right of $l$, the sites $p_5$ and $p_{10}$ finally become neighbors, and the corresponding edge in the dual of ConstrVD($S, L$) triangulates the corresponding graph. By construction, the endpoints of a given line segment $l$ are always neighbors in the new diagrams, and therefore, the line segments themselves are inserted.

Algorithms for computing the constrained Voronoi diagram VD($S, L$) and the constrained Delaunay triangulation $DT(S, L)$ have been proposed by several authors: Lee and Lin [Lee and Lin 86], Wang and Schubert [Wang and Schubert 87], Chew [Chew 89], and Wang [Wang 93]. Good implementation schemes can be found in [Seidel 88] and [Kao and Mount 92]. An application of $DT(S, L)$ to quality mesh generation can be found in [Chew 93].

## 6.3.3. Types of Generalizations

In this section, we simply list some of the most famous generalization schemes and show examples of some intuitive ones.

First, Voronoi diagrams can be considered in other spaces that use different *distance functions*. For the $L_1$ and the $L_2$ metric, similar construction schemes are available. The bisectors of two sites are no longer line segments (see Figure 6.13

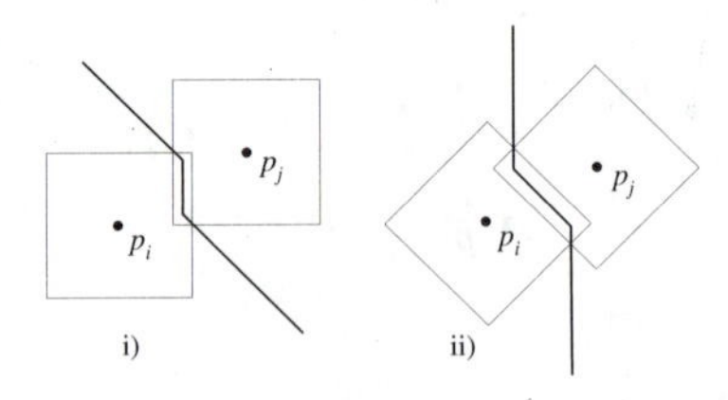


**Figure 6.13.** Bisectors under $L_1$ and $L_2$ distance functions. One can build the bisector by the intersections of scaled copies of the unit circle.

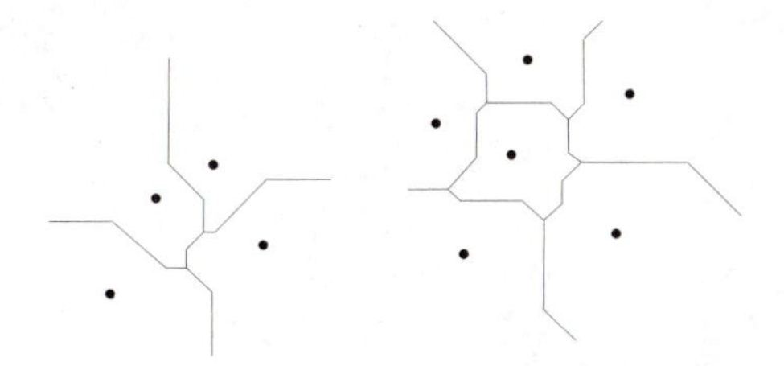

**Figure 6.14.** Voronoi diagrams under $L_1$ and $L_2$ distance functions.

for an example), but they consist of at most three segments and divide the plane into distinct half-spaces also. This means that in a brute force manner, we can simply compute the intersections of half-spaces for computing the Voronoi region of a single site. This results in a simple $O(n^3)$ construction algorithm. Optimal $O(n \log n)$ construction schemes for the $L_p$ norms were considered by several authors; see [Hwang 79, Lee 80, Lee and Wong 80]. Examples of Voronoi diagrams for $L_1$ and $L_2$ are shown in Figure 6.14.

More generally, we can consider *convex distance functions*. A convex distance function is defined by a convex set $C$ with fixed base point $O$. The distance between two points $p$ and $q$ is given by the following transformation (see also Figure 6.15). Translate $C$ onto $p$ and consider the ray through $q$ starting from $p$. The ray hits $C$ in a unique point $q'$, and the distance with respect to the convex set $C$ is defined by

$$d_C(p,q) := \frac{d(p,q)}{d(p,q')}.$$

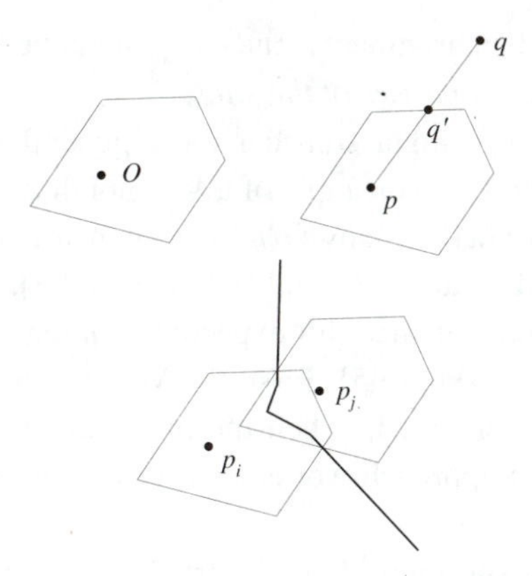


**Figure 6.15.** Bisectors under convex distance functions.

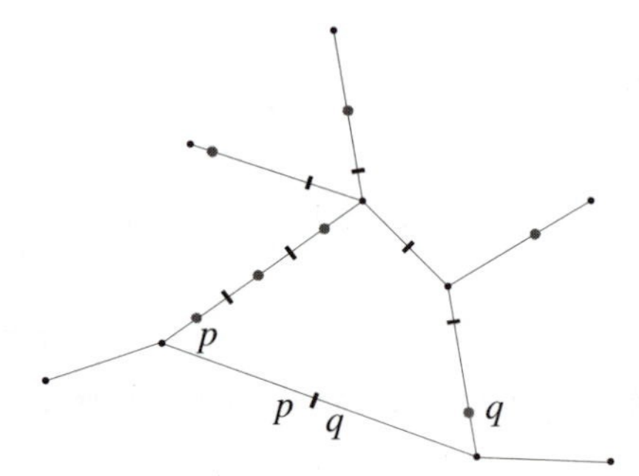


**Figure 6.16.** The Voronoi diagram of a set of sites on a graph.

This is also true for the unit circles of $L_1$ and $L_2$. The bisector of two points $p_i$ and $p_j$ is the locus of all points with the same distance to $p_i$ and $p_j$. Therefore, we can use scaled copies of $C$ translated to $p_i$ and $p_j$. The locus of all intersections of the scaled copies for all scale factors represents the bisector. Therefore, it can be easily shown that for a convex polygon $C$ the bisector of two points has complexity $O(|C|)$.

Furthermore, we can consider different *environments*. We have already seen that the Voronoi diagram can be generalized to 3D space. Many other environments may be used. For example, we can consider a set of sites on a graph $G = (V, D)$. The distance between two arbitrary points on the graph is given by the shortest distance along the graph. An example of the Voronoi diagram of a set of sites on a graph is shown in Figure 6.16. Construction and complexity results for various Voronoi diagrams on trees, and graphs can be found in [Hurtado et al. 04].

Beyond metrics and environments, one can generalize the concept of Voronoi diagram with respect to *properties of the sites.*

Straightforwardly, one might consider more general sites such as line segments or polygonal chains. An example of a Voronoi diagram of a set of line segments is given in Figure 6.17. Many $O(n \log n)$ construction schemes were presented; see [Yap 87a], [Fortune 87], and [Klein et al. 93]. A simple randomized incremental construction scheme with expected $O(n \log n)$ running time is presented in [Alt and Schwarzkopf 95]. First, the Voronoi diagram of the endpoints of the line segment is computed. Then the line segment bisectors are inserted successively. The given approach works also for curved objects with properties similar to line segments.

Furthermore, every site might have a certain weight that influences the distance function. That is, for two points $p_i$ and $p_j$ with real-valued weights $w(p_i)$

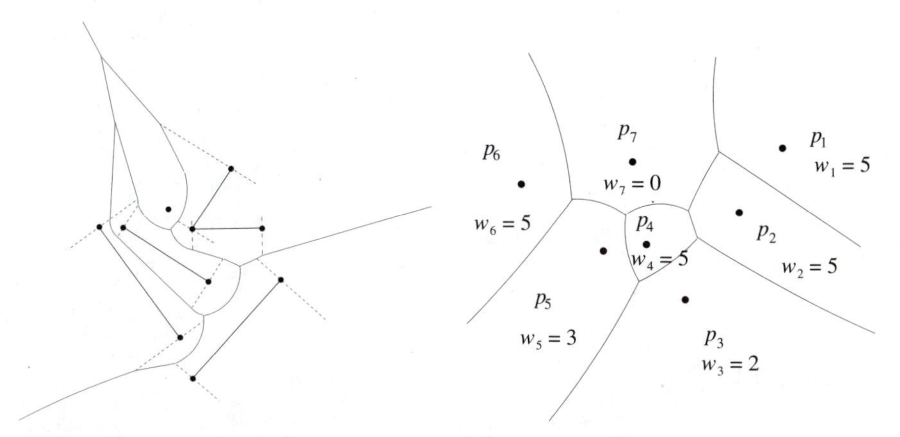


**Figure 6.17.** The Voronoi diagram of a set of non-intersecting polygonal objects.

**Figure 6.18.** The Voronoi diagram with additive weights.

and $w(p_j)$, the weighted bisector is defined to be the locus of points $q$ so that

$$w(p_i)d(p_i, q) = w(p_j)d(p_j, q)$$

if we apply the weights multiplicatively, or

$$w(p_i) + d(p_i, q) = w(p_j) + d(p_j, q)$$

if the weights are meant to be additive. An example of a Voronoi diagram with additive weights is given in Figure 6.18. For additive weights, the bisector is given by the arc of a hyperbola, because the difference of the distances $d(p_i, q) - d(p_j, q)$ is a constant $w(p_j) - w(p_i)$. Obviously, a big weight results in a smaller area.

Generalizations of Voronoi diagrams might consider also the *intention of the cell subdivision*. The *normal* Voronoi diagram subdivides the plane into cells of the same nearest neighborship. But we can ask also for a subdivision of the farthest neighborship. That is, a cell represents all points that belong to a site farther away than any other site. This subdivision is called the *farthest point Voronoi diagram*.

A more general concept of the neighborship subdivision is called the *kth order Voronoi diagram*. A $k$th order Voronoi diagram in 2D subdivides the plane into regions $R$ dedicated to a set of $k$ sites $p_{i_1}, p_{i_2}, \ldots, p_{i_k}$ such that for every point $p$ of a region $R$, the first $k$ nearest neighbors are $p_{i_1}, p_{i_2}, \ldots, p_{i_k}$. Figure 6.19 shows an example for $k = 3$. In this setting, the $(n-1)$ order Voronoi diagram of $n$ sites represents the farthest point Voronoi diagram.

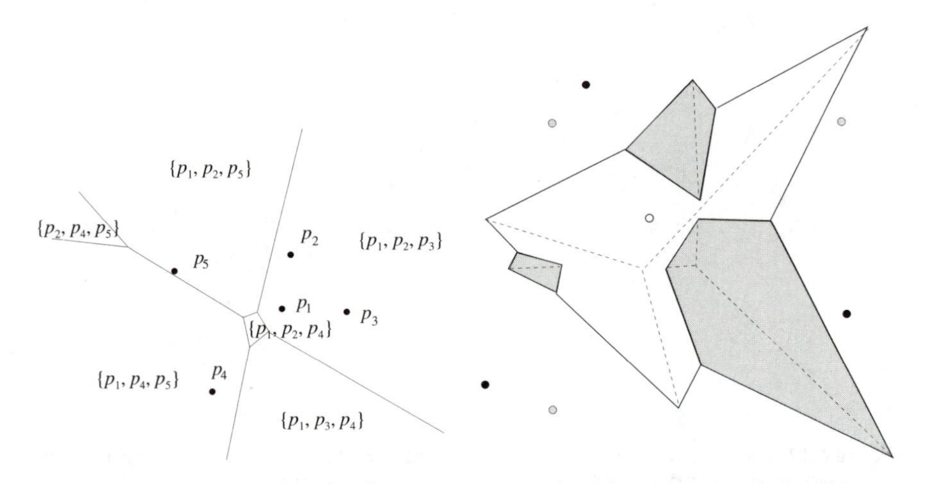


**Figure 6.19.** The 3-order Voronoi diagram.

**Figure 6.20.** The farthest color Voronoi diagram.

Furthermore, we can make use of colors for separating the point set $P$ into sets of points $P_1, P_2, \ldots P_k$. Every set $P_i$ has its own color, and $\cup_{i=1}^{k} P_i = P$ holds. The *color Voronoi diagram* subdivides the plane into regions of nearest neighborship with respect to the sets $P_i$. The region of $P_i$ contains all points $q$ that have a nearest neighbor stemming from $P_i$.

Additionally, we can combine the given concepts. For example, the *farthest color Voronoi diagram* represents a subdivision of the plane into regions with respect to sets of points $P_1, P_2, \ldots P_k$, so that the region of $P_i$ contains all points $q$ that have a farthest neighbor from $P_i$. An example of a farthest color Voronoi diagram is given in Figure 6.20 with three colors. Note that the regions are no longer connected and that a set might have an empty region. The diagram has complexity $O(nk)$ and can be computed in $O(nk \log n)$ using methods stemming from the famous book [Sharir and Agarwal 95] on *Davenport-Schinzel sequences*; see also [Abellanas et al. 01a] and [Abellanas et al. 01b].

# 6.4. Applications of the Voronoi Diagram

## 6.4.1. Nearest Neighbor or Post Office Problem

We consider the well-known post office problem. For a set $S$ of sites in the sphere and an arbitrary query point $q$, we want to compute the point of $S$ closest to $q$ efficiently.

In the field of computational geometry, there is a general technique for solving such query problems. One tries to decompose the query set into classes so that every class has the same answer. Then, for a single answer, we need to determine only its class. This technique is called the *locus approach*.

The Voronoi diagram represents the locus approach for the post office problem. The classes correspond to the regions of the sites. For a query point $q$, we want to determine its class/region and return its owner.

We present some simple ideas for the region query, which work in 2D and 3D.

**Decomposition of the diagram by slabs.** To solve the given task, the following simple slab method can be applied. The slab method was mentioned first in [Dobkin and Lipton 76]. It subdivides the cell decomposition into slabs $S_i$ and, in turn, every slab is subdivided by a set of segments $s_i$; see Figure 6.21.

The following is the slab method sketch:

- Draw a horizontal line through every vertex of the diagram and sort the lines in $O(n \log n)$ time; see Figure 6.21. The lines decompose the diagram into *slabs*;
- For every *slab*, sort the set of crossing edges of the Voronoi diagram. We must trace the DCEL of the Voronoi diagram and will find a sorted set of crossing segments in linear time.

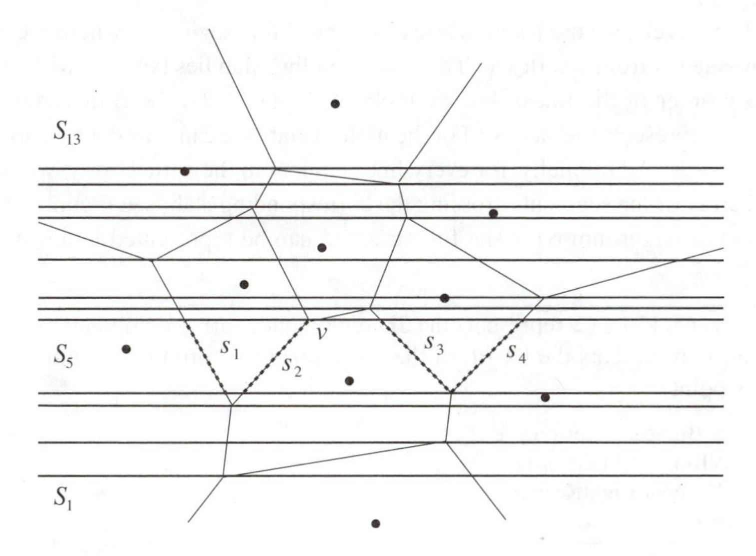


**Figure 6.21.** After constructing a sorted list of slabs and a sorted list of segments in every slab, a query point $q$ can be located quickly by a binary search.

SlabStructureConstr($T$) ($T$ represents the DCEL of the Voronoi diagram)

```
V := T.GetVertices
L := V.sortByY
S := new array(|L|, |2|)
i := 0
while L ≠ ∅ do
        S[i][1] := L.First
        L.DeleteFirst
        E_L := T.TraceLeft(S[i][1])
        E_R := T.TraceRight(S[i][1])
        EdgeArray := Concat(E_R, E_L)
        S[i][2] := EdgeArray
        i := i + 1
end while
return S
```

**Algorithm 6.6.** The construction of a slab search structure for the region query returns a 2D array $S$. For each vertical line, the array contains a sorted array of crossing segments.

The corresponding algorithm must return a data structure that supports a binary search. For example, we can insert the slabs into a binary tree. In turn, every tree node represents the segment subdivision of a slab by a binary tree representing segments.

Alternatively, we use a sorted array of vertical line segments, where every line entrance stems from a vertex $v$. The corresponding slab lies below $v$ with respect to the $y$ order of the lines. For example, in Figure 6.21, the fifth vertex from below, $v$, represents the slab $S_5$. For the highest slab, we can introduce an artificial vertex at $+\infty$. Additionally, for every line segment in the sorted array, we store a sorted array of the segments crossing the corresponding slab; see Figure 6.21 and compare to Algorithm 6.6. The full structure can be represented in an $(n \times n)$-

SlabQuery($S, V, q$) ($S$ represents the 2D region query array computed by Algorithm 6.6, $V$ represents the DCEL of the corresponding Voronoi diagram, and $q$ is a query point)

```
j := S.BinSearchAbove(q_y)
e := SBinSearchLeft(q_x)
p := V ReportRightRegion(e)
return p
```

**Algorithm 6.7.** The region query can be answered efficiently using a slab decomposition and a binary search.

dimensional 2D array $S$, where $S[i][1]$ denotes the Voronoi vertex and $S[i][2]$ contains an array of sorted segments. For example, in Figure 6.21 the slab $S_5$ is given by $S[5][]$ with $S[5][1] = v$ and $S[5][2] = [s_1, s_2, s_3, s_4]$.

Tracing through the DCEL as indicated by TraceLeft or TraceRight in Algorithm 6.6, compute the array of segments of the current slab. The trace can be done in linear time as follows. In the planar DCEL, we sucessively check the edges of the corresponding face of the current vertex and test for intersections. Thus, we will find the first intersection. Then we go on recursively with the next face using the corresponding pointers in the DCEL; see Section 9.1.1. See also Algorithm 6.8. Altogether, we need $O(n)$ time for computing a sorted array of line segments for a single slab, and $O(n^2)$ time for computing all slabs.

We can speed up the construction by the following idea. From one slab to the next slab, only a single vertex and its outgoing edges has to be considered, because the sorted array of segments can only change locally. If the sorted array of segments $S[i][2]$ for a slab $i$ is constructed, the next sorted array for vertex $S[i+1][1]$ is computed from $S[i][2]$ and the outgoing edges at $S[i+1][1]$. Thus, the next array of segments is constructed in time proportional to the degree of the corresponding node, which equals 3 in the Voronoi diagram.

Altogether, the incremental construction of the slabs takes $O(n)$ time but $O(n^2)$ space is required. In the corresponding algorithm, the operations $T$.TraceLeft($S[i][1]$) and TraceRight($S[i][2]$) will use the information of $S[i-1][2]$. Only the first slab has to be constructed from scratch. Note that presorting of the vertices and the construction of the Voronoi diagram result in $O(n \log n)$ construction time altogether.

A query is done efficiently as follows. For a query point $q$, we locate its *slab* in $O(\log n)$ time and then its region in $O(\log n)$ time, both by binary search.

For reporting the region of a query, it suffices to find the appropriate line or edge below $p$ or to the left of $q$, rather than the slab or the region itself. This is indicated in Algorithm 6.6 by BinSearchAbove, BinSearchLeft, and ReportRightRegion. For example, $S$.BinSearchAbove($q_y$) returns index $j$ if $q$ lies between the slabs $S[j][]$ and $S[j+1][]$, assuming that the slabs are ordered by increasing $y$-coordinates of the vertices.

**Theorem 6.8.** *Given a set $S$ of $n$ point sites in the plane, one can, within $O(n \log n)$ time and $O(n^2)$ storage, construct a data structure that supports nearest neighbor queries. For an arbitrary query point $q$, its nearest neighbor in $S$ can be found in time $O(\log n)$.*

It was shown already in [Cole 86] that we do not need to consider a unique data structure for every slab, because only one segment changes from one slab

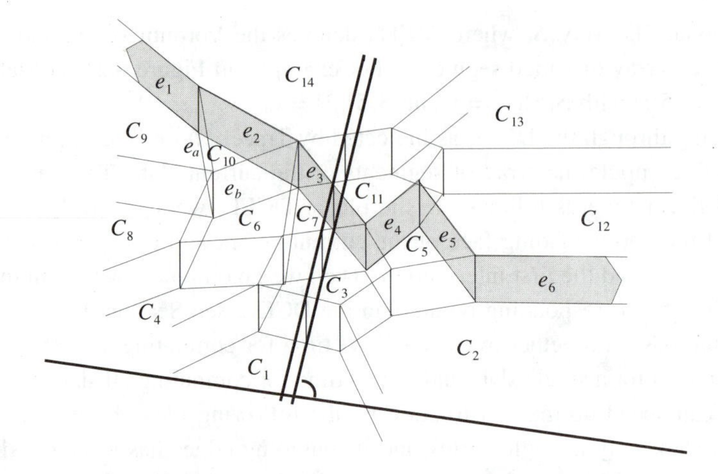


**Figure 6.22.** The separating surfaces for a 3D slab subdivision of convex cells. The separator $\sigma_{10}$ separates the cells $C_1, C_2, \ldots, C_{10}$ from $C_{11}, C_{12}, \ldots, C_{14}$.

to the other. This key idea was used in [Sarnak and Tarjan 86] for the *persistent search tree* model. In this model, we start with an empty search tree and insert the crossing segments sucessively. There is a linear size tree data structure that stores the changes over time and gives access to every structure in constant time. See also [Edelsbrunner 87].

The presented ideas can be extended to 3D. We can analogously subdivide the 3D diagram into slabs by horizontal planes at the vertices. Unfortunately, the given slabs will still contain convex 3D objects; see Figure 6.22 for a simple example of a slab. Therefore, we need a data structure for handling queries in 3D slab subdivisions, which is the subject of the next section. The methods of the next section can be applied also to the 2D Voronoi diagram and also to the full 3D Voronoi diagram.

**Separating surfaces for 2D and 3D point location in a convex cell decomposition.** Let us assume that we have a cell decomposition with convex cells as shown in Figure 6.22. The cells can be ordered as follows. We move a hyperplane perpendicular to the $x$-axis and parallel to the $y$-axis along the $x$-axis. Let $H_x$ denote the corresponding intersection with the given 3D slab at position $x$; see Figure 6.22. $H_x$ suggests a partial order of the neighboring cells as follows.

If $C$ and $D$ share a boundary face and $C$ lies below $D$ with respect to the corresponding hyperplanes $H_x$, then let $C \ll D$. The relation $\ll$ is acyclic because the cells are convex. Then we can easily extend the partial order to a total order

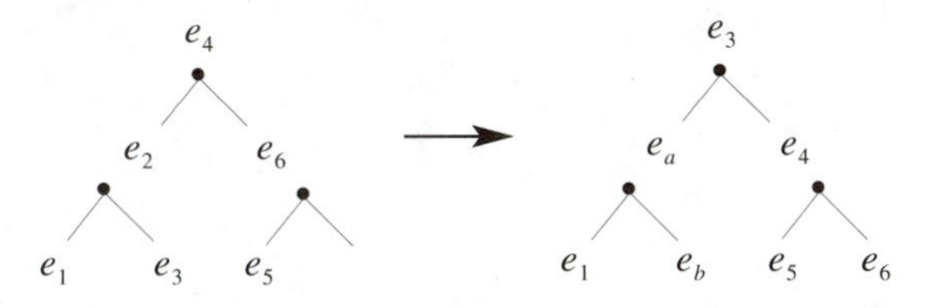


**Figure 6.23.** The separator $\sigma_{10}$ of Figure 6.22 is a binary tree of its edges. From $\sigma_{10}$ to $\sigma_9$, local updates are made.

of all cells. This means that we enumerate the cells by $C_1, C_2, \ldots, C_k$, so that $C_i \ll C_j$ implies $i < j$.

For every $i$, there is a sequence of surfaces that splits $C_1, C_2, \ldots, C_i$ from $C_{i+1}, C_{i+2}, \ldots, C_k$; see Figure 6.22 for an example for $i = 10$. Let $\sigma_i$ denote this *separator*; see also [Tamassia and Vitter 96] and [Edelsbrunner et al. 84]. The separators $\sigma_i$ and $\sigma_{i+1}$ differ in surfaces belonging to $C_i$. We can build a binary tree for the separators due to their numeration; see Figure 6.24.

The intention is that by a binary search we want to find the separators $\sigma_i$ and $\sigma_{i+1}$, so that the query point lies between them in a unique cell. In turn, for every separator, we should be able to decide whether the query point lies below or above the separator. This is very similar to the slab method. Fortunately, we can use the properties of the Voronoi diagram. The Voronoi cells were constructed by the intersections of hyperplanes. Therefore, they are convex, and every separator is an $x$-monotone chain by construction. This means that we can answer the above/below separator query by a binary search also. For the $x$-monotone separator, we can build a binary tree of its edges; see Figure 6.23 for the separator $\sigma_{10}$ in Figure 6.22.

Altogether, we have to construct a binary tree of separators, and for every separator, construct a binary tree of its surfaces. The binary tree of separators contains a single cell at every leaf of the tree. The path from the root to the leaf of the tree shows how this cell is enclosed by separators. If the order of the separators is given, we can build up the binary tree easily. For example, Figure 6.24 shows the binary separator tree for the cell decomposition of Figure 6.25.

Finally, we consider the construction of the separators. The presented idea works also for the Voronoi diagram in 2D and is more elegant as the simple slab subdivision. For convenience, we illustrate the construction in 2D; that is, surfaces are replaced with edges. Fortunately, a total order of the cells is obviously given by the $y$ order of the sites; see Figure 6.25. This is also true in 3D. We assume that no two points have the same $y$-coordinate. Otherwise, we can apply the methods presented in Section 9.5.

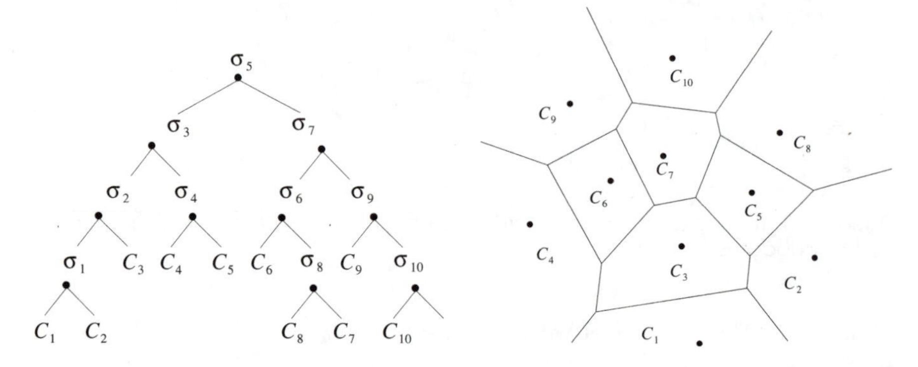


**Figure 6.24.** The binary separator tree for the cell decomposition of Figure 6.25. Every leaf represents a cell.

**Figure 6.25.** The $y$ order of the Voronoi sites gives the total order of the cell decomposition in 2D and 3D.

With the help of the Voronoi diagram, the separators are constructed incrementally. Let a total order $C_1, C_2, \ldots, C_k$, so that $C_i \ll C_j$ implies $i < j$ be given by increasing $y$-coordinates of the sites. For every $C_i$, the corresponding site $p_{j_i}$ is known.

The following is the separator construction sketch:

- The first $\sigma_k$ separator for $C_k$ and $C_1, C_2, \ldots, C_{k-1}$ is given by the boundary of $C_k$. The edges are given in $x$-monotone order. Build a binary tree of the edges and keep a reference to the DCEL of VD($S$) for every edge;
- Assume that the separator $\sigma_{i-1}$ is already constructed for $C_{i-1}, C_i, \ldots, C_k$ and $C_1, C_2, \ldots, C_{i-2}$, and a binary tree of the edges is given.
  - For $C_i$, delete the subchain of edges of $p_{j_i}$ which are already in the tree of $\sigma_{i-1}$;
  - Insert the subchain of edges of $p_{j_i}$ which were not in the tree of $\sigma_{i-1}$. The order of the whole chain and the balance of the tree must be maintained.

For example, the incremental reconstruction step in 3D from separator $\sigma_{10}$ to $\sigma_9$ in Figure 6.22 deletes the edge $e_2$ and inserts the chain $e_a$ and $e_b$; see Figure 6.23 for the corresponding trees.

The running time of the construction of the separators depends on the complexity of the cell decomposition. If the cell decomposition has $O(m)$ faces, then the trees are constructed in $O(m \log m)$ time due to the insertion and the deletion of $O(m)$ edges. The number of edges in all separators is obviously quadratic in $m$. A point location query for a point $p$ is answered by a binary search in $O(\log m)$ time as follows.

The following is the point location query sketch:

- Consider the root node $v$ of a binary tree of separators and the corresponding separator $\sigma$;
- If $v$ is a leaf, report the corresponding cell;
- Otherwise check whether $p$ lies below or above $\sigma$ as follows:
  - Consider the root $w$ of the binary tree of edges of $\sigma$ and the corresponding edge $e$;
  - Check whether $p$ lies in the corridor spanned by two vertical lines passing through the left and right endpoints of $e$. If this is true, check whether $p$ lies above or below $e$ and return this value;
  - Otherwise, if $p$ is not in the corridor of the left and right endpoints of $e$, branch to the child of $w$ of the segment tree that lies on the same side as $p$ with respect to the corridor at $e$. Repeat the process with this child and its associated edge (the new $w$ and $e$);
- If $p$ lies below $\sigma$, branch to the child of $v$ that contains the separator below $\sigma$. Otherwise, branch to the child of $v$ that contains the separator above $\sigma$. Repeat the process with this child (the new $v$).

The presented approach works in the same way in 3D and also on the full 3D Voronoi diagram also; that is, we can neglect the slab structure for 3D. In the point location query, the edges are replaced by surfaces and the corridor check is done for an area separated by three halfspaces. Altogether, the following result holds.

**Theorem 6.9.** *For a Voronoi diagram cell decomposition in* 3D *with $O(m)$ faces, there is a data point location structure based on separators that can be built up in $O(m \log m)$ time and $O(m^2)$ space. A point location query can be answered in $O(\log m)$ time.*

Finally, we present a very simple method for the nearest-neighbor query.

TraceLine($V, p_f, f, q$) ($V$ represents the DCEL of a Voronoi diagram, $q$ is a query point, and $p_f$ is an arbitrary point on a face $f$ of $V$. )

```
P := V.polyhedron(p_f, q)
(p_f, f) := V.TraceExit(P, p_f, q)
if p_f = Nil then
        return ReportSite(P)
else
        TraceLine(V, p_f, f, q)
end if
```

**Algorithm 6.8.** Answering a region query by tracing a line through the DCEL.

**Tracing a line in the diagram.** Another very simple method for the locus approach and the nearest neighbor problem is tracing a line from a specified vertex to the query point. This simple method works in arbitrary dimension and needs no preprocessing. Let the DCEL $V$ of the Voronoi diagram and a query point $q$ be given.

The following is the line tracing sketch:

- Choose an arbitrary point $p_f$ on a face $f$ of the Voronoi diagram $V$. Consider the line $l$ passing from $p_f$ to $q$. The line passes a convex polyhedron $P$ first;
- Move around $P$ using the DCEL until the exit point $p_f$ of $l$ for $P$ on a face $f$ of $P$ is found;
- If there is no such exit point, then report the site of $P$;
- Otherwise, go on with the line passing from $p_f$ to $q$.

It remains to describe the operation TraceExit in Algorithm 6.8. We simply trace the boundary of a polyhedron $P$ of $V$ that is crossed by the ray from $p_f$ to $q$ first. The face $f$ serves as the starting face. We are searching for the exit edge or face and its corresponding intersection point. In the planar DCEL, we successively check the edges of the corresponding polyhedron and test for intersection. In 3D, we also make use of the DCEL, and successively check all faces and test for intersection with the trace line.

The running time of the simple trace technique depends on the complexity of the Voronoi diagram and the chosen starting point. As noted earlier, in practical situations, the complexity of a Voronoi diagram in 3D often lies within $O(n)$ for $n$ sites. On the other hand, the expected number of steps in the tracing algorithm is given by $O(\log n)$ if the DCEL has linear size; see [Dwyer 91],

[Devroye et al. 98] and [Devroye et al. 04]. In the worst case, the presented algorithm needs $O(n^2)$ time. The trace method does not need a search query data structure.

**Theorem 6.10.** *Tracing a line through the Voronoi diagram of complexity $O(m)$ has time complexity $O(m)$ in the worst case, but expected time is $O(\log m)$.*

### 6.4.2. Other applications of the Voronoi Diagram in 2D and 3D.

With the help of the Voronoi diagram, we can answer nearest neighbor queries between the sites efficiently. The nearest neighbors from $S$ for a point $p_i \in S$ lie inside a neighbor region of $p_i$ in $\mathrm{VD}(S)$. This is easy to see by the following general argument, which holds in 2D and 3D.

Let us separate $S$ into two sets $S_1$ and $S_2$. The nearest neigbors of $S_1$ and $S_2$; that is, two points $p_1 \in S_1$ and $p_2 \in S_2$, where

$$d(p_1, p_2) = \min\{d(p_i, p_j) : p_i \in S_1, p_j \in S_2\}$$

share a surface in the Voronoi diagram of $S$ and, in turn, an edge in the Delaunay triangulation. This is easy to see by the following argument. If the line segment between $p_1$ and $p_2$ contains a point $c$ that belongs to a cell $p \neq p_1, p_2$, then we have $d(p, c) < d(p_1, c), d(p_2, c)$. Let $p \in S_1$. By the triangle inequality, we have

$$d(p, p_2) \leq d(p, c) + d(c, p_2) < d(p_1, c) + d(c, p_2) = d(p_1, p_2),$$

which gives a contradiction to the definition of $p_1$ and $p_2$.

Now, we can apply this argument for $S_1 = \{p_i\}$ and $S_2 = S \backslash S_1$.

**Theorem 6.11.** *Let $S$ be a set of points in 3D. For the nearest neighbor $p_j \in S$ of a point $p_i \in S$, there is an edge $p_i p_j$ in the Delaunay triangulation of $S$. All nearest neighbors can be found in time proportional to the complexity of the Voronoi diagram.*

Beyond nearest neighbor queries, there are many different geometrical applications of the Voronoi diagram and its dual. Here, we simply list a few of them, together with some performance results, provided that the diagram is given.

The Delaunay triangulation contains the *minimum spanning tree*. By definition, the minimum spanning tree is the smallest graph (with respect to overall edge length) that connects all sites. The algorithm of Kruskal [Kruskal, Jr. 56] computes a minimum spanning tree of an arbitrary graph $G = (V, E)$ in time $O(|E| \log |E|)$. Therefore, the minimum spanning tree of a set of points can be computed applying Kruskal's algorithm on the *sparse* Voronoi diagram.

**Theorem 6.12.** *The minimum spanning tree of a set of points in* 3D *is part of the Delaunay triangulation.*

*Proof:* We apply a divide-and-conquer argument. If the minimum spanning tree of $S$ is given, we can split the tree at every edge into two disjoint subsets of sites $S_1$ and $S_2$. As already shown, the edge connecting $S_1$ and $S_2$ has to be a Delaunay edge. □

The minimum spanning tree gives us a simple heuristic for the traveling salesman problem, TSP for short. The traveling salesman visits all sites and returns to its start point. We are interested in computing the shortest tour (with respect to edge length) that solves this problem. It was shown that the problem of computing the optimal TSP tour is NP-hard. If we follow the edges of the minimum spanning tree in a depth-first manner and return to the start, we have visited every edge exactly twice. The length of the minimum spanning tree is always smaller than a TSP tour. Thus, we can compute a TSP approximation in $O(|E| \log |E|)$ time, where $E$ denotes the set of edges in the Delaunay triangulation.

**Theorem 6.13.** *Following the minimum spanning tree of a set of sites in* 3D *in a depth-first manner gives a 2-approximation of the optimal TSP tour of the sites.*

The Voronoi diagram is helpful also for the subject of localization; see also [Hamacher 95]. Let us assume that the sites in 3D represent sources of danger, such as fire or earthquake. Then we would like to find a location inside a given convex area $A$ so that we have maximum security. In terms of the geometry, we are searching for the biggest empty ball with respect to the sites inside $A$. It can be shown that the center of this ball lies on the vertices of Voronoi diagram of the set of sites or on the boundary of the considered area $A$.

The geometric argument is as follows. For a possible center $c$, we enlarge the ball by moving $c$ until a site of $S$ is met. If this is only a single site, we can further enlarge the ball until a second site is met. Now we are on the surface of a Voronoi region. We move along this bisector and can enlarge the ball until three sites lie on the boundary of the ball. Thus, we are located on a Voronoi edge. Now, we can still enlarge the ball, moving along the edge until finally a fourth site is met.

Within this enlargment process, we might touch the boundary of the area $A$ at some stage. In this case, we have to find an optimal location on $A$. Either this location is given by an intersection with a surface or an edge of VD($S$), or $c$ is somewhere on $A$ if $A$ is met in the first stage of the enlargement. Since $A$ is convex, we will move the center toward a vertex of $A$ in order to enlarge the ball. Altogether, the following holds.

**Theorem 6.14.** *The biggest empty ball of a set of sites $S$ in 3D in a given convex area $A$ has its center $c$ either on a Voronoi vertex of* VD $S$ *or on the boundary of $A$. On the boundary of $A$, $c$ is either an intersection of $A$ with an edge or a surface of the diagram, or $c$ is a vertex of $A$.*

Additionally, for localization planning, the farthest color Voronoi diagram has some nice applications. Assume that we have subsets of sites $P_1, P_2, \ldots, P_k$ such that $P_i$ represents the same source; for example, $P_i$ might represent the location of all supermarkets or all hospitals. Now, we would like to choose a location such that one instance of every source set is nearby. Geometrically speaking, we are searching for the smallest ball that contains at least one instance of every set $P_i$ for $i = 1, \ldots, k$. Let us assume that every $P_i$ has its own color. Obviously, the optimal ball has four sites of different colors on its boundary and no other site with the same color inside the ball. Otherwise, we could shrink the ball. This means that for the center of this ball, the four sites represents the four farthest colors among all the colors. Thus, the center is a Voronoi vertex in the 3D farthest color Voronoi diagram. Altogether, we systematically check all *four-colored* Voronoi vertices in the given diagram.

**Theorem 6.15.** *Let $P_1, P_2, \ldots, P_k$ denote $k$ point sets in* 3D *such that each set has its own color. The smallest ball that contains at least one instance of every set $P_i$ for $i = 1, \ldots, k$ has its center on a four-colored Voronoi vertex of the farthest color Voronoi diagram.*

There are many other applications, such as motion planning or clustering; see the monographs mentioned at the beginning of this section. For example, using Voronoi diagrams for clustering of objects is shown in [Dehne and Noltemeier 85].

The complexity and the running time of the corresponding algorithms depend on the complexity of the Voronoi diagram. As we have already mentioned, the complexity of the diagrams in 3D is linear in many practical situations also. Thus, many of the presented problems can be solved in 3D with almost the same time bound as in the 2D case.

## 6.5. Voronoi Diagrams in Computer Graphics

In the following, we will present some applications in computer graphics that use Voronoi diagrams (or Delaunay graphs) to achieve an elegant or simple solution.

**Figure 6.26.** Roman mosaics from baths (Mosaic a, Mosaic b). (See Color Plate X.) ((left) Courtsey of D. Mesher at the SJSU Jewish Studies program. (right) Courtesy of the Hellenic Ministry of Culture.

## 6.5.1. Mosaics

One of the major subjects in computer graphics is the photorealistic rendering of scenes or environments modeled in the computer. However, a fairly recent topic in computer graphics is the so-called *non-photorealistic rendering* (NPR). For example, NPR might be used to produce mosaics.

Mosaics are made up of small pieces, such as stones, pebbles, or shells, that are closely set on a surface. The pieces are usually variously colored, so that the overall ensemble creates a picture when viewed from a distance. The individual pieces are small compared to the size of the mosaic and more or less congruent to each other. Examples are the wonderful Greek, Roman (see Figure 6.26), or Byzantine mosaics found in baths, houses, and temples.

In addition, a mosaic consisting of $N$ pieces can potentially carry more information than an image made up of $N$ pixels, because the pieces can carry extra information, such as shape and orientation. For instance, in the examples in Figure 6.26, one can see that the orientation of the pieces follows the edges, contours, or other features (e.g., medial axis) of the image. Because this is a very characteristic feature of mosaics, we want to achieve this effect in our computer-generated ones, too. However, we are now given two conflicting tasks: using only congruent pieces, we are to place them such that the area between them (the grout) is minimized, and such that the orientation of all of them (we assume the pieces have a unique "up" direction) is as close as possible to a prescribed one.

In the following, let us assume that the pieces are squares of equal size (later, we will see how to lift this restriction easily). At each point in the mosaic, we will prescribe the orientation of the pieces by a vector field $\phi$. A more formal statement is as follows.

**Problem 6.16.** Given a region $R \subset \mathbb{R}^2$ in the plane, and a vector field $\phi : R \to \mathbb{R}^2$, find $N$ sites $P_i \in R$ and $N$ orientations $o_i \in \mathbb{R}^2$, such that the following conditions hold: if $N$ squares of side length $s$ are placed on the sites, one at each $P_i$ and with orientation $o_i$, then

- they are disjoint;
- the area they cover is maximized;
- the sum of the angles between the $o_i$ and $\phi(P_i)$ is minimized.

The idea is that the region $R$ covers an image and the direction field is designed such that it aligns the squares with the edges in the image. When the $P_i$ have been found, we will color each square uniformly by the color found at position $P_i$ in the image.

Choosing exactly equal squares for the pieces of the mosaic is a simplification, but we can make the mosaic look more handcrafted after the sites have been found by distorting each square randomly by a small amount.

Actually, the problem of placing the sites $P_i$, as stated above, is the general problem of generating *low-discrepancy sampling patterns*, or *low-energy particle configurations*.[2]

**Placing the sites.** First, we will consider only the problem of placing the sites for equally aligned squares (i. e., we omit the orientation for the moment).

Imagine $N$ particles in region $R$, each one repelling its neighbors. When the particles come to a rest, we can compute the Voronoi diagram of the particle set, and this Voronoi will be a special variant, called the *centroidal Voronoi diagram* (CVD). In a CVD, we have the additional property that each site is located at the mass center (centroid) of its Voronoi region. Figure 6.27 shows an example; another example is a honeycomb.

More formally, let $V_i$ be the Voronoi region of site $P_i$. Then the *centroid* or *center of mass* is

$$C_i = \frac{\int_{V_i} x \, dx}{\text{Area}(A)}. \tag{6.1}$$

More generally, if we take a probability density function $\rho$ over $R$ into account, then the centroid is

$$C_i = \frac{\int_{V_i} x\rho(x) \, dx}{\int_{V_i} \rho(x) \, dx}. \tag{6.2}$$

[2]This problem occurs in many other areas, such as quantization (where a range of quantities is replaced by one representative from a code book), optimal placement of resources (where all resources should be placed such that the sum of the distances everyone has to travel to the nearest resource is minimized), or the modeling of the territorial behavior of animals (where each animal tries to maximize its own territory, ousting its neighbors as far as possible).

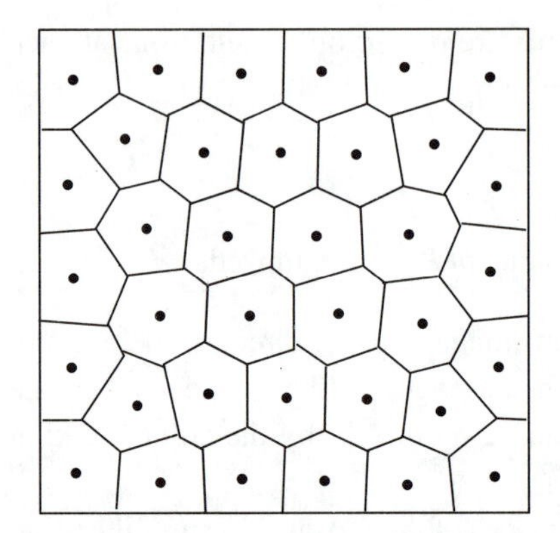

**Figure 6.27.** Example centroidal Voronoi diagram (CVD).

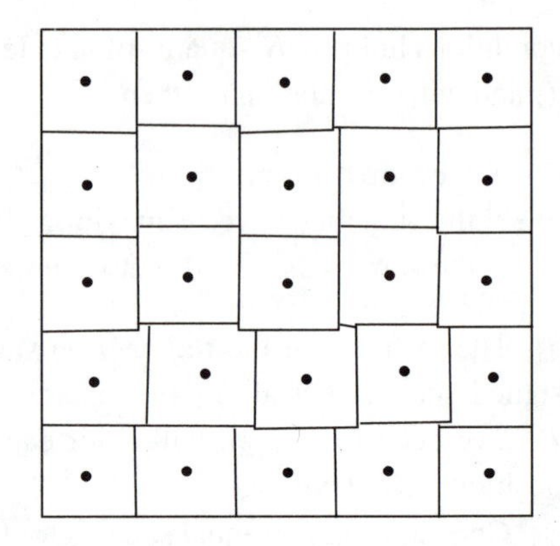

**Figure 6.28.** Example CVD with the Manhattan metric instead of the Euclidean metric.

Then the Voronoi diagram is a CVD if and only if

$$\forall i : C_i = P_i.$$

For a given Voronoi diagram $\{P_i, V_i\}_{i=1\ldots N}$, we define the *energy* (sometimes also called the cost) as

$$E(\{P_i, V_i\}_{i=1,\ldots,N}) = \sum_{i=1}^{N} \int_{V_i} \rho(\mathbf{x})|\mathbf{x} - C_i| \, \mathrm{d}x \,. \tag{6.3}$$

[Du et al. 99] proved that a necessary condition for $E$ to be minimal is that $\{P_i, V_i\}_{i=1\ldots N}$ is a CVD of $R$.

Observe that the CVDs produced so far look locally like hexagonal tilings. This is because in Equation (6.3), the Euclidean metric is used to measure distances and, thus, the range of influence of a Voronoi region. However, we would like more square-like tilings. Consequently, we need to choose a different metric, namely the $L_1$ metric, or Manhattan distance metric, $|x_1 - x_2| + |y_1 - y_2|$. Then the CVD becomes much more like a tiling with squares (see Figure 6.28).

In order to compute a CVD for a given region $R$, there are two well-known algorithms; Lloyd's method and MacQueen's method [Ju et al. 02]. While the latter is simpler to perform on a CPU, we will present the former, for reasons you will see shortly. In addition, Deussen et al. [Deussen et al. 00] report that it seems to be more stable than other particle spreading techniques (although, theoretically, its convergence in two or more dimensions has not been proven).

```
select an initial set of N random points P_i ∈ R
repeat
        construct the Voronoi regions V_i
        determine the centroids C_i of all V_i
        set P_i := C_i
until E({P_i, V_i}) is "small enough"
```

**Algorithm 6.9.** Lloyd's algorithm to compute a centroidal Voronoi diagram for a region $R$.

Lloyd's method is basically an iteration between computing Voronoi diagrams and mass centers.[3] Algorithm 6.9 outlines the procedure. There are two open issues: how to compute the Voronoi regions and how to determine the centroid.

Since an approximate CVD will serve our purpose perfectly, we can compute the Voronoi regions by the same method as in Section 5.1.2 [Hausner 01]: we place a right cone with its apex on each of the $P_i$, with its axis along the $z$-axis. Then we project all of them orthogonally on the $xy$-plane. This is exactly what happens when rendering all the cones into the frame buffer. For each pixel, the Z-buffer performs the nearest neighbor "search," because the pixel will get filled by that cone whose apex is nearest. For each cone, we choose a unique color, so that in the end, all pixels in the same Voronoi region will have the same ID (i. e., color).[4]

With this approach, we can even accommodate different metrics, because the shape of the cones embodies the metric. The function of a circular cone is $z = \sqrt{(x_1 - x_2)^2 + (y_1 - y_2)^2}$, which is exactly the distance of a point $\mathbf{x}_1$ from the apex $\mathbf{x}_2$. If we want to embody the Manhattan metric, we just render square cones whose function is $z = |x_1 - x_2| + |y_1 - y_2|$ (see Figure 6.29).

**Figure 6.29.** Different shapes of cones embody different metrics.

[3]Note that, in general, there is no unique CVD for a region $R$.

[4]Note that this technique does not compute the combinatorial version of the Voronoi diagram, i. e., it does not create combinatorial descriptions of the Voronoi regions, nor does it identify the neighbors of each region. But for Lloyd's algorithm, this is not needed.

What remains is to compute the centroids of the Voronoi regions (see Equation (6.1)). In a conventional CPU implementation, this would be done by a Monte Carlo–based integration technique [Ju et al. 02]. Here, we can simply scan the frame buffer and compute, for each color ID, the average of all $x$- and $y$-values, respectively, of pixels that have been colored with that ID. Again, this gives just an approximate centroid, but this is fine for our application.[5]

**Orienting the pieces.** So far, we have ignored the orientation of the pieces that are to be placed at the $P_i$. With circular cones, there is no distinguished orientation; and with square cones we have, so far, oriented them all in the same way.

The modification we have to make is quite simple [Hausner 01]: instead of rendering square cones with the same orientation, we just rotate the cones according to the vector field at the apex, $\phi(P_i)$.[6]

Here, we won't go into too much detail about how to obtain a suitable vector field $\phi$. One simple way is similar to the method in Section 5.1.2 of computing the Voronoi regions [Hausner 01]. We start with a set of curves in the region $R$, along which the mosaic pieces should be aligned. For each of the curves, we render (with orthogonal projection) a (tessellated) "curved mountain." Then, we read back the Z-buffer, which contains kind of a height field, and compute the discrete gradient at each pixel (see Figure 6.30). These gradients will be perpendicular to the curves at pixels close to the curve.

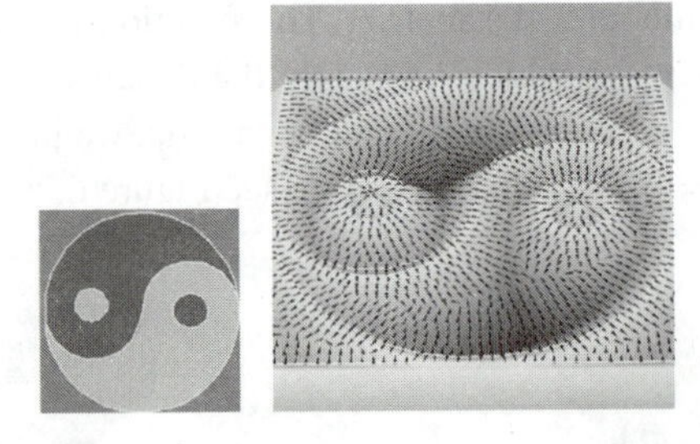

**Figure 6.30.** The vector field for orienting the mosaic pieces can be computed by discrete gradients from a set of curves [Hausner 01]. (Courtesy of Alejo Hausner and ACM.)

[5]The relative error of a centroid depends on the size of its Voronoi region, i.e., the number of pixels it occupies. In extreme cases, this could cause one centroid to vanish. However, this can be solved easily by rendering a small portion again with a higher resolution, or by removing the site and inserting a new random site in another, large Voronoi region.

[6]Of course, we now depart from the CVD, but we better accommodate the desired orientation of the pieces.

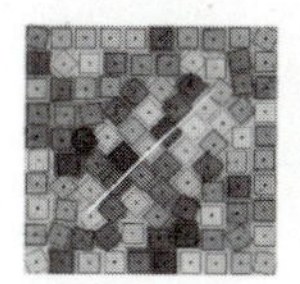 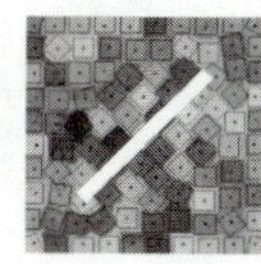 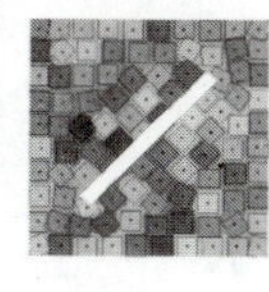 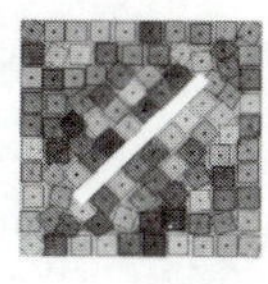 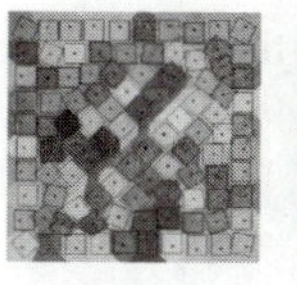 

**Figure 6.31.** Process to preserve edges of the original image in the final mosaic. (Hausner 01). (See Color Plate XI.) (Courtesy of Alejo Hausner and ACM.)

Edge avoidance. One prominent feature of mosaics is that the pieces usually do not straddle edges or contours of the image, to make them more pronounced in the mosaic. This is something we want to simulate in our computer-generated mosaics, too. This can be achieved by modifying the density function $\rho$ appropriately. Here, we should set $\rho$ to zero in a band around the edges, with the width of the band about the same as the size of the pieces.

Since we compute the new Voronoi region centroids by scanning the frame buffer for pixels with the region's color ID, all we need to do is to insert an additional step just before the scanning, in which we set all pixels inside the edge bands to a neutral color (e. g., white), so that they won't be counted for any Voronoi region. Thus, the Voronoi sites (i. e., centroids) will be pushed away from the edges. This process is illustrated in Figure 6.31.

Overall, the algorithm for computing a mosaic is shown in Algorithm 6.10. Compared to Algorithm 6.9, we might want to change the convergence criterion, e. g., we could stop if the energy does not decrease significantly anymore. The

---

1: select an initial set of $N$ random points $P_i \in R$
2: **repeat**
3: place a square pyramid at each site $P_i$, with apex at $P_i$ and axis along z-axis, and rotate it about z-axis to align it with the direction field $\phi(P_i)$
4: render all pyramids using orthogonal projection into the frame buffer, each with a unique color
5: draw edge bands over the discrete Voronoi diagram in frame buffer, using neutral color (white)
6: read color buffer back
7: scan buffer and compute average $x$ and $y$ values (centroids $C_i$) for each color "ID" $i$
8: set $P_i := C_i$
9: **until** convergence criterion is met
10: draw a square of size $s$ at each $P_i$ with uniform color found at position $P_i$ in the original image

---

**Algorithm 6.10.** Algorithm to generate a mosaic.

size $s$ of the final mosaic pieces can be estimated as

$$s = \delta \sqrt{\frac{S}{N}},$$

where $S$ is the number of pixels in the region $R$, and $\delta$ is a factor to account for the "dead space" between the squares (grout) due to different orientations.

**Varying the piece size.** Just as with the orientation, we can adapt the algorithm to vary the size of the mosaic pieces (squares) simply by changing the slope of the square cones (pyramids)

$$z = \alpha(|x_1 - x_2| + |y_1 - y_2|)$$

or even the aspect ratio of the squares

$$z = \beta|x_1 - x_2| + \frac{1}{\beta}|y_1 - y_2|.$$

This can be useful in areas of the image with fine detail.

Basically, this amounts to changing the probability density function $\rho$ in Equations (6.2) and (6.3). Accordingly, Lloyd's algorithm will pack the squares more densely in areas with high probability density.

**Figure 6.32.** Mosaics generated by randomly placing 6000 Voronoi sites on an image and coloring each Voronoi region uniformly by the color in the image. (See Color Plate XII.)

**Figure 6.33.** Two examples of mosaics (Hausner 01), generated with the CVD method. (See Color Plate XIII.) (Courtesy of Alejo Hausner and ACM.)

Examples. Figure 6.32 shows an example of a Voronoi diagram, where a number of sites have been placed randomly on an image, and each Voronoi cell is colored uniformly according to the color in the image at the Voronoi site. This is basically the technique presented in [Haeberli 90]. The example was generated using the software of [Hoff III et al. 99].

Figure 6.33 shows two examples generated with the method described above.

## 6.5.2. Natural Neighbor Interpolation

Interpolation is a very important technique that arises in many situations in computer graphics and other fields of engineering.

The problem statement can be formulated quite simply.

> Let $\mathcal{P} = \{P_1, \ldots, P_n\} \subset R^d$ be a set of fixed points in $d$-dimensional space. Sometimes, we call these points *data sites*. At each of these, we are given a real value $z_i$ (here, we will generalize this a bit). Now, the *interpolation problem* is to find a *good* (in some sense) function $\varphi$ such that
>
> $$\forall i = 1, \ldots, n \ : \ \varphi(P_i) = z_i.$$

An example is the interpolation of vectors in space, e. g., normals, which are given only at a discrete set of points, e. g., the vertices of a model or a point cloud obtained from a scanner (see Figure 7.19).

There are almost infinite ways to solve this problem. Here, we will describe one method that utilizes Voronoi diagrams in a neat way.

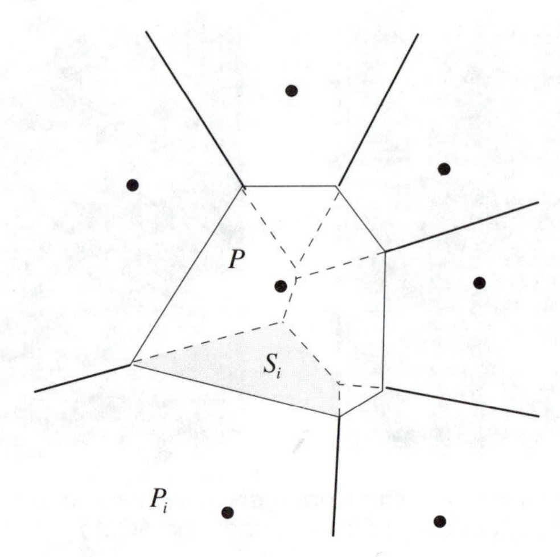


**Figure 6.34.** Inserting $P'$ changes the Voronoi diagram. The thick and dashed lines together constitute the Voronoi diagram over $\mathcal{P}$, while the thick and the thin lines show the Voronoi diagram over $\mathcal{P}'$. The thin lines alone represent the Voronoi region for $P$, the point at which the interpolation is to be computed.

An extremely simple method is the following. Given a point $P$, determine the nearest neighbor of $P$, $P_{i^*} \in \mathcal{P}$. Set $\varphi(P) := z_{i^*}$. Finding the nearest neighbor amounts to identifying the Voronoi region in which $P$ is located. However, an explicit construction of the Voronoi diagram is not needed (although it could be used to accelerate the search). Obviously, this interpolant is a piecewise constant function and, thus, not continuous.

The remedy is, more or less, clear: we need to involve the "neighbors" of $P$ or $P_{i^*}$, respectively. This was proposed by [Sibson 80, Sibson 81]. The idea is to utilize the Voronoi diagram for determining the neighbors of $P$ *and* for determining their "influence" on $P$.

More precisely, we first construct the Voronoi diagram over $\mathcal{P}$, $\mathcal{V}(\mathcal{P})$. Voronoi regions that share a $(d-1)$-dimensional face[7] are called *neighbors*.

Second, we construct the Voronoi diagram over $\mathcal{P}' := \mathcal{P} \cup P$, $\mathcal{V}(\mathcal{P}')$. This is equivalent to inserting $P$ into the Delaunay diagram over $\mathcal{P}$ (see Section 6.2), and then carrying the changes over to the Voronoi diagram. Note that the changes will usually be only local (see Figure 6.34).

Let us denote the region of a site $P_i$ in $\mathcal{V}(\mathcal{P})$ by $R_i$, the region of the same site in $\mathcal{V}(\mathcal{P}')$ by $R_i'$, and the region of site $P$ by $R$. The *natural neighbors* of $P$ are defined as the points $P_i$ that are neighbors of $P$ in $\mathcal{V}(\mathcal{P}')$.

[7] In 2D, these are common edges in the diagram; in 3D, these are the common polygons.

Next, we define the influence of a natural neighbor. This is done by further partitioning region $R$ into subregions

$$S_i := R_i \cap R.$$

From Figure 6.34, it should be obvious that $P_i$ is a natural neighbor of $P$ if and only if $S_i \neq \emptyset$. Let $\mathrm{Vol}(S_i)$ denote the volume of these subregions (or, more precisely, the Lebesgue measure in $d$ dimensions). Now, the *natural coordinate* of $P$ associated to $P_i$ is defined by

$$s_i(P) = \frac{\mathrm{Vol}(S_i)}{\sum_j \mathrm{Vol}(S_j)}. \tag{6.4}$$

(Note that $\sum_j \mathrm{Vol}(S_j) = \mathrm{Vol}(R)$.) Sometimes, these are also called *Sibson's coordinates*.

Straightforwardly, the interpolant now is

$$\varphi(P) = \sum_{i=1}^{n} z_i s_i(P). \tag{6.5}$$

The natural coordinates have three important properties [Sibson 80, Sibson 81], as follows.

1. The $s_i(P)$ are well-defined if and only if $P$ is an inner point of the convex hull.

2. $s_i(P)$ is a continuous function of $P$, and is continuously differentiable *except* at the data sites in $\mathcal{P}$. More precisely, [Piper 93] $\varphi(P)$ is

   (a) $C^0$, but not necessarily $C^1$ if $P$ coincides with some data site $P_i$;

   (b) $C^1$, but not necessarily $C^2$ if $P$ lies on one of the Delaunay spheres of the data sites;

   (c) $C^\infty$ everywhere else.

3. The $s_i(P)$ satisfy the identity

$$\sum_i s_i(P) P_i = P; \tag{6.6}$$

   i. e., $P$ is a convex combination of its neighbors.[8]

[8] Here, we have, somewhat laxly, identified the points with their position vectors.

Thus, the interpolant $\varphi$ is $C^\infty$ almost everywhere *except* on the data sites and a number of spheres.

Clearly, all $s_j(P) \to \delta_{ij}$ (Kronecker delta) as $P \to P_i$. So, $\varphi(P) \to z_i$ as $P \to P_i$, no matter, from which direction $P$ approaches $P_i$. However, this can happen with different "slopes," depending on the direction from which we approach $P_i$, because the overlap of the Voronoi regions can decrease with different "speeds." Something similar happens across the borders of the Delaunay spheres, because there, the set of natural neighbors of $P$ changes: as long as $P$ stays outside a Delaunay sphere, that sphere remains a valid Delaunay sphere; however, as soon as $P$ enters the sphere, it is no longer valid and must be replaced by (usually three) other Delaunay spheres. And, in that event, $P$ "steals" some of the space of the Voronoi region of the farthest site of that (former) Delaunay sphere.

The reason the $s_i(P)$ and, thus, $\varphi$ are well-defined only inside the convex hull of $\mathcal{P}$ is that if $P$ is outside, then $P$ becomes itself a member of the convex hull. Consequently, its Voronoi region is unbounded and, thus, some of the $s_i(P)$ become unbounded. So, if we want to interpolate outside the convex hull of $\mathcal{P}$, too, we need to introduce sort of "sentinel points;" i. e., we have to add points to $\mathcal{P}$ on a sufficiently large bounding box around $\mathcal{P}$.

Instead of attaching fixed values to each data site, we can attach functions. So, assume that to each $P_i$ is attached a continuously differentiable function $h_i \in C^1$ from $\mathbb{R}^d$ to $\mathbb{R}$ satisfying $h_i(P_i) = 0$ (for instance, the distance from $P_i$). So, the more general interpolation now is $\varphi(P) = \sum_i^n s_i(P)h_i(P)$.

One question remaining is: how can we increase continuity of the interpolant? One way was proposed by [Hiyoshi and Sugihara 00].[9] However, in the following, we will present a simpler way [Boissonnat and Cazals 01, Boissonnat and Cazals 00].

We just define the natural neighbor interpolation as

$$h(P) = \sum_{i=1}^{n} s_i^{1+\omega}(P)h_i(P), \tag{6.7}$$

with some arbitrarily small $\omega > 0$.

**Lemma 6.17.** *$h(P)$ interpolates the $h_i$, and $h \in C^1$, provided $h_i(P_i) = 0$.*

*Proof:* As we stated earlier, the $s_i$ and (by definition) the $h_i$ are continuous over $\mathbb{R}^d$ and, thus, $h$ is continuous over $\mathbb{R}^d$. Since $s_j(P_i) = \delta_{ij}$, we have $h(P_i) = h_i(P_i)$. (It happens, in our case, that $\forall i : h(P_i) = 0$.)

[9] In fact, they can achieve arbitrary continuity. However, the method becomes more and more computationally expensive as continuity increases.

```
generate set S_1 of random points inside bounding box of 𝒫
compute S' := {X ∈ S_1 | P is nearest neighbor to X}
for all P_i do
    compute S_i := {X ∈ S' | p_i is nearest neighbor to X} s_i ← |S_i|/|S'|
end for
```

**Algorithm 6.11.** Simple Monte Carlo–like method to compute natural neighbors.

Since the $s_i$ are in $C^1$ at all $P \notin \mathcal{P}$, we can differentiate $h$ yielding

$$\frac{\partial h}{\partial x^j}(P) = \sum_{i=1}^{n} s_i^{1+\omega}(P)\, h_i(P) + (1+\omega)\, s_i^{\omega}(P)\, \frac{\partial s_i}{\partial x^j}(P)\, h_i(P), \qquad \square$$

where $x^j$ denotes the $j$th Cartesian coordinate.

As $P \to P_k$, we have $h_k(P) \to 0$, $s_i(P) \to \delta_{ik}$. So the second term inside the sum above vanishes for all $i$, as $P$ approaches $P_i$, and from the first term only $s_k$ remains. Thus, $\frac{\partial h}{\partial x^j}(P) \to \frac{\partial h_k}{\partial x^j}(P_k)$, which is, by prerequisite, in $C^1$.

The final question remaining is: how do we actually compute the subregions $S_i$ of the Voronoi region $R$ of $P$ (the "new" point)? Actually, we need to compute only Vol($S_i$).

This latter observation will help to achieve a fairly fast way to compute the natural coordinates. There are two possibilities. One is to *simulate* an insertion of $P$ into the Delaunay triangulation of $\mathcal{P}$ [Boissonnat and Cazals 00]. This is basically the algorithm described in Section 6.2, except that we identify only the conflicting tetrahedra, and then walk around them to accumulate the volumes. So, the complexity of computing the natural coordinates for one $P$ using this approach is the same as the one for Delaunay insertion.

Another possibility is to use randomized sampling to compute the volumes in a Monte-Carlo-like fashion. Actually, we don't even need to compute the volumes themselves, but just the *ratios*. So we just generate a number of random points inside the bounding box of $\mathcal{P}'$. For each of them, we compute its nearest neighbor with respect to $\mathcal{P}'$, and we throw away those points for which $P$ is not the nearest neighbor (we are only interested in random points that are located inside the Voronoi region of $P$). For each of the random points remaining, we determine the nearest neighbor with respect to $\mathcal{P}$. Now, the natural coordinate of $P$ with respect to $P_i$ is just the ratio of the respective numbers of points. This simple algorithm is summarized in Algorithm 6.11.

**Plate I.** A superposition of a terrain and a TIN of its height field (Wahl et al. 04). (See Figure 1.3.)

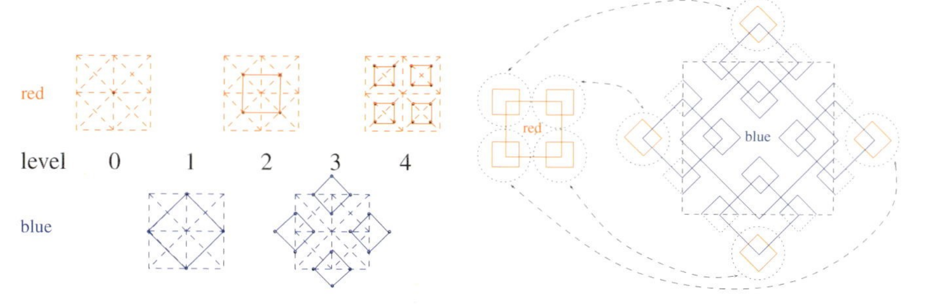


**Plate II.** The 4-8 subdivision can be generated by two interleaved quadtrees. The solid lines connect siblings that share a common father. (See Figure 1.6.)

**Plate III.** The red quadtree can be stored in the unused "ghost" nodes of the blue quadtree. (See Figure 1.7.)

**Plate IV.** Some results of the texture synthesis algorithm (Wei and Levoy 00). In each pair, the image on the left is the original one, the one on the right is the (partly) synthesized one. (See Figure 2.14.)

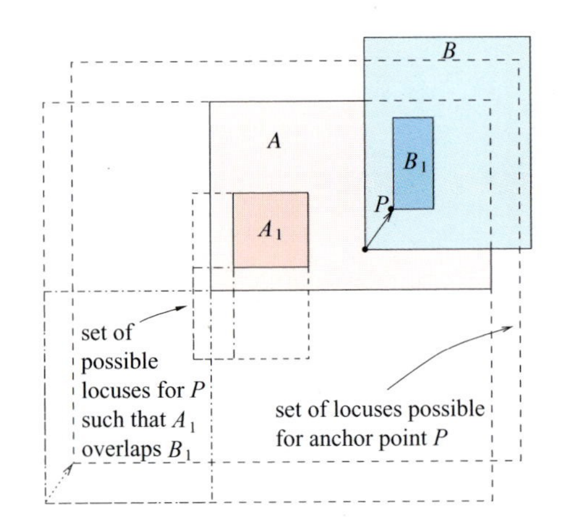


**Plate V.** By estimating the volume of the Minkowski sum of two BVs, we can derive an estimate for the cost of the split of a set of polygons associated with a node. (See Figure 4.8.)

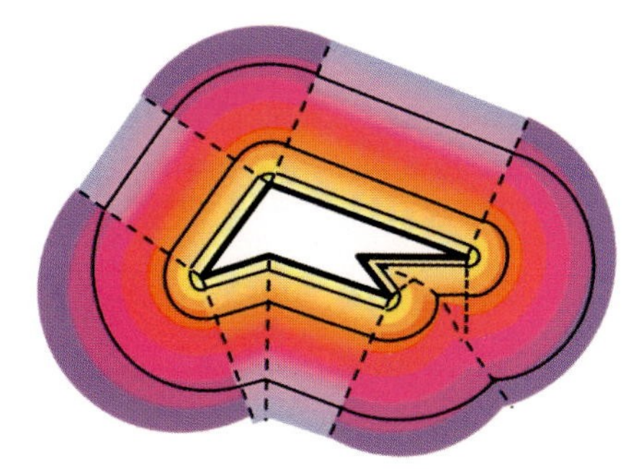

**Plate VI.** Example of a distance field in the plane for a simple polygon (thick black lines). The distance field inside the polygon is not shown. The dashed lines show the Voronoi diagram for the same polygon. The thin lines show isosurfaces for various isovalues. (See Figure 5.1.)

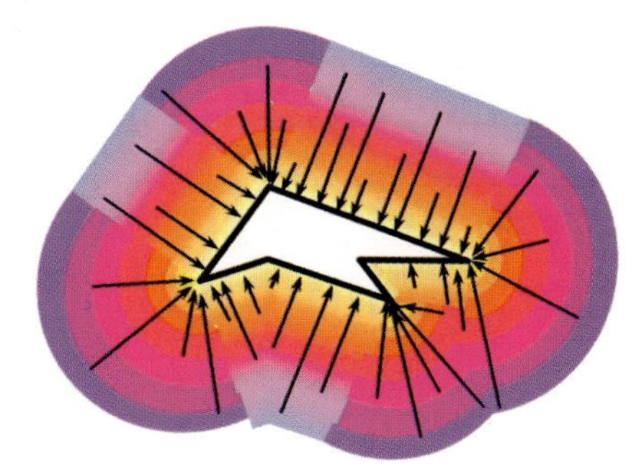

**Plate VII.** The vector distance field for the same polygon as shown on the left. Only a few vectors are shown, although (in theory) every point of the field has a vector. (See Figure 5.2.)

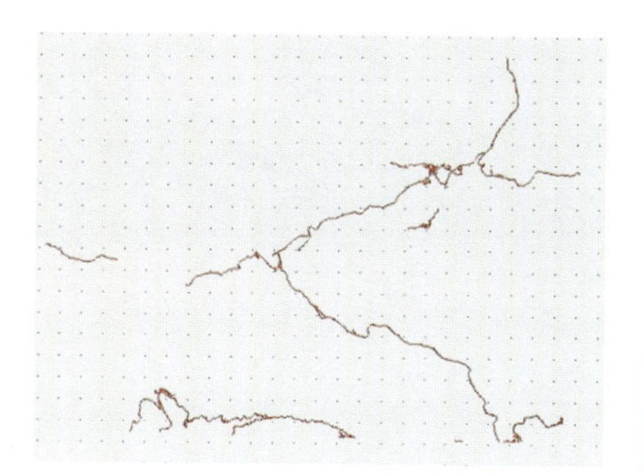

**Plate VIII.** A network of roads described in the plane by a set of edges (left) and the distance map of these roads. The distance of each point in the plane from a road is color-coded. It could be used, for example, to determine the areas where new houses can be built. (See Figure 5.3.)

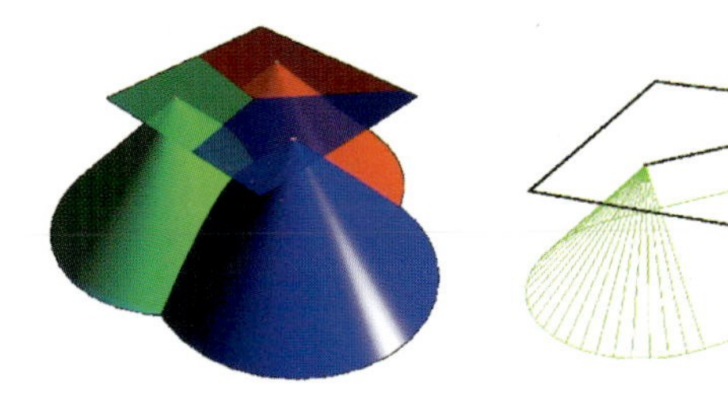

**Plate IX.** (left) The distance function of a point site in the plane is a cone. (middle) More complex sites have a bit more complex distance functions. (right) The distance function of sites in 3D is different for the different slices of the volume (Hoff III et al. 99). (See Figure 5.6.) (Courtesy of K. Hoff, T. Culver, J. Keyser, M. Lin, and D. Manocha, Copyright ACM.)

**Plate X.** Roman mosaics from baths (Mosaic a, Mosaic b). See Figure 6.26.)

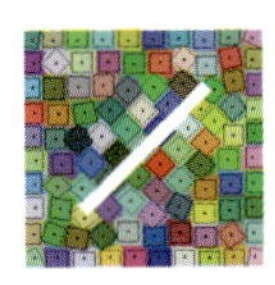

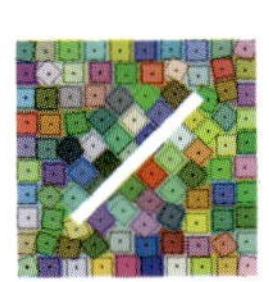

**Plate XI.** Process to preserve edges of the original image in the final mosaic. (Hausner 01). (See Figure 6.31.)

**Plate XII.** Mosaics generated by randomly placing 6000 Voronoi sites on an image and coloring each Voronoi region uniformly by the color in the image. (See Figure 6.32.)

**Plate XIII.** Two examples of mosaics (Hausner 01), generated with the CVD method. (See Figure 6.33.)

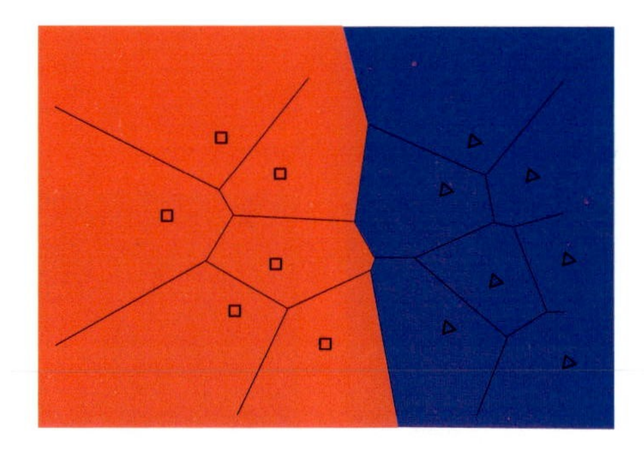

**Plate XIV.** Conceptually, the nearest-neighbor classifier partitions the feature space into Voronoi cells. (See Figure 7.14.)

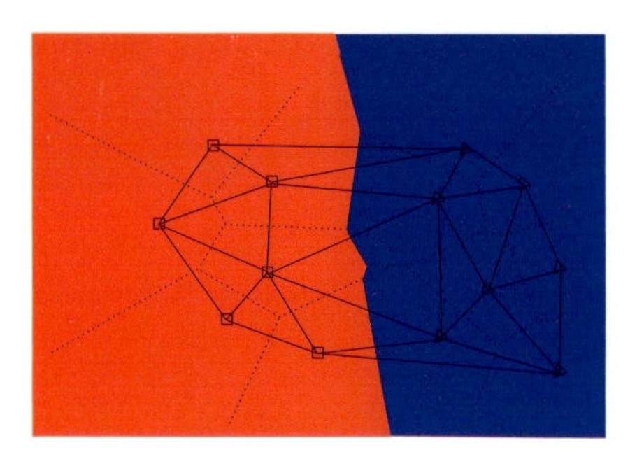

**Plate XV.** The decision boundary of the nearest-neighbor classifier runs between Delaunay neighbors with different colors. (See Figure 7.15.)

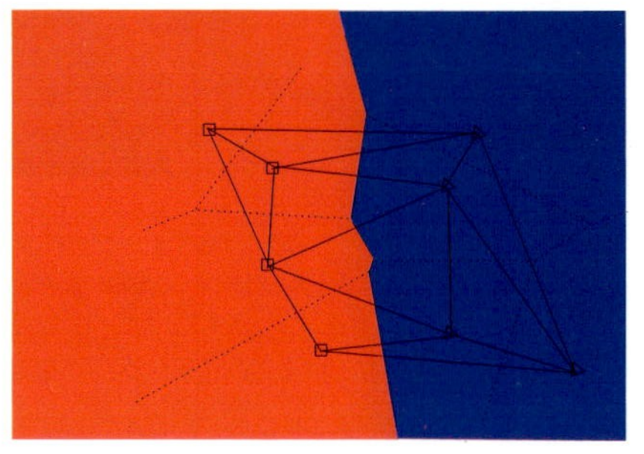

**Plate XVI.** The edited training set has the same decision boundary, but fewer nodes. (See Figure 7.16.)

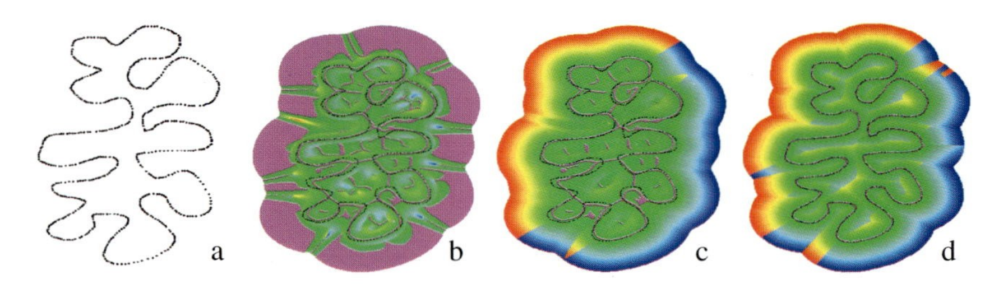


**Plate XVII.** Visualization of the implicit function $f(\mathbf{x})$ over a 2D point cloud. Points $\mathbf{x} \in \mathbb{R}^2$ with $f(\mathbf{x}) \approx 0$, i.e., points on or close to the surface, are shown magenta. Red denotes $f(\mathbf{x}) \gg 0$, and blue denotes $f(\mathbf{x}) \ll 0$. (a) point cloud; (b) reconstructed surface using the definition of (Adamson and Alexa 03); (c) utilizing the centered covariance matrix produces a better surface, but it still has several artifacts; (d) surface and function $f(\mathbf{x})$ based on the SIG. (See Figure 7.21.)

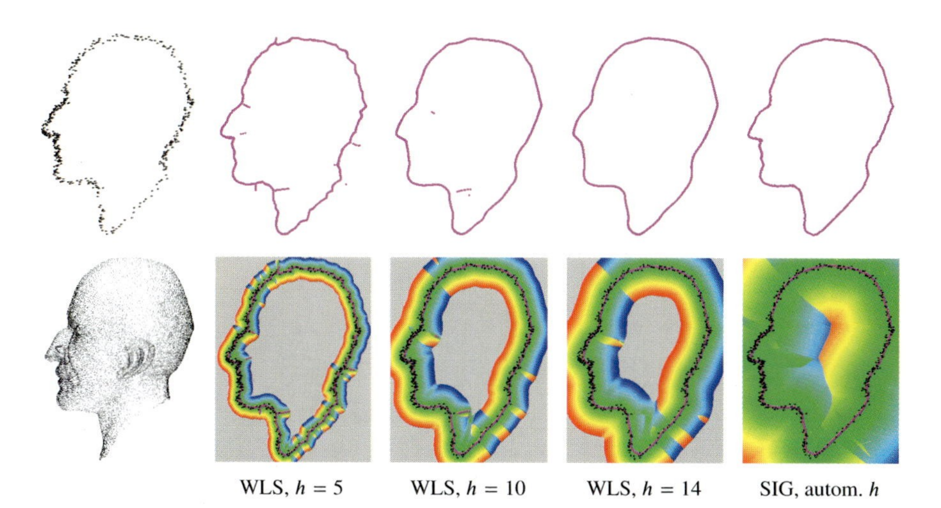


**Plate XVIII.** Reconstructed surface based on simple WLS and with proximity graph (rightmost) for a noisy point cloud obtained from the 3D Max Planck model (leftmost). Notice how fine details, as well as sparsely sampled areas, are handled without manual tuning. (See Figure 7.23.)

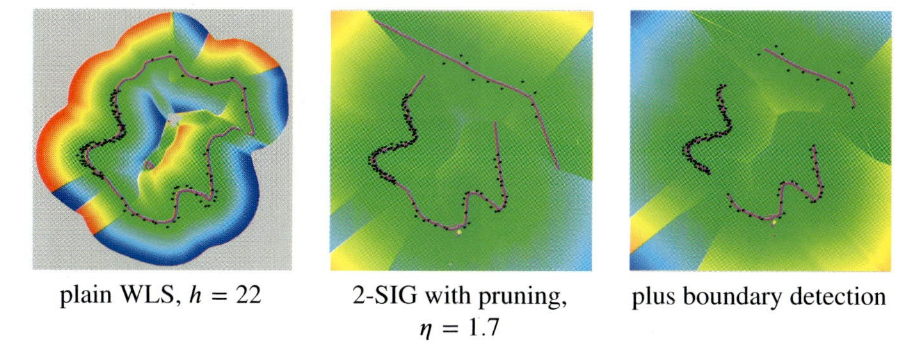


**Plate XIX.** Automatic sampling density estimation for each point allows you to determine the bandwidth automatically and independently of scale and sampling density (middle), and to detect boundaries automatically (right). (See Figure 7.24.)

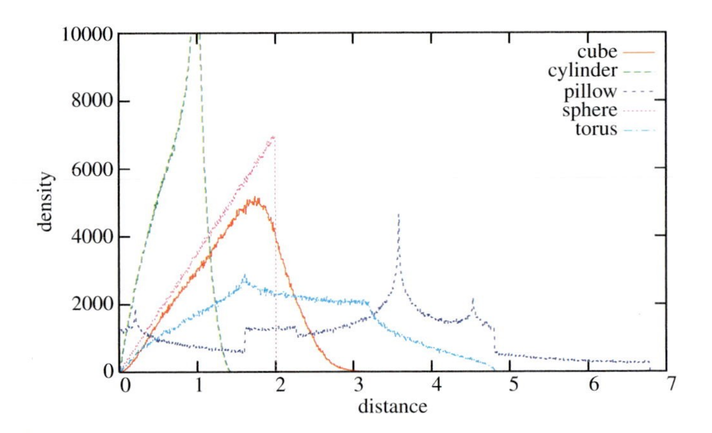


**Plate XX.** The shape distributions of a number of different simple objects. (See Figure 2.15.)

**Plate XXI.** One of the test objects for Figure 7.30. (See Figure 7.31.)

# 7

# Geometric Proximity Graphs

In Chapter 6, we studied a data structure that provides a notion of proximity among its sites, without explicitly saying so, namely Delaunay diagrams. In this chapter, we will learn about other graphs that are defined based on the notion of proximity (with different concrete meaning).

Interest in geometric proximity graphs (sometimes also referred to as *neighborhood graphs*) has been at a high level for about 20 years, in many diverse areas such as computational geometry, theoretical computer science, and graph theory. In a sense, polygonal meshes, which are an extremely common boundary representation of graphical objects, are a special kind of proximity graph.

These graphs can serve as a powerful tool to capture the *structure* or *shape* of otherwise unstructured point sets. Therefore, they have numerous applications

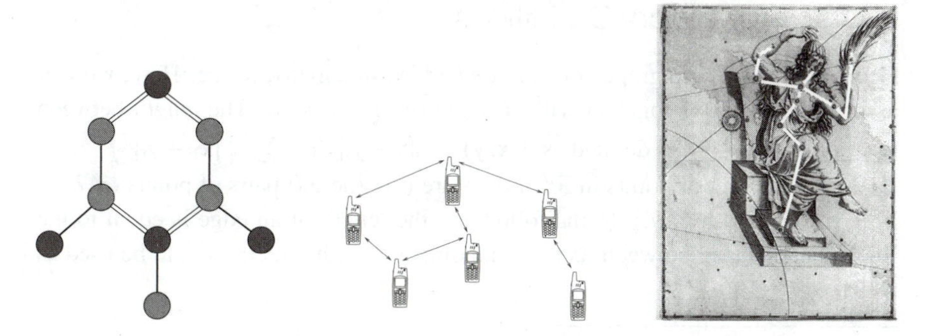

**Figure 7.1.** Identifying all pairs of close nodes in a set can reveal significant structure, such as a molecule or a mobile ad-hoc network. However, care must be exercised when establishing that proximity is indeed relevant (Cassiopeia courtesy of [RAS]).

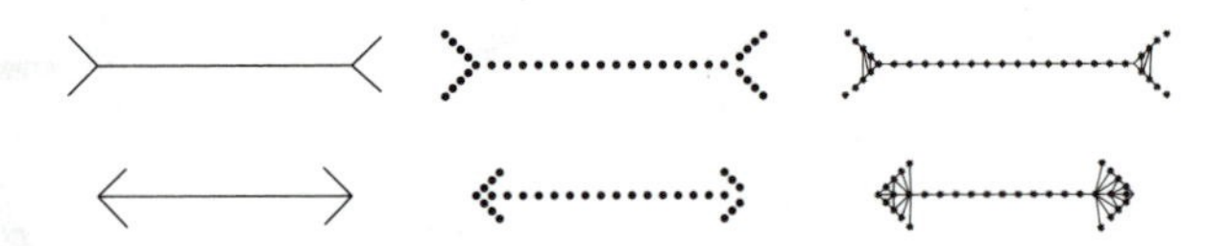

**Figure 7.2.** Proximity graphs can even help in psychology to explain some optical illusions.

in areas such as computer graphics, computer vision, geography, information retrieval, routing in ad-hoc networks, and computational biology, among many others (see Figure 7.1).

Even in psychology, they can help to explain some optical illusions [Sattar 04]. For instance, the well-known Mueller-Lyer illusion [Coren and Girgus 78] consists of two arrows, one with inward pointing arrowheads and the other with outward pointing arrowheads (see Figure 7.2).

In this chapter, we will present a small number of neighborhood graphs (other than polygonal meshes) and a few applications in computer graphics, where they can help to detect structure in point clouds.

## 7.1. A Small Collection of Proximity Graphs

In this section, we define a number of common proximity graphs and highlight some of their properties. This will usually be done by introducing a *neighborhood* that defines exactly when two nodes (i. e., points) are neighbors of each other. The following definitions and discussions build on [Jaromczyk and Toussaint 92].

### 7.1.1. Preliminary Definitions

Geometric graphs are graphs that are *embedded* in a metric space. Here, we will assume the space $\mathbb{R}^d$ together with an $L_p$ norm, $1 \leq p \leq \infty$. The *length* between two points $\mathbf{x}, \mathbf{y} \in \mathbb{R}^d$ is defined as $d(\mathbf{x}, \mathbf{y}) := \|\mathbf{x} - \mathbf{y}\|_p = \left(\sum_{i=1}^{d} |x_i - y_i|^p\right)^{1/p}$.

Let $V$ be a set of points in $\mathbb{R}^d$. Edges are (unordered) pairs of points $(p, q) \in V \times V$, denoted by $pq$.[1] In the following, the length of an edge is equal to the Euclidean distance between its two endpoints (any other metric could be used as well).

[1]Technically, one should distinguish between the combinatorial structure of the graph, given by $V$ and the set of edges $E \subset V \times V$, and the *geometrical realization*, given by points that have an actual location in space and edges that are straight line segments connecting the points. In the following, we will not differentiate between those two.

Proximity graphs are geometric graphs where the edges connect points that are in *proximity* to each other (or where at least one point is close to the other). If $pq$ is an edge in such a proximity graph, then we say that $p$ is a *neighbor* of $q$ (and vice versa).

Exactly how proximity is defined depends on the type of neighborhood graph. It is always a geometric property, which, at least, involves the two points that are neighbors of each other (and, therefore, connected by an edge).

This property often involves spheres, so we define the sphere with center $\mathbf{x}$ and radius $r$ as $S(\mathbf{x}, r) := \{\mathbf{y} \in \mathbb{R}^d \mid d(x, y) = r\}$. Analogously, we define the (closed) *ball* $B(\mathbf{x}, r) := \{\mathbf{y} \in \mathbb{R}^d \mid d(x, y) \leq r\}$.

### 7.1.2. Definitions of Some Proximity Graphs

**Unit disk graph.** This is probably the neighborhood graph with the simplest definition. The set of edges of the unit disk graph, $UDG(V)$, is defined as

$$E := \{pq \mid d(p, q) \leq 1\} ;$$

i. e., two nodes are connected by an edge iff their distance is at most 1.

This definition is motivated by mobile ad-hoc networks, where each node (e. g., cell phone) can reach only other nodes within a certain radius (assuming that all phones have the same transmission power).

**Relative neighborhood graph.** We define a *lune* $L(p, q) := B(\mathbf{p}, d) \cap B(\mathbf{q}, d)$, where $d = \|\mathbf{p} - \mathbf{q}\|$ (see Figure 7.3).

The relative neighborhood graph (RNG) of $V$, $RNG(V)$, is defined by the set of edges

$$E := \{pq \parallel L(p, q) \cap V = \varnothing\} .$$

In other words,

$$pq \in E \Leftrightarrow \nexists v \in V : d(p, v) < d(p, q) \ \wedge \ d(q, v) < d(p, q).$$

Or, in yet other words,

$$pq \in E \Leftrightarrow \forall v \in V : d(p, q) \leq \max\{d(p, v), d(q, v)\} .$$

**Gabriel graphs.** The neighborhood here is a so-called *diameter sphere* $G(p, q) := B\left(\frac{\mathbf{p}+\mathbf{q}}{2}, \frac{d}{2}\right)$, where $d = \|\mathbf{p} - \mathbf{q}\|$ (see Figure 7.4).

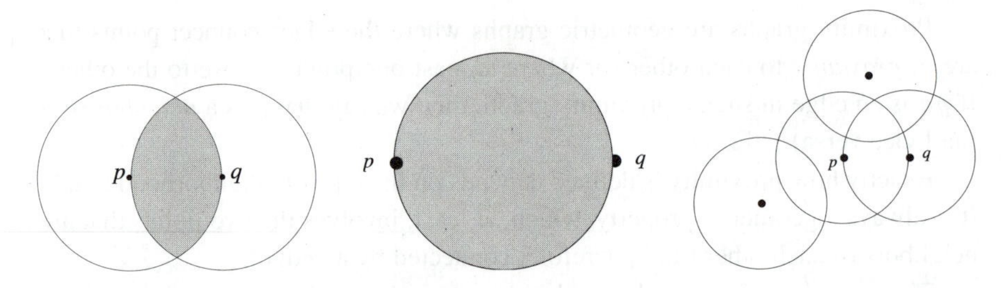


**Figure 7.3.** The *lune* is the defining neighborhood of the relative neighborhood graph.

**Figure 7.4.** The defining neighborhood for the Gabriel graph is the *diameter sphere.*

**Figure 7.5.** The sphere-of-influence graph is defined by *spheres of influence* around each point.

The Gabriel graph (GG) over $V$, $GG(V)$, is defined by the set of edges

$$E := \{pq \parallel G(p,q) \cap V = \varnothing\}.$$

In other words,

$$pq \in E \Leftrightarrow \forall v \in V : d(p,q) \le \sqrt{d^2(p,v) + d^2(q,v)}.$$

**$\beta$-skeletons.** This is a family of neighborhood graphs, parameterized by $\beta$, $1 \le \beta < \infty$.

For a fixed $\beta$, the neighborhood is the intersection of two spheres:

$$U_\beta(p,q) := B\left((1-\frac{\beta}{2})\mathbf{p} + \frac{\beta}{2}\mathbf{q}, \frac{\beta}{2}d\right) \cap B\left((1-\frac{\beta}{2})\mathbf{q} + \frac{\beta}{2}\mathbf{p}, \frac{\beta}{2}d\right),$$

where $d = \frac{\beta}{2}\|\mathbf{p}-\mathbf{q}\|$. The $\beta$-skeleton over $V$, $BG_\beta(V)$, is defined by the set of edges

$$E := \left\{pq \parallel U_\beta(p,q) \cap V = \varnothing\right\}.$$

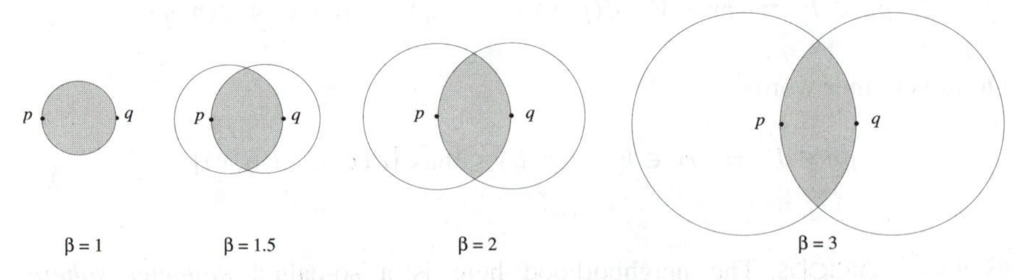


**Figure 7.6.** Examples of the family of lunes that define the $\beta$-skeleton.

It is easy to see that $RNG(V) = BG_2(V)$ and $GG(V) = BG_1(V)$. Moreover, this family of proximity graphs is monotonic with respect to $\beta$, i. e., $\beta_1 > \beta_2 \Rightarrow BG_{\beta_1} \subset BG_{\beta_2}$. In other words, lower values of $\beta$ give denser graphs.

**Sphere-of-influence graph.** The sphere-of-influence graph (SIG) seems to be less well-known [Michael and Quint 03, Boyer et al. 00, Jaromczyk and Toussaint 92].

While in the previous two graphs, the neighborhood is defined between *pairs* of points, here, the neighborhood, its "sphere of influence," is defined for each point individually. More precisely, for each point $p \in V$, the radius $r_p$ to its nearest neighbor is determined. Then the SIG over $V$, $SIG(V)$, is defined by the set of edges

$$E := \left\{ pq \parallel d(p,q) \leq r_p + r_q \right\} .$$

The SIG tends to connect points that are "close" to each other relative to the local point density. In contrast, the RNG and the GG tend to connect points such that the overall edge length is small (think of cities and highways: it makes sense to connect two cities directly by a highway, *unless* there is a third one close by that could be connected as well by a small detour). The extreme along that line is the minimum spanning tree (see below) that has the least overall edge length.

Another difference is that the SIG, unlike the RNG or GG, can be disconnected (see Figure 7.7). This may or may not be desired, depending on the application.

A straightforward way to reduce the likeliness or the number of "gaps" in the SIG is an extension, the $r$-SIG, $r \in \mathbb{N}$ [Klein and Zachmann 04b]. Here, the

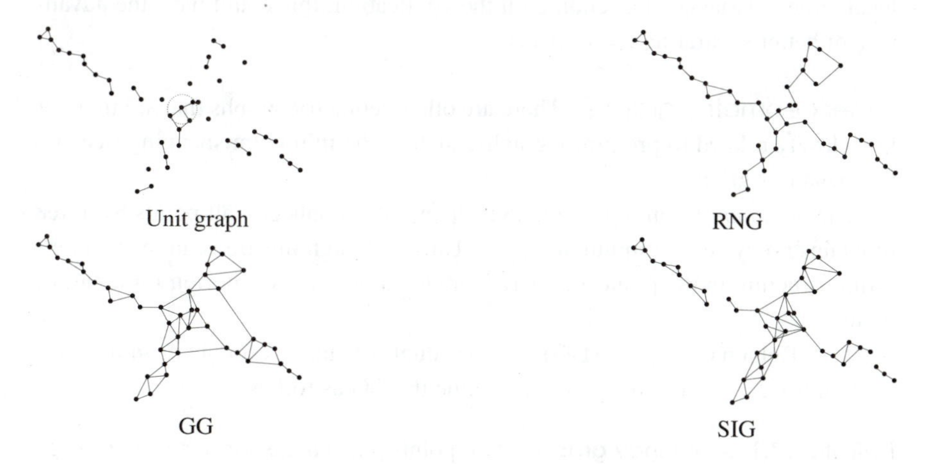


**Figure 7.7.** Examples of proximity graphs.

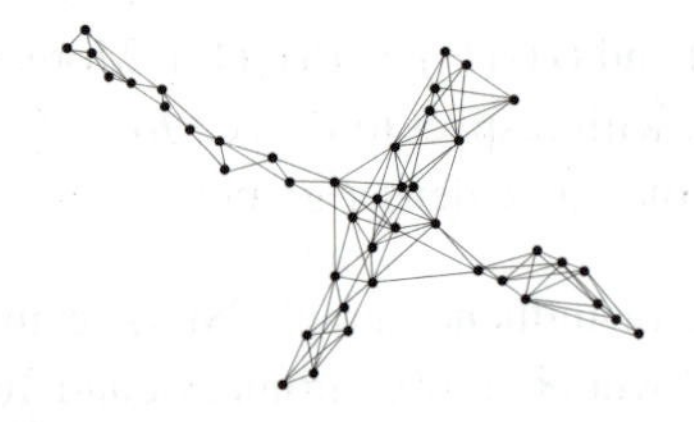

**Figure 7.8.** The 3-SIG over the same point cloud as in Figure 7.7.

**Figure 7.9.** The 3-SIG with additional pruning.

sphere of influence is not determined by the nearest neighbor, but by the $r$th nearest neighbor. Obviously, the larger $r$, the more points that are directly connected by an edge (see Figure 7.8).

A further extension can be applied to the $r$-SIG, if it is too dense or contains edges that are too long. This is *pruning* of edges based on a statistical outlier detection method over the lengths of edges [V. Barnett 94]. In statistics, an outlier is a single observation that is far away from the rest of the data. One definition of "far away" in this context is "larger than $Q_3 + 1.5 \cdot IQR$," where $Q_3$ is the third quartile ($Q_2$ would be the median), and $IQR$ is the interquartile range $Q_3 - Q_1$. Our experiments showed that best results are achieved by pruning edges with length of at least $Q_3 + IQR$ [Klein and Zachmann 04b]. Figure 7.9 shows the pruned 3-SIG of the example point set.

Another extension could be to use ellipsoids instead of spheres [Klein and Zachmann 04b]. Then, we have the additional degree of freedom of the orientation of the ellipsoids. For instance, we could orient them along the direction of local largest variance. Depending on the application, this could have the advantage of better separating close sheets.

Other geometric graphs. There are other geometric graphs that are more or less closely related to proximity graphs, such as the minimum spanning tree and the Delaunay graph.

The *minimum spanning tree* (MST) spans (i.e., connects) all points by a tree (contains no cycles) of minimal length. Thus, although this tree can reveal interesting structure in the point set, there is no local proximity criterion for a pair of points.

The Delaunay graph (DG) is the dual of the Voronoi diagram (see Section 6.1.2). Alternatively, we can define the DG as follows.

**Definition 7.1. (Delaunay graph.)** Two points $p$ and $q$ are connected by an edge if they satisfy the *empty circle property*. The points $p$ and $q$ satisfy the empty

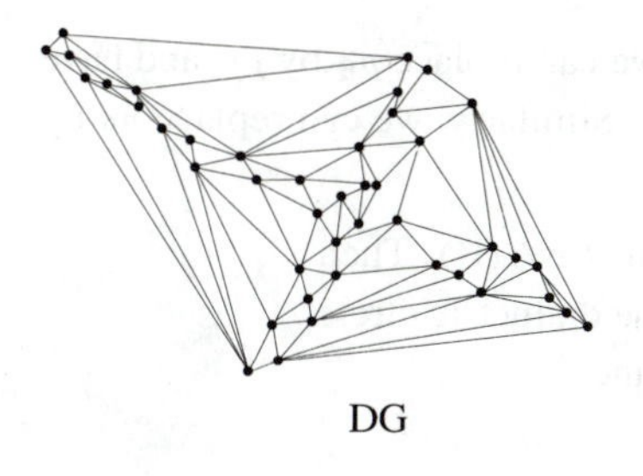


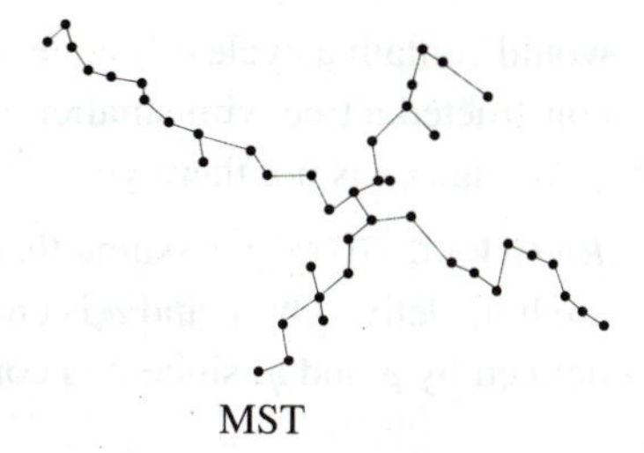


**Figure 7.10.** Example DG and MST for the same set of points as in Figure 7.7.

circle property iff there is a circle (or, in $\mathbb{R}^d$, a hypersphere) such that $p$ and $q$ are on its boundary and no point of $V$ is in the interior of this circle.[2]

Actually, this definition is equivalent to the one given in Section 6.1.2. For the sake of uniqueness, we assume that all points are in *general position*, meaning that in $\mathbb{R}^d$, no $d+1$ points lie on a common hyperplane, and no $d+2$ points lie on a common hypersphere. Figure 7.10 shows the DG and the MST for the same point set as in Figure 7.7.

### 7.1.3. Inclusion Property

Here, we demonstrate the following:

**Lemma 7.2.** *For a given set of points $V$,*

$$MST(V) \subseteq RNG(V) \subseteq GG(V) \subseteq DG(V).$$

This implies, in particular, that $RNG(V)$ and $GG(V)$ are connected, and that $|MST(V)| \leq |RNG(V)| \leq |GG(V)| \leq |DG(V)|$. Furthermore, in $\mathbb{R}^2$, the number of edges is linear in the number of points.

*Proof:* The proof is fairly simple. Assume for the moment that $V \subseteq \mathbb{R}^2$.

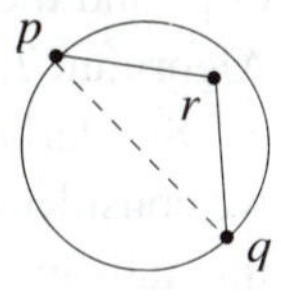


$MST(V) \subseteq RNG(V)$**:** Assume, for the sake of contradiction, that some edge $pq$ is in the MST but not in the RNG. Then, the lune formed by $p$ and $q$ is not empty, i. e., there is some other point $r \in L(p,q)$. But then, $d(p,r) < d(p,q)$ and $d(q,r) < d(p,q)$. As a consequence, the original MST is not minimal, which can be seen as follows. The original MST cannot contain both eges $pr$ and $qr$ (otherwise, it

[2]Then, there is also a maximal empty hypersphere that has exactly $d+1$ points on its boundary, two of which are $p$ and $q$.

would contain a cycle). If edge $pr$ is not there, we can replace $pq$ by $pr$, and have constructed a tree with smaller total edge length. Similarly, we can replace $pq$ by $qr$ if edge $qr$ is not there yet.

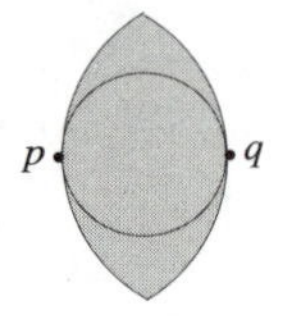


**$RNG(V) \subseteq GG(V)$:** Assume that $pq$ is an edge in $RNG(V)$. Then the lune defined by $p$ and $q$ is empty, and so is the diametric circle defined by $p$ and $q$, since it is contained in the lune.

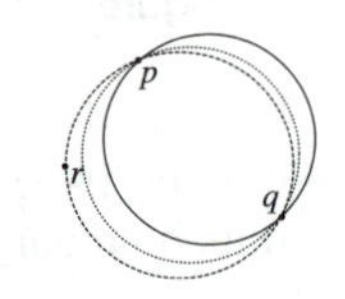


**$GG(V) \subseteq DG(V)$:** Assume that $pq$ is an edge in $GG(V)$. Then, the diametric circle defined by $p$ and $q$ is empty. Thus, the pair $(p, q)$ also satisfies the empty circle property as defined by Definition 7.1. Usually, the DG has more edges than the GG. This can be seen as follows. Imagine that the circle can slide to the left or right, such that its diameter increases just so that $p$ and $q$ always stay on its boundary (of course, it won't be a diametric any more). Because of the definition of the GG, we can slide the circle, at least a little bit, and at least to one side, without hitting another point. Now, imagine that we slide the circle until it hits a third point, $r$. So, we have now found two more pairs of points, $(p, r)$ and $(q, r)$, that satisfy the empty circle property.[3] □

In higher dimensions, the proofs work just the same (with hyperspheres instead of circles).

### 7.1.4. Construction Algorithms

GG and RNG. We start with a simple algorithm for constructing the GG over a set $V$ of points. From Lemma 7.2, we know that we can start with $DG(V)$, and then remove those edges that are not neighbors according to the diameter circle property. In a brute-force algorithm, we could check this property for an edge $pq$ of $DG(V)$ by just testing each point of $V$ to see whether it is inside the diameter circle of $pq$. A more efficient algorithm is to test only the neighbors of $p$, and those of $q$, against a candidate diameter circle. This is summarized in Algorithm 7.1.

Similarly, we can construct the $RNG(V)$ by starting with the $DG(V)$. Then we consider each edge $pq$ of it in turn and check if there is any other point $r \in V$ that falls inside the lune made by $p$ and $q$; i. e., if

$$d(p,r) < d(p,q) \ \wedge \ d(q,r) < d(p,q). \tag{7.1}$$

[3] Because we always assume $V$ to be in general position (see Definition 7.1).

```
construct DG(V), set GG := DG(V)
for all edges pq ∈ GG do
        for all neighbors r of p or q do
                if r inside diameter circle of p and q then
                        delete edge pq from GG
                end if
        end for
end for
```

**Algorithm 7.1.** Simple algorithm to construct the GG, starting from the Delaunay graph of a set of points.

Another way to construct the GG would be by using the brute-force approach. We can consider each potential pair of points $(p, q)$ (there are $O(n^2)$ of them). For each of them, we check whether there is any other point $r$ (there are $O(n)$) that is inside their diametric sphere. This test essentially involves two distance computations; i. e.,

$$\|p, r\|^2 + \|q, r\|^2 < \|p, q\|^2,$$

which takes time $O(d)$ in $d$-dimensional space. Overall, this brute-force approach would take $O(dn^3)$.

We can easily improve on this by the following observation. When checking the pair $(p, q)$ to see whether they are neighbors in the GG, we must test whether any other point $r$ is inside the diametric sphere (see Figure 7.11, left image). At the same time, we can check whether $r$ is in the right half-space of the plane $H_{qp}$ that is orthogonal to the diameter $pq$ and goes through $q$. If it is, then it cannot

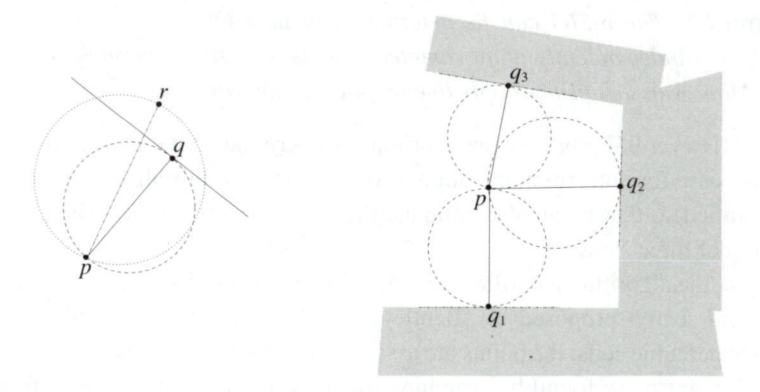


**Figure 7.11.** A simple heuristic leads to an $O(dn^2)$ algorithm to construct the GG.

```
for all p ∈ V do
    N_p ← V \ p
    for all q ∈ N_p do
        for all rinV do
            if Eq. 7.1 true then
                consider next q
            else
                if r is to the right of H_qp then
                    remove r from N_p
                end if
            end if
        end for
    end for
end for
```

**Algorithm 7.2.** Simple algorithm and heuristic to build the GG.

be a Gabriel neighbor to point $p$ (because $q$ would be inside the diametric sphere of $pr$).

So a better algorithm is the following. For each point $p$, we keep a list of neighbor candidates, $N_p$ (initially, this is the whole point set). Then, as we test $q_i \in N_p$ for being a real neighbor, we remove all $r$ from $N_p$ that are to the right of $H_{q_i p}$ (see Figure 7.11, right image). This reduces the average compexity to $O(dn^2)$. Algorithm 7.2 summarizes this heuristic.

The algorithm to construct the relative neighborhood graph is analogous to the one for constructing the GG.

SIG. Here, we demonstrate the following.

**Lemma 7.3.** *The $r$-SIG can be determined in time $O(n)$ on average for uniformly and independently point-sampled models with size $n$ in any fixed dimension. Moreover, it consumes only linear space in the worst case.*

*Proof:* [Dwyer 95] proposed an algorithm to determine a SIG in linear time in the average case for uniform point clouds. As $r$ is constant, this algorithm can easily be modified so that it can also compute the $r$-SIG in linear time. The algorithm consists of three steps.

First, the algorithm identifies the $r$-nearest neighbors of each point by utilizing the *spiral search* proposed by [Bentley et al. 80]: the space is subdivided into $O(n)$ hypercubic cells, the points are assigned to cells, and the $r$-nearest neighbors of each point $p$ are found by searching the cells in increasing distance from the cell containing $p$. As $O(1)$ cells are searched for each point on average and a

```
initialize grid with n cells
for all p ∈ V do
    assign p to its grid cell
end for
for all p ∈ V do
    find rth nearest neighbor to p by searching the grid cells in spiral order around
        p with increasing distance
end for
for all p ∈ V do
    for all cells around p that intersect the sphere of influence around p (in spiral
        order) do
        assign p to cell
    end for
end for
for all cells in the grid do
    for all pairs p_i, p_j of points assigned to the current cell do
        if spheres of influence of p_i and p_j intersect then
            create edge p_i p_j
        end if
    end for
end for
```

**Algorithm 7.3.** Simple algorithm to compute the $r$-SIG in $O(n)$ time on average.

single query can be done in $O(1)$ [Bentley et al. 80], this first step can be done in time $O(n)$.

Second, each point is inserted into every cell that intersects the $r$-nearest-neighbor sphere. On average, most spheres are small so that each point is inserted into a constant number of cells, and a constant number of points is inserted into each cell.

Finally, within each cell, all pairs of points that have been assigned to this cell are tested for intersection of their spheres of influence. Because each cell contains only a constant number of points, this can also be done in time $O(n)$.

[Avis and Horton 85] have shown that the 1-SIG has at most $c \cdot n$ edges where $c$ is a constant. This $c$ is always bounded by 17.5 [Avis and Horton 85, Edelsbrunner et al. 89]. [Guibas et al. 92a] extended this result to the $r$-SIG over a point cloud from $\mathbb{R}^d$ and showed that the number of edges is bounded by $c_d \cdot r \cdot n$, where the constant $c_d$ depends only on the dimension $d$. That means the $r$-SIG consumes $O(n)$ space in the worst case. □

Moreover, as mentioned in [Toussaint 88], ElGindy has observed that the line-segment intersection algorithm introduced [Bentley and Ottmann 79] can be used to construct a SIG in the plane in $O(n \log n)$ time in the worst case. The algorithm of [Guibas et al. 92a] constructs the $r$-SIG in time

$$O(n^{2-\frac{2}{1+\lfloor (d+2)/ \rfloor}+\epsilon} + rn \log^2 n),$$

for any $\epsilon > 0$ in the worst case.

# 7.2. Classification

## 7.2.1. Problem Statement

Classification is a fundamental class of techniques that has applications in a huge number of areas, such as pattern recognition, machine learning, and robotics.

It is used to partition a "universe" of all possible objects, each described as a bunch of data, into a set of *classes*. It is mainly characterized by its *decision rule*. This rule determines, for a given object, which class it belongs to.

There are two fundamental classes of decision rules: *parametric* and *non-parametric* rules.

Parametric decision rules make the membership classification based on the knowledge of the a priori probabilities of occurrence of objects belonging to some class $C_i$. This is captured by the probability density functions, $p(X|C_i)$, which characterize how probable the measurement $X$ is should the membership class be $C_i$.

Very often, however, it is difficult to describe these probability density functions a priori, and often they are just unknown.

In such cases, non-parametric decision rules are attractive, because they do not require such a priori knowledge. Instead, these rules rely directly on the *training set* of objects. For this set, the class membership is known precisely for each object in this set. This is a priori knowledge, too, but much less difficult to acquire. (For instance, it could be provided by a human.) Therefore, this is also called *supervised learning*. The idea is that a "teacher" feeds a number of example objects to the "student" and provides the correct answer for each of them. After the learning phase, the student tries to determine the correct answer for unseen objects based on the examples it has seen so far.

Usually, objects are represented by a set of *features*, each of which can usually be represented by a real number. Thus, objects can be represented as points in the so-called *feature space*, and classes are subsets of this space. Since we work with

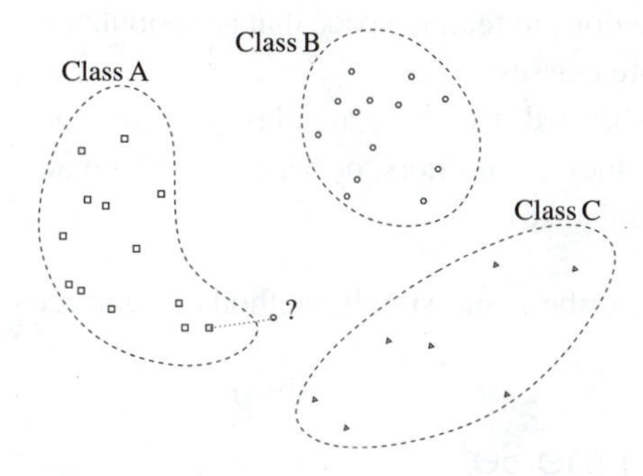


Figure 7.12. The nearest-neighbor yields a simple classifier.

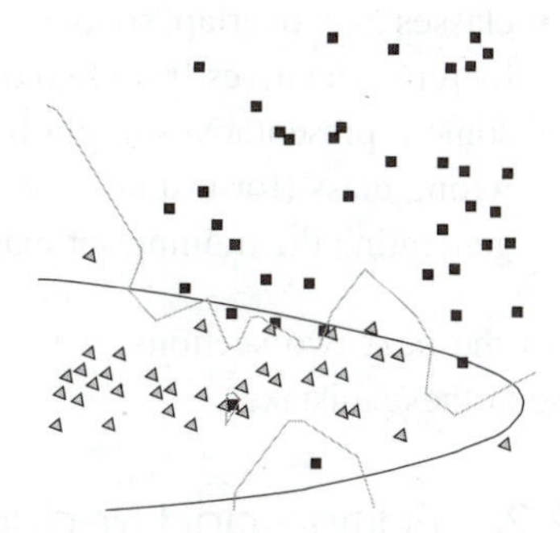

Figure 7.13. Comparison of the decision boundaries induced by the (optimal) Bayes classifier (squares), and the nearest-neighbor classifier (triangles).

points, we can usually define a measure of *distance* between points (which might be more or less well adapted to the objects).

A very simple, non-paramtric decision rule is the *nearest-neighbor classifier*. As its name suggests, it is based on the distance measure, and it assigns an unknown object to the same class as the nearest object from the (known) training set (see Figure 7.12). Most often, the Euclidean metric is used as a distance measure, although it is not always clear that this is best suited for the problem at hand.

Although this rule is extremely simple, it is fairly good (see Figure 7.13). More precisely, as the size of the training set goes to infinity, the asymptotic probability of error for the nearest-neighbor rule, $P_e^{NN}$, can be bounded by

$$\frac{P_e^{\mathrm{NN}}}{P_e^{\mathrm{opt}}} < 2 - P_e^{\mathrm{opt}} \frac{N}{N-1},$$

where $P_e^{\mathrm{opt}}$ is the optimal, so-called Bayes probability of error, and $N$ is the size of the training set [Cover and Hart 67]. In other words, the NN error is never two times worse than the optimal error.

Although the nearest-neighbor rule offers remarkably good performance (largely due to its simplicity), it is by no means a silver bullet. Some of its problems are:

- large space requirements (to store the complete training set);
- in high dimensions, it is very difficult to find the nearest neighbor in time better than $O(N)$ (this is called the *curse of dimensionality*).

In addition, there are two more problems that all classification methods must deal with:

- classes may overlap, so there are regions in feature space that are populated by representatives from two or more classes;
- some representatives might be mislabeled, i. e., they are classified into the wrong class (for instance, because they are outliers, or because the human generating the training set made a mistake).

In the next two sections, we will describe some simple methods to correct some of these mistakes.

### 7.2.2. Editing and Reducing the Set

Imagine that the classes correspond to colors, i. e., each point in the training set is labeled with a color. The nearest-neighbor rule induces a partitioning of the feature space into a number of *cells*, each of which belongs to exactly one point of the training set. These cells are exactly the Voronoi cells of the Voronoi diagram induced by the training set (see Chapter 6 and Section 6.4.1).

So a class is exactly the union of all Voronoi cells that have the same color (see Figure 7.14). The *decision boundary* of a decision rule are those points in feature space that separate classes. More precisely, a point is on the decision boundary iff any arbitrarily small, closed ball centered on this point contains points of two different classes. With the nearest-neighbor rule partitioning feature space into a set of Voronoi cells, the decision boundaries are exactly those boundaries of the Voronoi cells that lie between cells of different colors.

In practice, the Voronoi diagram and the decision boundary are rarely computed explicitly, especially in higher dimensions. But it is clear that the decision boundary alone is sufficient for classification. So, we can reduce the training set by removing those points that won't change the decision boundary. This is called *editing* the training set.

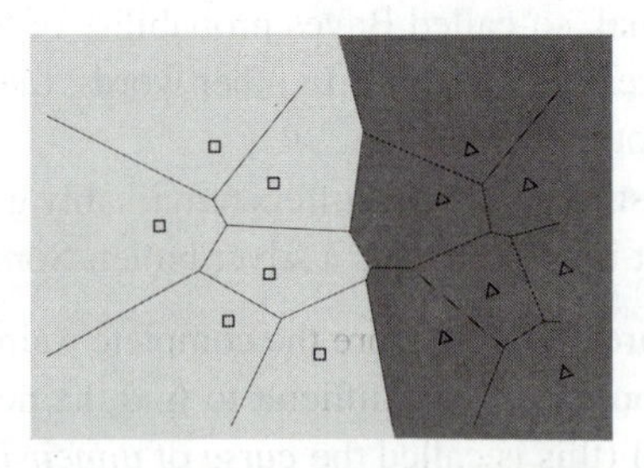

**Figure 7.14.** Conceptually, the nearest-neighbor classifier partitions the feature space into Voronoi cells. (See Color Plate XIV.)

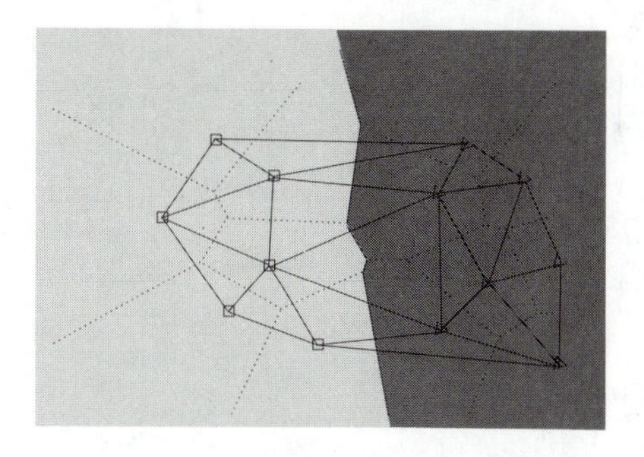

**Figure 7.15.** The decision boundary of the nearest-neighbor classifier runs between Delaunay neighbors with different colors. (See Color Plate XV.)

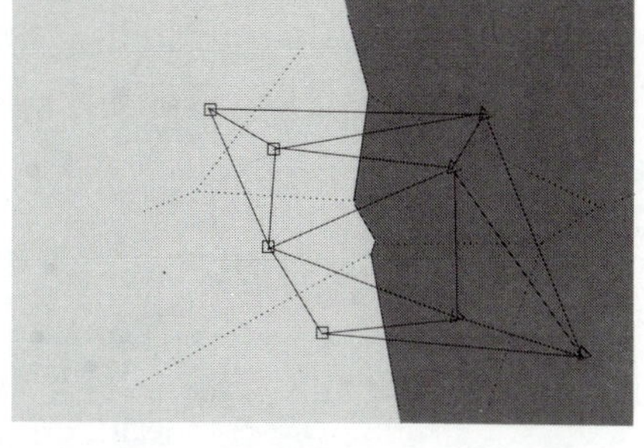

**Figure 7.16.** The edited training set has the same decision boundary but fewer nodes. (See Color Plate XVI.)

Instead of thinking in terms of the Voronoi diagram and Voronoi cells, we can think in terms of the DG (see Figure 7.15). The decision boundary then runs between neighbors in this graph that have different colors.

With the DG, it is very simple to edit the training set: we just remove all those nodes (i. e., points) whose neighbors all have the same color. This will change the DG and the Voronoi diagram, but it will not change the decision boundary (see Figure 7.16).

## 7.2.3. Proximity Graphs for Editing

One problem with Delaunay editing is that it can still leave too many points (see Figure 7.17). These are usually points that are well separated from other classes and contribute only portions of the decision boundary very remote from the cluster of class representatives. Thus, these points are usually less important for the classification of unknown points, which can be expected to come from the same distribution as the training set.

Another problem is that computing the DG takes, in the worst case, $\Theta(n^{\lceil d/2 \rceil})$, which is prohibitively expensive in higher dimensions (sometimes even in three dimensions).

So, it makes sense to pursue the following, approximate solution [Bhattacharya et al. 81]: compute a proximity graph over a given training set, which can be achieved more efficiently, edit the training set according to this graph, and then classify unknown points using this edited training set. Of course, the decision boundary will be different, but hopefully not too much, so that classification of unknown points will still yield (mostly) the same result.

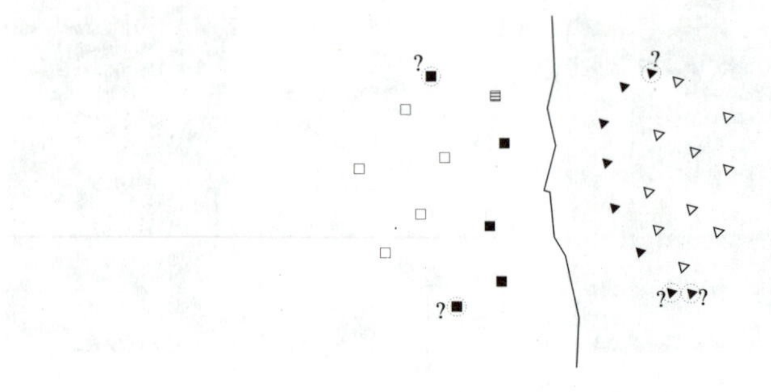

**Figure 7.17.** Delaunay editing often leaves too many points in the edited set. The unfilled points will be removed by Delaunay editing, but the contribution of the circled points is questionable.

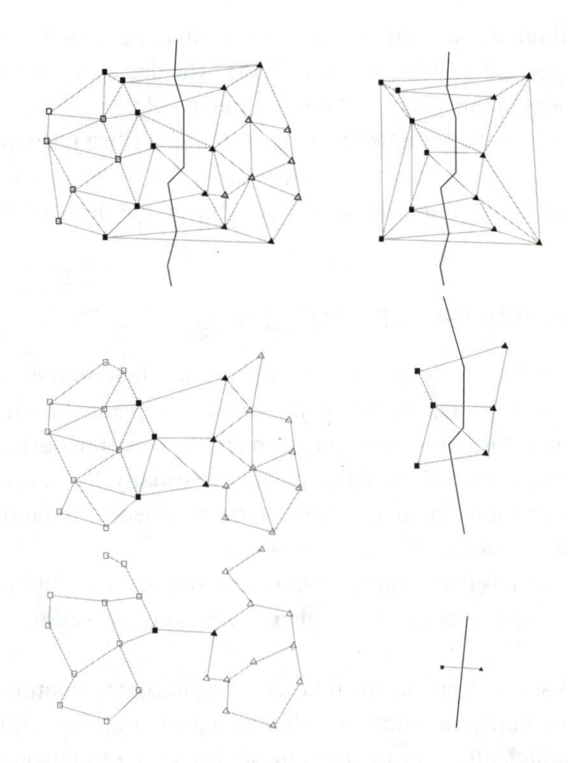

**Figure 7.18.** Different proximity graphs yield different edited (i. e., reduced) training sets and, thus, different approximations to the decision boundary.

```
build the proximity graph of the entire training set
for all samples in the set do
    find all its neighbors according to the proximity graph
    determine the most represented class (the "winning" class)
    mark the sample if the winning class is different from the label of the sample
end for
delete all marked samples from the set
```

**Algorithm 7.4.** Simple algorithm to clean (edit) a training set from incorrectly labeled samples.

Figure 7.18 shows the same training set edited with the DG, GG, and RNG (see Section 7.1.2). Because the GG contains (usually) fewer edges than the DG, more nodes are marked having neighbors of the same color and are, thus, removed. So the GG usually yields a smaller training set than the DG. A similar relation holds between RNG and GG.

Consequently, the decision boundary changes, because it is, being induced by the nearest-neighbor rule, the boundary between Voronoi cells of different colors in the Voronoi diagram over the edited training set. As you can see in Figure 7.18, the GG does not change the decision boundary much, but the RNG can change it quite substantially.

## 7.2.4. Cleaning the Training Set

In general, editing can serve several purposes. One of them is to reduce the size of the training set. This is the meaning we have understood so far. Another one is to remove samples from the training set that have been mislabeled, that overlap with other classes (see Figure 7.13), or that are just outliers.

This task can be solved by utilizing proximity graphs, too. A very simple method will be described in the following [Sanchez et al. 97].

The basic idea is to identify "bad" samples by examining their neighborhood. If most of the samples are correctly labeled, and the training set is sufficiently dense, then an incorrectly labeled sample will likely have a large number of differently labeled neighbor samples. This idea is summarized in Algorithm 7.4.

This algorithm can be refined straightforwardly a bit by modifying the rule when to mark a sample. For instance, we could mark a sample only if the number of neighbors in the winning class is at least 2/3 of the number of all neighbors. In addition, this method can be applied repeatedly to the training set.

**Figure 7.19.** 3D scanners (left) produce large point clouds (right) from real objects (middle).

## 7.3. Surfaces Defined by Point Clouds

In the past few years, point clouds have had a renaissance caused by the widespread availability of 3D scanning technology (see Figure 7.19). Such devices produce a huge number of points, each of which lies on the surface of the real object in the scanner (see Figure 7.19). These sets of points are called *point clouds*. For one object, there can be tens of millions of points. Usually, they are not ordered (or only partially ordered) and contain some noise.

In order to render [Pfister et al. 00, Rusinkiewicz and Levoy 00, Zwicker et al. 02, Bala et al. 03] and interact [Klein and Zachmann 04a] with objects represented as point clouds, one must define an appropriate surface (even if it is not explicitly reconstructed).

This definition should produce a surface as close to the original surface as possible while being robust against noise (introduced by the scanning process). At the same time, it should allow the fastest possible rendering and interaction with the object.

In the following, we present a fairly simple definition [Klein and Zachmann 04b].

### 7.3.1. Implicit Surface Model

The surface definition begins with *weighted least squares interpolation* (WLS).

Let a point cloud $\mathcal{P}$ with $N$ points $\mathbf{p}_i \in \mathbb{R}^d$ be given. Then an appealing definition of the surface from $\mathcal{P}$ is the zero set $S = \{\mathbf{x} | f(\mathbf{x}) = 0\}$ of an implicit function [Adamson and Alexa 03]

$$f(\mathbf{x}) = \mathbf{n}(\mathbf{x}) \cdot (\mathbf{a}(\mathbf{x}) - \mathbf{x}), \tag{7.2}$$

where $\mathbf{a}(\mathbf{x})$ is the weighted average of all points $\mathcal{P}$,

$$\mathbf{a}(\mathbf{x}) = \frac{\sum_{i=1}^{N} \theta(\|\mathbf{x} - \mathbf{p}_i\|)\mathbf{p}_i}{\sum_{i=1}^{N} \theta(\|\mathbf{x} - \mathbf{p}_i\|)}. \tag{7.3}$$

Usually, a Gaussian *kernel* (weight function),

$$\theta(d) = e^{-d^2/h^2}, \quad d = \|\mathbf{x} - \mathbf{p}\|, \tag{7.4}$$

is used, but other kernels work as well.

The bandwidth of the kernel, $h$, allows us to tune the decay of the influence of the points. It should be chosen such that no holes appear [Klein and Zachmann 04a].

Theoretically, $\theta$'s support is unbounded. However, it can be safely limited to the extent where it falls below the machine's precision, or some other suitably small threshold $\theta_\varepsilon$. Alternatively, one could use the cubic polynomial [Lee 00]

$$\theta(d) = 2\left(\frac{d}{h}\right)^3 - 3\left(\frac{d}{h}\right)^2 + 1,$$

or the tricube weight function [Cleveland and Loader 95]

$$\theta(d) = \left(1 - \left|\frac{d}{h}\right|^3\right)^3,$$

or the Wendland function [Wendland 95]

$$\theta(d) = \left(1 - \frac{d}{h}\right)^4 \left(4\frac{d}{h} + 1\right),$$

all of which are set to 0 for $d > h$ and, thus, have compact support (see Figure 7.20 for a comparison). However, the choice of kernel function is not critical [Härdle 90].

The normal $\mathbf{n}(\mathbf{x})$ is determined by weighted least squares. It is defined as the direction of smallest weighted covariance, i. e., it minimizes

$$\sum_{i=1}^{N} \left(\mathbf{n}(\mathbf{x}) \cdot (\mathbf{a}(\mathbf{x}) - \mathbf{p}_i)\right)^2 \theta(\|\mathbf{x} - \mathbf{p}_i\|) \tag{7.5}$$

for fixed $\mathbf{x}$ and under the constraint $\|\mathbf{n}(\mathbf{x})\| = 1$.

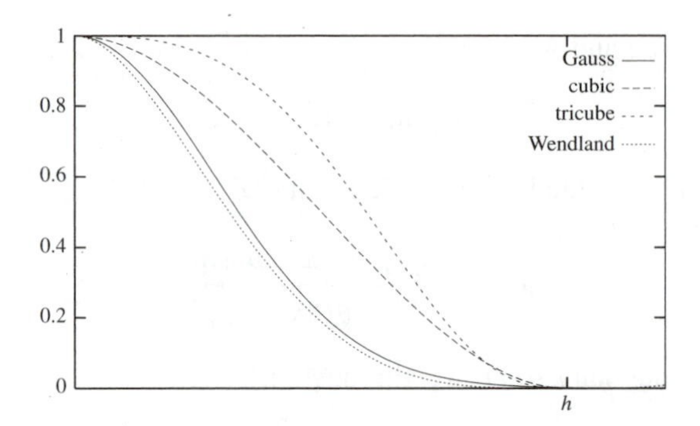


**Figure 7.20.** Different weight functions (kernels). Note that the horizontal scale of the Gauss curve is different, so that the qualitative shape of the curves can be compared better.

Note that, unlike [Adamson and Alexa 03], we use $\mathbf{a}(\mathbf{x})$ as the center of the *principal component analysis* (PCA), which makes $f(\mathbf{x})$ much more well behaved (see Figure 7.21). Also, we do not solve a minimization problem like [Levin 03, Alexa et al. 03], because we are aiming at an extremely fast method.

The normal $\mathbf{n}(\mathbf{x})$ defined by Equation (7.5) is the smallest eigenvector of the centered covariance matrix $\mathbf{B} = (b_{ij})$ with

$$b_{ij} = \sum_{k=1}^{N} \theta(\|\mathbf{x} - \mathbf{p}_k\|)(p_{k_i} - a(\mathbf{x})_i)(p_{k_j} - a(\mathbf{x})_j). \tag{7.6}$$

There are several variations of this simple definition, but for sake of clarity, we will stay with this basic one.

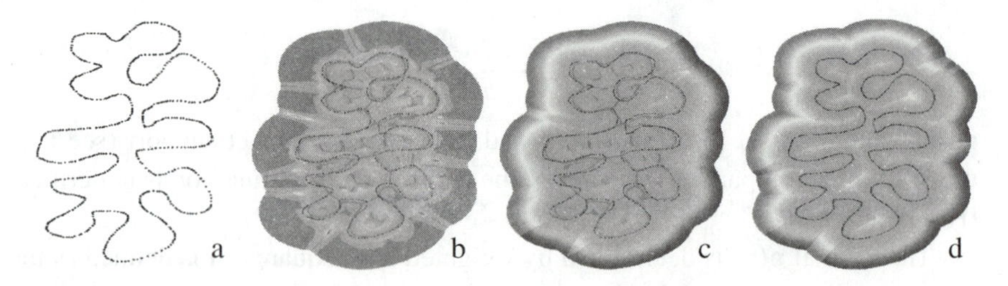


**Figure 7.21.** Visualization of the implicit function $f(\mathbf{x})$ over a 2D point cloud. Points $\mathbf{x} \in \mathbb{R}^2$ with $f(\mathbf{x}) \approx 0$, i.e., points on or close to the surface, are shown magenta. Red denotes $f(\mathbf{x}) \gg 0$, and blue denotes $f(\mathbf{x}) \ll 0$. (a) point cloud; (b) reconstructed surface using the definition of [Adamson and Alexa 03]; (c) utilizing the centered covariance matrix produces a better surface, but it still has several artifacts; (d) surface and function $f(\mathbf{x})$ based on the SIG. (See Color Plate XVII.) (Courtesy of Elsevier.)

### 7.3.2. Euclidean Kernel

This simple defintion (and many of its variants) has several problems. One problem is that the Euclidean distance $\|\mathbf{x}-\mathbf{p}\|$, $\mathbf{p} \in \mathcal{P}$, can be small, while the distance from $\mathbf{x}$ to the closest point on $S$ and then along the shortest path to $\mathbf{p}$ on $S$ (the geodesic) is quite large.

This can produce artifacts in the surface $S$ (see Figure 7.21). Two typical cases are as follows. First, assume $\mathbf{x}$ is halfway between two (possibly unconnected) components of the point cloud. Then it is still influenced by *both* parts of the point cloud, which have similar weights in Equations (7.3) and (7.5). This can lead to an *artificial* zero subset $\subset S$, where there are no points from $\mathcal{P}$ at all. Second, let us assume that $\mathbf{x}$ is inside a cavity of the point cloud. Then $\mathbf{a}(\mathbf{x})$ gets "drawn" closer to $\mathbf{x}$ than if the point cloud was flat. This makes the zero set *biased* towards the "outside" of the cavity, away from the true surface. In the extreme, this can lead to cancellation near the center of a spherical point cloud, where all points on the sphere have a similar weight.

This thwarts algorithms based solely on the point-cloud representation, such as collision detection [Klein and Zachmann 04a] or ray tracing [Adamson and Alexa 04].

The problems just mentioned could be alleviated somewhat by restricting the surface to the region $\{\mathbf{x} : \|\mathbf{x}-\mathbf{a}(\mathbf{x})\| < c\}$ (since $\mathbf{a}(\mathbf{x})$ must stay within the convex hull of $\mathcal{P}$). However, this does not help in many cases involving cavities.

### 7.3.3. Geodesic Distance Approximation

As illustrated in the preceding section, the main problems are caused by a distance function that does not take the topology of $S$ into account. Therefore, we will try to approximate the geodesic distances on the surface $S$. Unfortunately, we do not have an explicit reconstruction of $S$, and in many applications, we do not even want to construct one. Instead, we use a geometric proximity graph where the nodes are points $\in \mathcal{P}$.

In principle, we could use any proximity graph, but the SIG seems to yield the best results.

We define a new distance function $d_{\text{geo}}(\mathbf{x}, \mathbf{p})$ as follows. Given some location $\mathbf{x}$, we compute its nearest neighbor $\mathbf{p}_1^* \in \mathcal{P}$. Then we compute the closest point $\hat{\mathbf{p}}$ to $\mathbf{x}$ that lies on an edge adjacent to $\mathbf{p}_1^*$. Now, conceptually, the distance from $\mathbf{x}$ to any $\mathbf{p} \in \mathcal{P}$ could be defined as

$$d_{\text{geo}}(\mathbf{x}, \mathbf{p}) = \min_{\mathbf{n} \in \{\mathbf{p}_1^*, \mathbf{p}_2^*\}} \{d(\mathbf{n}, \mathbf{p}) + \|\hat{\mathbf{p}} - \mathbf{n}\|\},$$

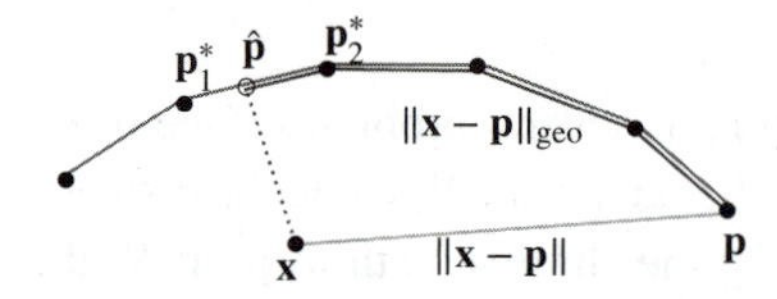

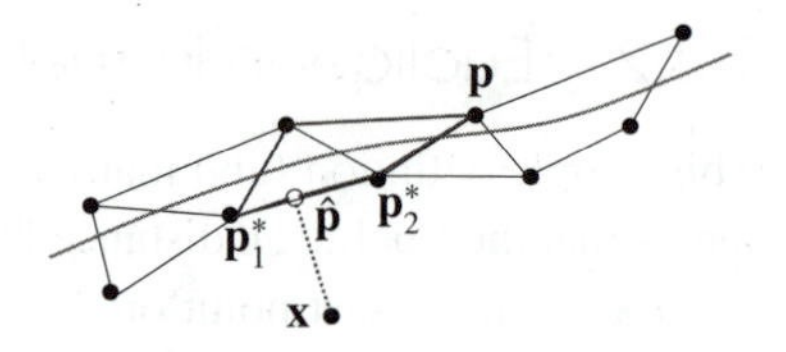


**Figure 7.22.** A proximity graph can help to improve the quality of the implicit surface by approximation of the geodesic distance instead of the Euclidean distance. (Courtesy of Elsevier.)

where $d(\mathbf{p}^*, \mathbf{p})$ for any $\mathbf{p} \in \mathcal{P}$ is the accumulated length of the shortest path from $\mathbf{p}^*$ to $\mathbf{p}$, multiplied by the number of "hops" along the path (see Figure 7.22, left image). However, it is not obvious that this is always the desired distance. In addition, it is desirable to define the distance function with as few discontinuities as possible. Therefore, we take the weighted average (see Figure 7.22, right image)

$$\begin{aligned} d_{\text{geo}}(\mathbf{x}, \mathbf{p}) = \; & (1-a)\big(d(\mathbf{p}_1^*, p) + \|\hat{\mathbf{p}} - \mathbf{p}_1^*\|\big) \\ & + a\big(d(\mathbf{p}_2^*, p) + \|\hat{\mathbf{p}} - \mathbf{p}_2^*\|\big), \end{aligned} \tag{7.7}$$

with the interpolation parameter $a = \|\hat{\mathbf{p}} - \mathbf{p}_1^*\|$.

Note that we do *not* add $\|\mathbf{x} - \hat{\mathbf{p}}\|$. That way, $f(\mathbf{x})$ is non-zero even far away from the point cloud.

Of course, there are still discontinuities in $d_{\text{geo}}$, and thus in function $f$. These can occur at the borders of the Voronoi regions of the cloud points, in particular at borders where the Voronoi sites are far apart from each other, such as points close to the medial axis.

The rationale for multiplying the path length by the number of hops is the following: if an (indirect) neighbor $\mathbf{p}$ is reached by a shortest path with many hops, then there are many points in $\mathcal{P}$ that should be weighted much more than $\mathbf{p}$, even if the Euclidean distance $\|\mathbf{p}^* - \mathbf{p}\|$ is small. This is independent of the concrete proximity graph used for computing the shortest paths.

Overall, when computing $f$ by Equations (7.2) to (7.6), we use $d_{\text{geo}}$ in Equation (7.4).

## 7.3.4. Automatic Bandwidth Computation

Another problem of the simple surface definition (without proximity graph) is the bandwidth $h$ in Equation (7.4), which is a critical parameter.

On the one hand, if $h$ is chosen too small, then the variance may be too large, i. e., noise, holes, or other artifacts may appear in the surface. On the other hand, if it is chosen too large, then the bias may be too large, i. e., small features in the surface will be smoothed out. To overcome this problem, [Pauly et al. 02] proposed to scale the parameter $h$ adaptively.

Here, we can use the proximity graph (e. g., a SIG) to estimate the local sampling density, $r(\mathbf{x})$, and then determine $h$ accordingly. Thus, $h$ itself is a function $h = h(\mathbf{x})$.

Let $r_1$ and $r_2$ be the lengths of the longest edges incident to $\mathbf{p}_1^*$ and $\mathbf{p}_2^*$, respectively (see Figure 7.22). Then, we set

$$r(\mathbf{x}) = \frac{1}{r} \cdot \frac{\|\hat{\mathbf{p}} - \mathbf{p}_2^*\| \cdot r_1 + \|\hat{\mathbf{p}} - \mathbf{p}_1^*\| \cdot r_2}{\|\mathbf{p}_2^* - \mathbf{p}_1^*\|}, \tag{7.8}$$

$$h(\mathbf{x}) = \frac{\eta\, r(\mathbf{x})}{\sqrt{-\log \theta_\varepsilon}}, \tag{7.9}$$

where $\theta_\varepsilon$ is a suitably small value (see Section 7.3.1), and $r$ is the number of nearest neighbors that determine the radius of each sphere of influence. (Note that $\log \theta_\varepsilon < 0$ for realistic values of $\theta_\varepsilon$.) Thus, $p_i$ with distance $\eta\, r(\mathbf{x})$ from $\hat{\mathbf{p}}$ will be assigned a weight of $\theta_\varepsilon$ (see Equation (7.4)).

We have now replaced the scale- and sampling-dependent parameter $h$ by another one, $\eta$, that is *independent* of scale and sampling density. Often, this can just be set to 1, or it can be used to adjust the amount of "smoothing." Note that this automatic bandwidth detection works similarly for many other kernels as well (see Section 7.3.1).

Depending on the application, it might be desirable to involve more and more points in Equation (7.2), so that $\mathbf{n}(\mathbf{x})$ becomes a least-squares plane over the complete point set $\mathcal{P}$ as $\mathbf{x}$ approaches infinity. In that case, we can just add $\|\mathbf{x} - \hat{\mathbf{p}}\|$ to Equation (7.8).

Figure 7.23 shows that the automatic bandwidth determination allows the WLS approach to handle point clouds with varying sampling densities without any manual tuning. The smoothing of the different sampling densities is very similar, compared to the "scale" (i. e., density). Notice the fine detail in the range from the tip of the nose throughout the chin, and the relatively sparse sampling in the area of the skull and at the bottom.

### 7.3.5. Automatic Boundary Detection

Another benefit of the automatic sampling density estimation is a very simple boundary detection method. This method builds on the one proposed by [Adamson and Alexa 04].

The idea is simply to discard points $\mathbf{x}$ with $f(\mathbf{x}) = 0$, if they are "too far away" from $\mathbf{a}(\mathbf{x})$ *relative* to the sampling density in the vicinity of $\mathbf{a}(\mathbf{x})$. More precisely, we define a new implicit function

$$\hat{f}(\mathbf{x}) = \begin{cases} f(\mathbf{x}), \text{ if } |f(\mathbf{x})| > \varepsilon \vee \|\mathbf{x} - \mathbf{a}(\mathbf{x})\| < 2r(\mathbf{x}), \\ \|\mathbf{x} - \mathbf{a}(\mathbf{x})\|, \text{ otherwise.} \end{cases} \quad (7.10)$$

Figure 7.24, right image, shows that this simple method is able to handle different sampling densities fairly well.

### 7.3.6. Complexity of Function Evaluation

The geodesic kernel needs to determine the nearest neighbor $\mathbf{p}^*$ of a point $\mathbf{x}$ in space (see Section 6.4.1). Using a simple kd-tree, an approximate nearest neighbor in 3D can be found in $O(\log^3 N)$ [Arya et al. 98].[4]

[Klein and Zachmann 04b] shows that all points $\mathbf{p}_i$ influencing $\mathbf{x}$ can be determined in constant time by a depth-first or breadth-first search through the graph.

Overall, $f(\mathbf{x})$ can be determined in $O(\log^3 N)$.

To achieve a fast practical function evaluation, we also need to compute the smallest eigenvector quickly [Klein and Zachmann 04a]. First, we compute the smallest eigenvalue $\lambda_1$ by determining the three roots of the cubic characteristic polynomial $\det(\mathbf{B} - \lambda\mathbf{I})$ of the covariance matrix $\mathbf{B}$ [Press et al. 92]. Then, we compute the associated eigenvector using the Cholesky decomposition of $\mathbf{B} - \lambda_1\mathbf{I}$.[5]

In our experience, this method is faster than the Jacobi method by a factor of 4, and it is faster than singular value decomposition by a factor 8.

## 7.4. Intersection Detection between Point Clouds

The surface defintion described in the previous sections is well suited for quick rendering, for instance, by ray tracing. In addition, it can also be used to determine the intersection between two point-cloud objects [Klein and Zachmann 05]. This will be explained in the following sections.

The problem is the following. Given two point clouds $A$ and $B$, the goal is to determine whether there is an intersection, i. e., a common root $f_A(x) = f_B(x) =$

[4]Under mild conditions, nearest neighbor can be done in $O(\log N)$ time by using a Delaunay hierarchy [Devillers 02], but this may not always be practical.

[5]The second step is possible because $\mathbf{B} - \lambda\mathbf{I}$ is positive semi-definite. Let $\lambda(A) = \{\lambda_1, \lambda_2, \lambda_3\}$ with $0 < \lambda_1 \leq \lambda_2 \leq \lambda_3$. Then, $\lambda(\mathbf{B} - \lambda\mathbf{I}) = \{0, \lambda_2 - \lambda_1, \lambda_3 - \lambda_1\}$. Let $\mathbf{B} - \lambda\mathbf{I} = \mathbf{USU}^\mathsf{T}$ be the singular value decomposition. Then, $x^\mathsf{T}(\mathbf{B} - \lambda\mathbf{I})x = x^\mathsf{T}\mathbf{USU}^\mathsf{T}x = y^\mathsf{T}\mathbf{S}y \geq 0$. Thus, the Cholesky decomposition can be performed if full pivoting is done [Higham 90].

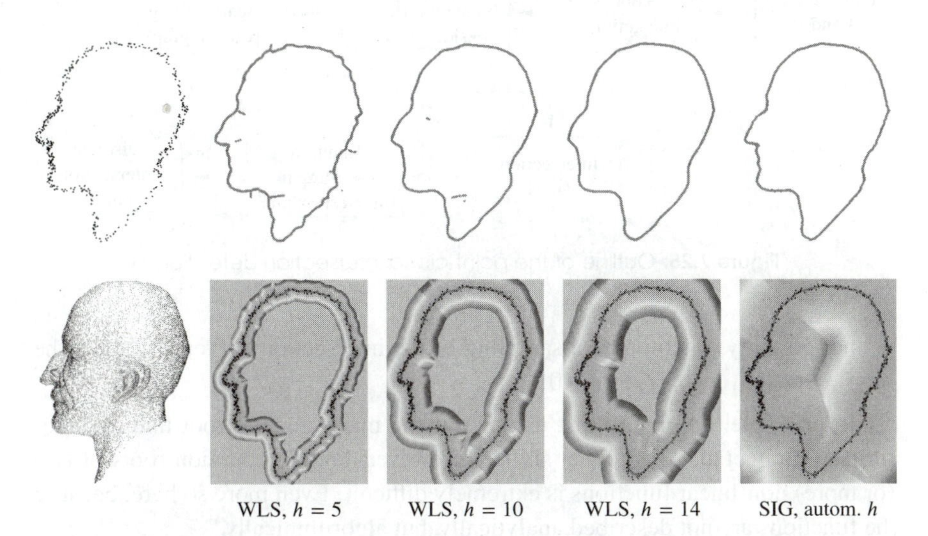


**Figure 7.23.** Reconstructed surface based on simple WLS and with proximity graph (rightmost) for a noisy point cloud obtained from the 3D Max Planck model (leftmost). Notice how fine details, as well as sparsely sampled areas, are handled without manual tuning. (See Color Plate XVIII.) (Courtesy of Elsevier.)

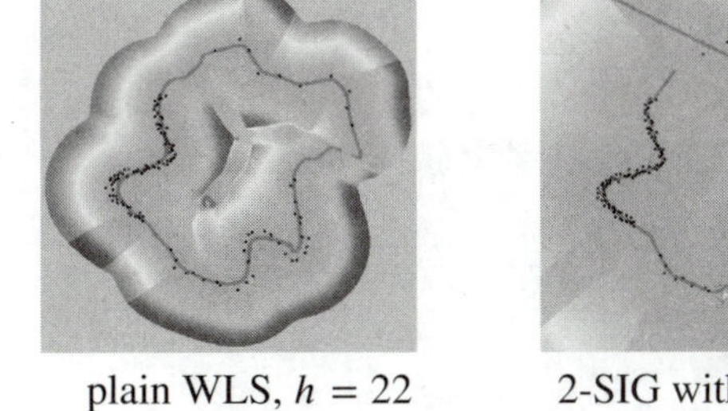


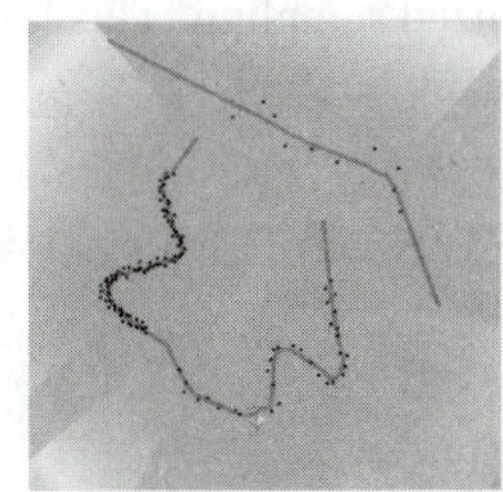


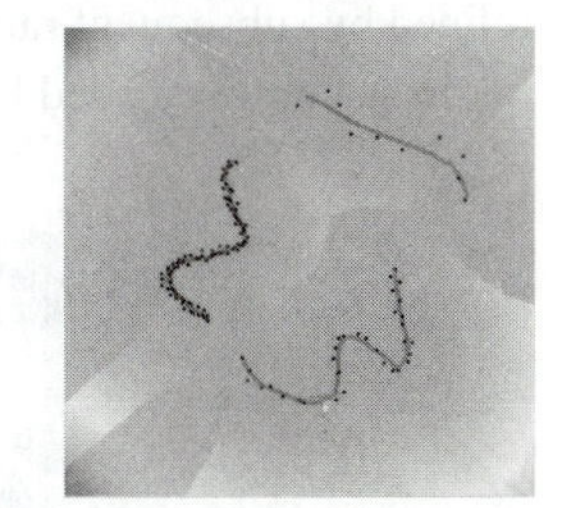


**Figure 7.24.** Automatic sampling density estimation for each point allows you to determine the bandwidth automatically and independently of scale and sampling density (middle), and to detect boundaries automatically (right). (See Color Plate XIX.) (Courtesy of Elsevier.)

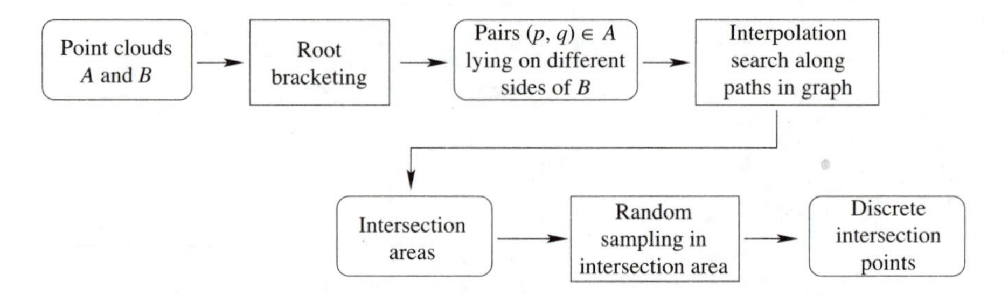


**Figure 7.25.** Outline of the point-cloud intersection detection.

0, and, possibly, to compute a sampling of the intersection curve(s), i. e., of the set $\mathcal{Z} = \{x \mid f_A(x) = f_B(x) = 0\}$.

In principle, one could just use one of the many general root-finding algorithms [Pauly et al. 03, Press et al. 92]. However, finding common roots of two (or more) non-linear functions is extremely difficult. Even more so here, because the functions are not described analytically, but algorithmically.

Fortunately, here we can exploit the proximity graph we already have in the surface definition of the objects and, thus, achieve much better performance.

First, the algorithm tries to bracket intersections by two points on one surface and on either side of the other surface (see Figure 7.25). Second, for each such bracket, it finds an approximate point in one of the point clouds that is close to the intersection (see Figure 7.26). Finally, this approximate intersection point is refined by subsequent randomized sampling. This last step is optional, depending on the accuracy needed by the application.

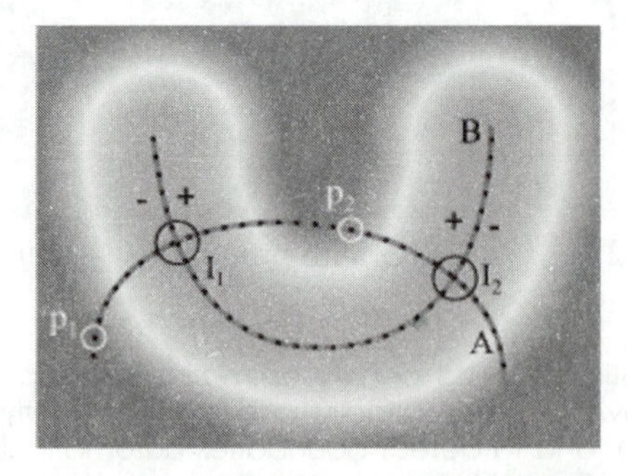


**Figure 7.26.** Two point clouds $A$ and $B$ and their intersection spheres $I_1$ and $I_2$. The root-finding procedure, when initialized with $p_1, p_2 \in A$, will find an approximate intersection point inside the *intersection sphere* $I_1$.

Steps 1 and 3 will be solved by a randomization approach. Both steps 1 and 2 will utilize the proximity graph. In the following, we describe each step in detail.

### 7.4.1. Root Bracketing

As mentioned earlier, the algorithm starts by constructing random pairs of points on different sides of one of the surfaces. The two points should not be too far apart, and the pairs should evenly sample the surface.

An exhaustive enumeration of all pairs is, of course, prohibitively expensive. Therefore, we pursue the following randomized (sub-)sampling procedure.

Assume that the implicit surface is conceptually(!) approximated by surfels (2D discs) of equal size [Pfister et al. 00, Rusinkiewicz and Levoy 00]. Let $\text{Box}(A, B) = \text{Box}(A) \cap \text{Box}(B)$ and $\overline{A} = A \cap \text{Box}(A, B)$. Then we want to randomly draw points $p_i \in \overline{A}$ such that each surfel $s_i$ gets occupied by at least one $p_i$; here, "occupied by $p_i$" means that the projection of $a(p_i)$ along the normal $n(p_i)$ onto the supporting plane of $s_i$ lies within the surfel's radius.

For each $p_i$, we can easily determine another point $p_j \in \overline{A}$ in the neighborhood of $p_i$ (if any), so that $p_i$ and $p_j$ lie on different sides of $f_B$. We represent the neighborhood of a point $p_i$ by a sphere $C_i$ centered at $p_i$.

An advantage of this approach is that the application can specify the density of the intersection points that are to be returned by our algorithm. From these, it is fairly easy to construct a discretization of the complete intersection curves (e. g., by using randomized sampling again).

Note that we never need to actually construct the surfels or assign the points from $A$ explicitly to the neighborhoods, which we describe in the following. Section 7.4.2 describes how to choose the radius of the spheres $C_i$.

In order to find a $p_j \in A \cap C_i$ on the "other side" of $f_B$, we use $f_B(p_i) \cdot f_B(p_j) \leq 0$ as an indicator. This, of course, is reliable only if the normals $n(x)$ are consistent throughout space. If the surface is manifold, this can be achieved by a method similar to [Hoppe et al. 92]: utilizing a proximity graph (such as the SIG), we can propagate a normal to each point $p_i \in A$. Then, when defining $f(x)$, we choose the direction of $n(x)$ according to the normal stored with the nearest neighbor of $x$.[6]

In order to sample $A$ such that each (conceptual) surfel is represented by at least one point in the sample, we use the following.

---

[6]Surprisingly, the direction of $n(x)$ is consistent over fairly large volumes without any preconditioning.

**Lemma 7.4.** *Let $A$ be a uniformly sampled point cloud. Further, let $S_A$ denote the set of conceptual surfels approximating the surface of A inside the intersection volume of $A$ and $B$, and let $a = |S_A|$. Then we can occupy each surfel with at least one point with probability $p = e^{-e^{-c}}$, where c is an arbitrary constant, just by drawing $n = O(a \ln a + c \cdot a)$ random and independent points from $\overline{A}$. These points will be denoted as $A'$.*

*Proof:* see [Klein and Zachmann 05].

For instance, if we want $p \geq 97\%$, we must choose $c = 3.5$, and if $a = 30$, then we must draw $n \approx 200$ random points.

The next section will show how to choose an appropriate size for the neighborhoods $C_i$. After that, Section 7.4.3 will propose an efficient way to determine the other part $p_j$ of the root brackets given a point $p_i \in A'$. □

## 7.4.2. Size of Neighborhoods

The radius of the spherical neighborhoods $C_i$ must be chosen so that, on the one hand, all $C_i$ cover the whole surface defined by $A$. On the other hand, the intersection with each adjoining neighborhood of $C_i$ has to contain at least one point in $A'$ so as to not miss any collisions lying in the intersection of two neighborhoods. The situation is illustrated in Figure 7.27.

To determine the minimal radius of a spherical neighborhood $C_i$, we introduce the notion of *sampling radius*.

**Definition 7.5. (Sampling radius.)** Let a point cloud $A$ as well as a subset $A' \subseteq A$ be given. Consider a set of spheres, centered at $A'$, that cover the surface de-

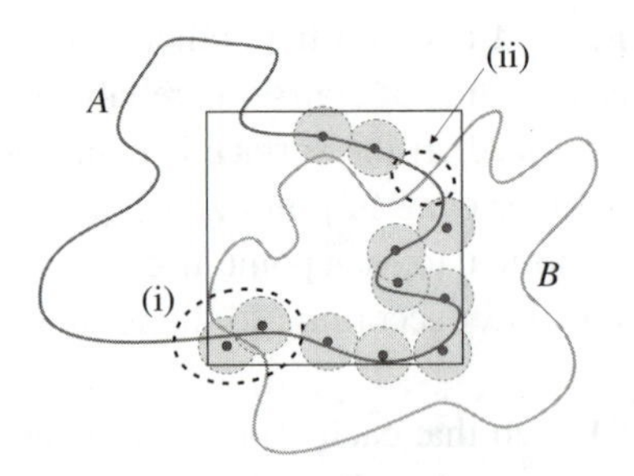


**Figure 7.27.** If the spherical neighborhoods $C_i$ (filled disks) are too small, not all collisions can be found. (i) Adjoining neighborhoods might not overlap sufficiently, so their intersection contains no randomly chosen cloud point. (ii) The surface might not be covered by neighborhoods $C_i$.

fined by $A$ (not $A'$), where all spheres have equal radius. We define the sampling radius $r(A')$ as the minimal radius of such a sphere covering.

The sampling radius $r(A')$ can obviously be estimated as the radius $r$ of a surfel $s_i \in S_A$.

Let $F_A$ denote the surface area of the implicit surface over $\overline{A}$. Then the surfel radius $r$ can be determined by

$$\frac{F_A}{a} = \pi r^2 \Rightarrow r = \sqrt{\frac{F_A}{a\pi}}.$$

Assume that the implicit surface over $\overline{A}$ can also be approximated by surfels of size $r(A)$. Then, $F_A$ can be estimated by

$$F_A = |\overline{A}| \cdot \pi r(A)^2.$$

Overall, $r(A')$ can be estimated by

$$r(A') = r(A) \cdot \sqrt{\frac{|\overline{A}|}{a}} \approx r(A) \cdot \sqrt{\frac{\mathrm{Vol}(A,B)}{\mathrm{Vol}(A) \cdot a} \cdot |A|}.$$

The size of $\overline{A}$, $|\overline{A}|$, can easily be estimated as $|A|\frac{\mathrm{Vol}(A)}{\mathrm{Vol}(A,B)}$, and the sampling radius $r(A)$ can easily be determined in the preprocessing.

### 7.4.3. Completing the Brackets

Given a point $p_i \in A'$, we need to determine other points $p_j \in A \cap C_i$ on the other side of $f_B$ in order to bracket the intersections. From a theoretical point of view, this could be done by testing $f_B(p_i) \cdot f_B(p_j) \leq 0$ for all points $p_j \in A' \cap C_i$ in time $O(1)$ because $|A'|$ can be a chosen constant. In practice, however, the set $A' \cap C_i$ cannot be determined quickly. Therefore, in the following, we propose an adequate alternative that works in time $O(\log\log N)$.

We observe that $A' \cap C_i \approx A' \cap A_i$, where $A_i := \{x \mid 2r(A') - \delta \leq \|x - p_i\| \leq 2r(A')\}$ is an *anulus* around $p_i$ (or, at least, these are the $p_j$ that we need to consider to ensure a certain bracket density). By construction of $A'$, $A' \cap A_i$ has a similar distribution as $A \cap A_i$. Observe further, that we don't necessarily need $p_j \in A'$.

Overall, the idea is to construct a random sample $B_i \subset A \cap C_i$ such that $B_i \subset A_i$, $|B_i| \approx |A' \cap A_i|$, and such that $B_i$ has a similar distribution as $A' \cap A_i$.

This sample $B_i$ can be constructed quickly by the help of Lemma 7.4. We just choose randomly $O(b \ln b)$ many points from $A \cap A_i$, where $b := |A' \cap C_i|$.

We can describe the set $A \cap A_i$ very quickly if the points in the CPSP (close-pairs shortest-path) map stored with $p_i$ are sorted by their geodesic[7] distance from $p_i$. Then we just need to use interpolation search to find the first point with distance $2r(A') - \delta$ and the last point with distance $2r(A')$ from $p_i$. This can be done in time $O(\log \log |A \cap C_i|)$ per point $p_i \in A'$. Thus, the overall time to construct all brackets is in $O(\log \log N)$.

### 7.4.4. Interpolation Search

Having determined two points $p_1, p_2 \in A$ on different sides of surface $B$, the next goal is to find a point $\hat{p} \in A$ "between" $p_1$ and $p_2$ that is "as close as possible" to $B$. In the following, we will call such a point the *approximate intersection point* (AIP). The true intersection curve $f_B(x) = f_A(x) = 0$ will pass close to $\hat{p}$ (usually, it does not pass through any points of the point clouds).

Depending on the application, $\hat{p}$ might already suffice. If the true intersection points are needed, then we refine the output of the interpolation search by the procedure described in Section 7.4.6.

Here, we can exploit the proximity graph. We just consider the points $P_{12}$ that are on the shortest path between $p_1$ and $p_2$, and we look for $\hat{p}$ that assumes $\min_{p \in P_{12}} \{|f(p)|\}$.

Let us assume that $f_B$ is monotonic along the path $\overline{p_1 p_2}$. Then, instead of doing an exhaustive search along the path, we can utilize interpolation search to look for $\hat{p}$ with $f(\hat{p}) = 0$.[8] This makes sense here, because the "access" to the key of an element, i. e., an evaluation of $f_B(x)$, is fairly expensive [Sedgewick 89]. The average runtime of the interpolation search is in $O(\log \log m)$, $m$ = number of elements.

Algorithm 7.5 for the interpolation search assumes that the shortest paths are precomputed and stored in a map.

However, in practice, the memory consumption, albeit linear, could be too large for huge point clouds. In that case, we can compute the path $P$ on the fly at runtime by Algorithm 7.6. Theoretically, the overall algorithm is now in linear time. However, in practice, it still behaves sublinearly because the reconstruction of the path is negligible compared to evaluating $f_B$.

If $f_B$ is not monotonic along the paths between the brackets, but the sign of $f_B(x)$ is consistent, then we can use a binary search to find $\hat{p}$. (The complexity in that case is $O(\log m)$.)

---

[7]By using the geodesic distance (or, rather, the approximation thereof), we basically impose a different topology on the space where $A$ is embedded, but this is actually desirable.

[8]In practice, the interpolation search will never find exactly such a $\hat{p}$, but instead a pair of adjacent

```
l, r = 1, n
d_{l,r} = f_B(P_1), f_B(P_n)
while |d_l| > ϵ and |d_r| > ϵ and l < r do
        x = l + ⌈ (−d_l / (d_r − d_l)) (r − l) ⌉ {*}
        d_x = f_B(P_x)
        if d_x < 0 then
                l, r = x, r
        else
                l, r = l, x
        end if
end while
```

**Algorithm 7.5.** Pseudocode for the root-finding algorithm based on interpolation search. $P$ is an array containing the points of the shortest path from $p_1 = P_1$ to $p_2 = P_n$, which can be precomputed. $d_i = f_B(P_i)$ approximates the distance of $P_i$ to object $B$. (*) Note that either $d_l$ or $d_r$ is negative.

```
q.insert(p_1);  clear P
repeat
        p = q.pop
        P.append( p )
        for all p_i adjacent to p do
                if d_geo(p_i, p_2) < d_geo(p_1, p_2) then
                        insert p_i into q with priority d_geo(p_i, p_2)
                end if
        end for
until p = p_2
```

**Algorithm 7.6.** This algorithm can be used to initialize $P$ for Algorithm 7.5 if precomputing and storing all shortest paths in a map is too expensive. ($q$ is a priority queue.)

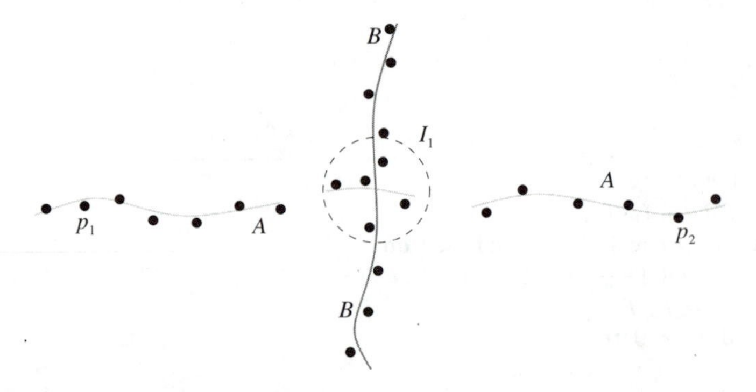


**Figure 7.28.** Models with boundaries can cause errors ($I_1$ could remain undetected), which can be avoided by "virtual" edges in the proximity graph.

### 7.4.5. Models with Boundaries

If the models have boundaries and the sampling rate of the root-bracketing algorithm is too low, not all intersections will be found (see Figure 7.28). In that case, some AIPs might not be reached, because they are not connected through the proximity graph.

Therefore, we modify the $r$-SIG a bit for our purpose. After constructing the graph, we usually prune away all "long" edges by an outlier detection algorithm (see Section 7.1.2). Now, we only mark these edges as "virtual." Thus, we can still use the $r$-SIG for defining the surface as before. For our interpolation search, however, we can also use the virtual edges so that small holes in the model are bridged.

### 7.4.6. Precise Intersection Points

If two point clouds are intersecting, the interpolation search computes a set of AIPs. Around each of them, an *intersection sphere* of radius $r = \min(\|x - \hat{p}_1\|, \|x - \hat{p}_2\|)$ contains a true intersection point, where

$$x = \frac{1}{d_1 + d_2}(d_2 p_1 + d_1 p_2),$$

the $\hat{p}_i$ have been computed by the interpolation search, lying on different sides of surface $B$, and $d_i = f_B(p_i)$. This idea is illustrated in Figure 7.29. So if the AIPs are not precise enough, we can sample each such sphere to get more accurate (discrete) intersection points.

---

points on the path that straddle $B$.

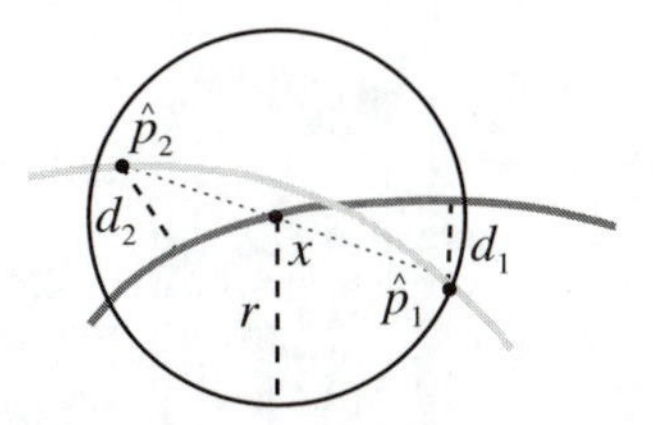


**Figure 7.29.** An intersection sphere centered at an AIP $p_i$. Its radius $r$ can be determined approximately with the help of a second point on the other side of $B$.

More precisely, if we want a precise collision point's distance from the surfaces to be smaller than $\epsilon_2$, we cover a given intersection sphere by $s$ smaller spheres with *diameter* $\epsilon_2$ and sample that volume by $s \ln s + cs$ many points so that each of the $s$ spheres gets a point with high probability. For each of these, we just determine the distance to both surfaces.

[Rogers 63] showed that a sphere with radius $a{\cdot}b$ can be covered by at most $s = \lceil \sqrt{3}\, a \rceil^3$ smaller spheres of radius $b$. Since we would like to cover the intersection sphere by spheres with radius $b = \epsilon_2/2$, we need to choose $a = 2r/\epsilon_2$, so that $a \cdot b = r$. As a consequence,

$$s = \lceil \sqrt{3}\,\frac{2r}{\epsilon_2} \rceil^3.$$

For example, to cover an intersection sphere with spheres of *radius* $\epsilon$, then $\epsilon_2 = 2\epsilon$ and $s = \lceil \sqrt{3}\, r/\epsilon \rceil^3$.

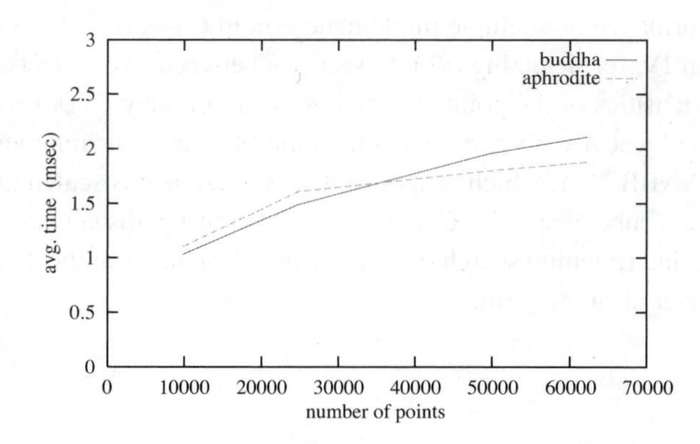


**Figure 7.30.** The plot shows the running time depending on the size of the point clouds for two objects (see Figures 7.19 and 7.31). The time is the average of all intersection detection times for distances between 0 and 1.5 and a lot of different orientations for two identical copies of each object.

**Figure 7.31.** One of the test objects for Figure 7.30. (See Color Plate XXI.)

### 7.4.7. Running Time

Under some mild assumptions about the point cloud, the running time of the algorithm is in $O(\log \log N)$, where $N$ is the number of points [Klein and Zachmann 05].

Empirically, it was found that about $n = O(a \ln a) = 200$ samples in phase 1 of the algorithm yields a sufficient bracket density so that the error is only about 0.1%.

The performance of a simple implementation in C++, running on a 2.8 GHz Intel Pentium IV, for detecting *all* intersections between two objects, depending on different densities of the point clouds, is shown in Figure 7.30. (We define the density of an object $A$ with $N$ points as the ratio of $N$ over the number of volume units of the AABB of $A$, which is at most 8, as each object is scaled uniformly so that it fits into a cube of size $2^3$. Computing the shortest paths between $(p_i, p_j)$ on the fly during interpolation search (see Algorithm 7.6) incurs at most an additional 10% of the overall running time.

# 8

# Kinetic Data Structures

In the previous chapters, we discussed hierarchical geometric data structures that are mainly of a static nature. However, in computer graphics (and many other fields), objects move constantly. So, how can we design data structures that allow efficient updating in the presence of an input set, the location of which in space changes continuously?

We can make two observations that help with designing such data structures:

- When an object (i.e., a line, a point, a polygon, etc.) moves by just $\epsilon$, then a conventional, static data structure usually remains valid;
- If an object moves farther than some $\epsilon$ and the data structure has to be updated, then this update is (usually) local, i.e., the data structures over the old configuration and over the new one, respectively, differ only by a small amount.

From these observations arise the following questions.

- How can we *determine* when the data structure has to be updated?
- How can we *update* it efficiently?

Both of these questions must be answered with less effort than just rebuilding the static data structure every time an object has moved and a query occurs.

The solution is a very generic algorithm technique called *kinetic data structure* (KDS). It makes a few fairly mild assumptions about the objects it can deal with:

- All objects (points, lines, line segments, polygons, etc.) follow a known (often called *published*) flight path;
- The flight path is an algebraic function in $t$ (i.e., it is a function containing only the four basic operators plus square root; the reason for this is that

```
initialize DS (often like static algorithm)
initialize certificates
compute failure time of all certificates
init priority queue over failure times
loop
    retrieve earliest certificate from queue
    update KDS (if it's an external event, then this will result in a new attribute)
    compute new certificates and/or remove some
    compute failure time of new certificates
    update queue
end loop
```

**Algorithm 8.1.** Basic "main loop" of most kinetic data structures.

we want to exclude flight paths that can go back and forth arbitrarily often within a time period $\epsilon$);

- Flight paths can be changed at discrete times, i.e., they must not change arbitrarily often. Usually, the application wants to change the flight path, if, for instance, two objects collide. As you will see in this chapter, such a change incurs additional work.

The "main loop" for virtually all KDSs looks basically as shown in Algorithm 8.1.

## 8.1. General Terminology

In this section, we will introduce some terminology related to KDSs [Basch et al. 97, Guibas 98]. For the moment, this terminology might remain a bit abstract, but we will explain each of these terms again with a concrete example.

**Attribute.** Denotes the output of the KDS. Examples: convex hull of a set of points, closest pair among a set of points. The attribute usually is part of the KDS. Sometimes, the KDS itself is the attribute (e. g., BSP).

**Combinatorial structure.** The KDS without any coordinates (and probably any other "float-like" values). Examples: a graph, a tree, pointers. The combinatorial structure is usually part of the KDS.

**Certificate.** A predicate over a number of objects from the input and the data structure. Examples: $\mathbf{pn} < 0$ (point below/on/above plane), $\begin{vmatrix} a_x - c_x & a_y - c_y \\ b_x - c_x & b_y - c_y \end{vmatrix}$

(point **c** is left/on/right of line through points **a**, **b**). The KDS maintains a number of certificates.

**Proof.** A set of certificates with the following property: the combinatorial structure together with the current concrete values (coordinates) remains valid for the given set of inputs, as long as all certificates in the proof are true. Obviously, we will try to determine the minimal size of a proof for a KDS.

**Failure time.** As the objects involved in a certificate move through their published flight path, that certificate eventually becomes false. The earliest time this happens is called the *failure time* or *death time* of the certificate. Since all flight paths are published, this can be computed, basically, by finding the smallest root of an algebraic function (which could become very complex).

Obviously, the KDS remains valid until the earliest failure time among all certificates in its proof.

**Event.** This is another term for the failure of a certificate. There can be two reasons why a certificate fails, which are called *external* and *internal* events.

An external event occurs whenever the attribute (i.e., the output) of the KDS changes. At this time, one or more of the certificates must fail.

An event is called internal if a certificate has failed but the attribute of the KDS does not change.

Speaking of events only makes sense if the attribute of the KDS is unique. (For instance, the convex hull is unique, but a BSP is not.)

**Efficient.** When designing a KDS one should strive to minimize the number of internal events. A KDS is called efficient if the ratio $\frac{\text{num. external event}}{\text{num. internal events}} \in O(\log^x n)$ in the worst case.

**Response time.** This is the time (worst case or average case) needed to update the KDS whenever a certificate fails. If the response time is in $O(\log^x n)$, then the KDS is called *responsive*.

**Compact.** A KDS is called compact, if the number of certificates in its proof is linear (or almost linear) in the size of the input.

**Local.** A KDS is called local, if each object of the input participates only in a small (i.e., polylog) number of certificates. This is important because it ensures that the KDS can be updated efficiently if an object changes its flight path.

As a simple example of a KDS, we will present a kinetic BSP in the following example. It will be introduced in three steps:

1. static segment tree;
2. kinetic segment tree;
3. kinetic BSP tree.

This is adapted from [de Berg et al. 01].

## 8.2. Static Segment Tree

This is a quick recap, so if you are familiar with this classic data structure, you can proceed to the next section.

Given a set $S = \{s_1, \ldots, s_n\}$ of intervals $s_i = [a_i, b_i] \subseteq \mathbb{R}$, and a query "point" $q \in \mathbb{R}$, we are looking for the subset $S' = \{s \in S | q \in s\}$. This is often called a *stabbing query* and can be extended trivially to $\mathbb{R}^d$.

Before describing the data structure, we define an *elementary interval* (EI). Let $X = \{a, b | a \in s \in S, b \in s \in S\} = \{x_i\}$ denote the set of *endpoints*, sorted by increasing value in a list. Then the interval $[x_i, x_{i+1})$ is called an EI. For convenience, we also include the intervals $(-\infty, x_0)$ and $[x_{2n}, \infty)$.

The segment tree is a balanced binary tree over the set of EIs, $X$. Each leaf $v_i$ stores one EI $\text{Int}(v_i) := [x_i, x_{i+1})$. We define the interval of an inner node $v$ as $\text{Int}(v) := \text{Int}(v_1) \cup \text{Int}(v_2)$, where $v_{1,2}$ are the two children of $v$. Each node stores the so-called *canonical subset* $S(v) := \{s \in S | \text{Int}(v) \subseteq s, \text{Int}(\text{parent}(v)) \nsubseteq s\}$ (in other words, each interval $s \in S$ is stored with several nodes $v_k$ in the tree, such that $\bigcup_k \text{Int}(v_k) = s$, and such that $s$ is stored as high up as possible). An example is shown in Figure 8.1.

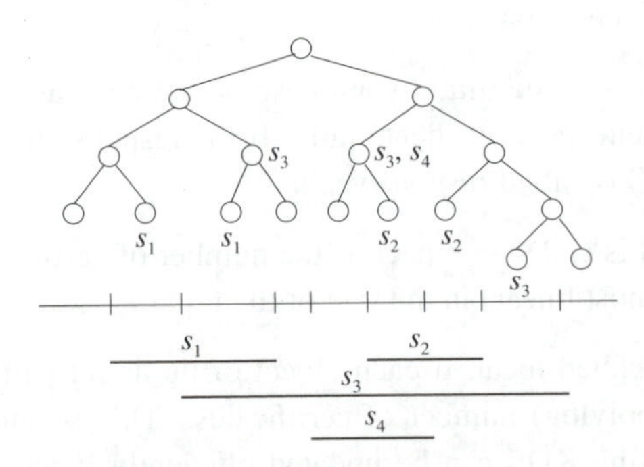


**Figure 8.1.** Example of a segment tree.

```
query(v, q)
output S(v)
if v is inner node then
    if q ∈ Int(v_1) then
        query(v_1, q)
    else
        query(v_2, q)
    end if
end if
```

**Algorithm 8.2.** Simple algorithm for answering a stabbing query.

The construction of a segment tree is very easy: generate and sort $X$; construct a skeleton binary tree over $2n + 1$ nodes; compute $\text{Int}(v)$ for all nodes in the tree; and sift each segment $s$ top-down through the tree.

The algorithm for answering a stabbing query is quite simple, as shown in Algorithm 8.2.

It is easy to see that such a segment tree needs $O(n \log n)$ space and $O(n \log n)$ construction time, and that a query can be answered in time $O(\log n + k)$, where $k$ is the size of the output.

## 8.3. Kinetic Segment Tree

By turning the static segment tree into a kinetic one, we are moving one step closer to our final goal of a kinetic BSP [de Berg et al. 01].

Now segments (i.e., intervals on $\mathbb{R}$) can move freely—they can swap positions, they can shorten or lengthen, etc. Observe, however, that the situation is not as difficult as it might seem, because of the following:

1. During the construction of the static segment tree, we never really needed to know the exact values of the endpoints of the intervals; it was completely sufficient to know their position in the sorted list of all endpoints;

2. For each segment $s$, there is a certain amount $\epsilon > 0$ by which we can move $s$ without invalidating the segment tree;

3. Whatever the position of a certain number segments, the skeleton of the segment tree is always the same. The only thing that changes is the labeling of the nodes with segment IDs.

These observations give rise to the following modifications of the original segment tree data structure:

- Segments are defined as $s_i := (a_i, b_i)$, $a_i, B_i \in \mathbb{N}$, $1 \leq a_i, b_i \leq 2n$. $a_i, b_i$ are called the *rank* of the endpoints;
- An array $R[1, \ldots, 2n]$ stores the current values of the endpoints (so the interval for $s_i$ is given by $\left[R_{a_i}, R_{b_i}\right]$). This array is kept sorted by value;
- With each segment $s$, we maintain a *fragment list* $\mathcal{L}(s) := \{v | s \in S(v)\}$. $\mathcal{L}$ is kept sorted from "left to right;"
- For each elementary interval $[i, i+1]$, we store a pointer to its associated leaf $v_i$.

The certificates in this KDS are $Z_i := R_i < R_{i+1}$, $i = 1, \ldots, 2n-1$. The proof remains valid until two endpoints swap position in array $R$. In that case, exactly two certificates are affected. The ranks of all other interval endpoints remain the same, so they are associated with the same nodes in the tree (in other words, their fragment lists remain the same).

Let $s$ be one of the two intervals affected by an endpoint swap.[1] There are four possible cases for which endpoints have swapped and how. Consider the one where before $s = [i, j]$ and after the swap $s = [i, j+1]$ (and analogously for the other segment).

The naïve update algorithm would just delete both segments from all nodes (with the help of array $\mathcal{L}$), and then sift them top-down through the tree. This would yield an $O(\log n)$ complexity.

But we can do better. The idea is to "augment" the segment $s = [i, j]$ by just the EI $[j, j+1]$, and then just update the tree starting with $v_j$, which is the leaf for $[j, j+1]$, and which is now covered by $s$, too. The pseudocode shown in Algorithm 8.3 gives more details.

1: $v \leftarrow v_j$ {leaf of $[j, j+1]$}
2: **repeat**
3: $\mu \leftarrow$ sibling of $v$
4: **if** $s \in S(\mu)$ **then**
5: delete $s$ from $S(\mu)$
6: $v \leftarrow$ parent($v$)
7: **end if**
8: **until** $s \notin S(\mu)$
9: insert $s$ into $S(v)$

**Algorithm 8.3.** Updating a kinetic segment tree.

[1] We omit the case where both endpoints belong to the same interval, as this is fairly easy to deal with.

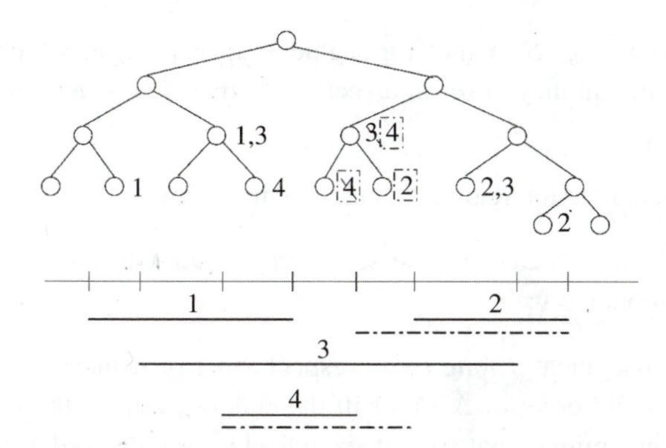


**Figure 8.2.** An example of a segment tree over four segments. The dashed segments show the segments 4 and 2, respectively, after two of their endpoints have swapped position. The changes in the tree are highlighted with dashed boxes.

Analogously, we can traverse the segment tree bottom-up in order to augment the other segment by an EI. Note that the test in Step 4 is necessarily false if $v$ is a *left* child.

The running time of Algorithm 8.3 is $O(h)$, where $h$ is the height of the node $v$ that gets $s$ attached, i.e., $[j, j+1] \in \text{Int}(v)$ and $s \in S(v)$ (after the update). This is made possible by the array $\mathcal{L}$, which allows us to perform Step 4 in Algorithm 8.3 in time $O(1)$. When starting with the leaf of the EI, we just need to inspect the front/back of $\mathcal{L}$, and then we work our way through $\mathcal{L}$.

**Lemma 8.1.** *Let $S$ be a set of $n$ moving segments on the real line $\mathbb{R}$. There is a kinetic segment tree with $O(n \log n)$ space, which can be updated in expected time $O(1)$ when an event occurs (i.e., two endpoints swap position). The worst-case time is still $O(\log n)$. The KDS is local and efficient.*

Locality and efficiency are given, since any segment participates in at most four certificates, and there are no internal events. The $O(1)$ time bound can be proven fairly easily by showing that the average height in a balanced tree is $O(1)$.

Figure 8.2 shows an example how the labeling changes as a result of an event.

## 8.4. Kinetic BSP in the Plane

We can now proceed to tackle our goal of designing a kinetic BSP. Here, we will restrict our discussion to the 2D case as presented in [de Berg et al. 01], but there are similar variants for the 3D case [Agarwal et al. 97, Agarwal et al. 98, Comba 99].

We are given a set $S$ of moving segments in the plane, which we want to restrict further in that they never intersect each other. Let us first introduce some notation:

- Let $\Delta(v)$ denote the region of node $v$ in the plane;
- Let $S(v) := S \cap \Delta(v)$ be the set of fragments that must be stored in the subtree rooted at $v$;
- We call a segment $s$ *long (with respect to v)* iff it intersects both the left and the right border of $\Delta(v)$, or iff the endpoints lie on these borders (you will see in a minute that $\Delta(v)$ does indeed have such borders);
- The *rank* of an endpoint is just the rank of its $x$-coordinate;
- Let $x_l(v), x_r(v)$ denote the smallest and largest, respectively, ranks of all endpoints in $S(v)$ (relative to *all* other $x$-coordinates!).

Now, we will describe the construction of the (static) BSP. Basically, this amounts to consider three cases:

1. $S(v) = \varnothing$: $v$ is a leaf;
2. $S(v)$ does not contain long segments: split the set of fragments by a vertical splitting line $x_{\text{mid}} = \lfloor \frac{x_l + x_r}{2} \rfloor$; such a node is called a *vertical node*. The line $x_{\text{mid}}$ will pass through an endpoint, but not necessarily one of $S(v)$. In fact, it could happen that all fragments in $S(v)$ are completely to one side of $x_{\text{mid}}$; such a node can cause fragmentation;
3. $S(v)$ contains $l$ long segments: we can sort these from bottom to top, since there are no intersections. They will be used as splitting lines. We create a *multi-way* node with $l + 1$ children, each of which corresponds to a region between two consecutive long segments. Each child gets those fragments assigned that are within such a region. Another name for such a node is a *horizontal node*; it does not cause fragmentation.

By construction, the regions have four borders: two vertical (parallel) ones on the left and right side, and two more or less horizontal ones at the top and bottom. This is called a *trapezoid*.[2] See Figure 8.3 for an illustration of how such a BSP looks.

[2]Trapezoid partitionings of the plane have applications for many other algorithms in computational geometry. Sometimes, they are also called *vertical decomposition*. They can be generalized to 3D, where they are sometimes called *cylindrical decomposition*.

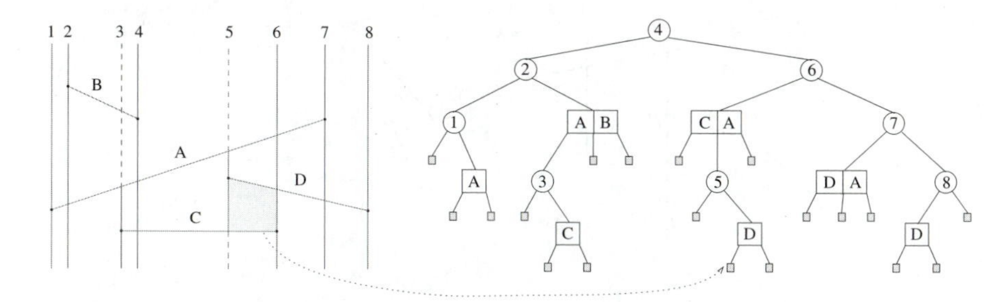


**Figure 8.3.** An example of a kinetic BSP tree. On the left, four segments are shown, together with the vertical splitting lines. At the top of each vertical line is its ID, which happens to also be the endpoint's rank in the situation shown. On the right is the corresponding BSP tree; round nodes are vertical splitting nodes, while square nodes are multi-way (i.e., horizontal) nodes. Leaves are small grey nodes.

Before proceeding, we would like to point out the following properties of this tree:

- The tree without the multi-way nodes can be considered a segment tree over the $x$-coordinates of all endpoints;
- The parent of a multi-way node is always a vertical node;
- Any region $\Delta(v)$ has at most two neighbor regions adjacent to its left side and to its right side, respectively, and one above and below, respectively;
- As soon as a segment cannot cause fragmentation within a region (because it has become a long segment with respect to that region), it will be used as a splitting line.

In order to obtain a KDS, we must throw in a few more ingredients, most of which we already used for the kinetic segment tree:

- array $R[1, \ldots, 2n] : \mathbb{N} \rightarrow \mathbb{R}$, just as before,
- a fragment list $\mathcal{L}$, as before,
- for each endpoint, a pointer to the neighbor region,
- for each leaf, four pointers to the neighbor regions.

Since segments are not allowed to intersect, it is obvious that the combinatorial structure changes only when the $x$-coordinates of two endpoints swap ranks. This yields the certificates, which are exactly the same as for the kinetic segment tree.

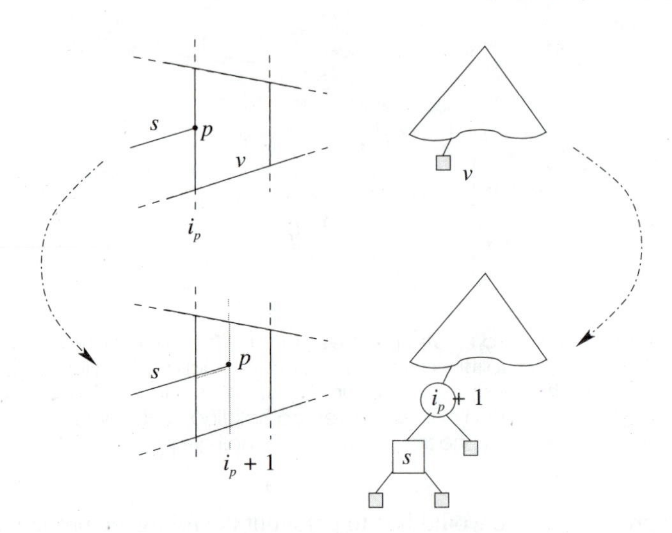


**Figure 8.4.** This is the first case of leaf operations to be performed on a kinetic BSP when an event happens. The leaf $v$ is replaced by a small subtree. Its root is a vertical node with a splitting line through $p$ (the endpoint that has swapped ranks with another endpoint).

For describing the algorithm, we assume that the rank of the right endpoint of segment $s$ has incremented by 1. This is an event in which $s$ has to augmented by one elementary fragment (EF). All other cases are analogous. The algorithm will proceed in two steps: update leaf and then reestablish the tree properties.

Step 1. Let $i_p$ = rank of endpoint $p$ before the event (afterwards it is $i_p + 1$). There are two cases.

In the first case, $p$ is on the *left* border of the region $\Delta(v)$ of leaf $v$ before the event, and after the event it is somewhere in the interior (i.e., $x_r(v) > i_p + 1$). In that case, we just replace $v$ by a little subtree, as shown in Figure 8.4.

The second case is when $p$ is on the left border before the event, and *on* the right border after the event (i.e., $x_r(v) = i_p + 1$). Again, we replace the leaf $v$ by something else, but now this depends on whether $v$'s parent is a vertical node or a multi-way (horizontal) node. In the first case, this distinction was not necessary because the root of the new subtree was a vertical node. See Figure 8.5 for the exact operations.

Step 2. After the leaf has been modified in Step 1, we obtain a new BSP that is valid and produces the new partitioning of the plane. However, it does not necessarily possess the properties outlined above. In particular, the horizontal

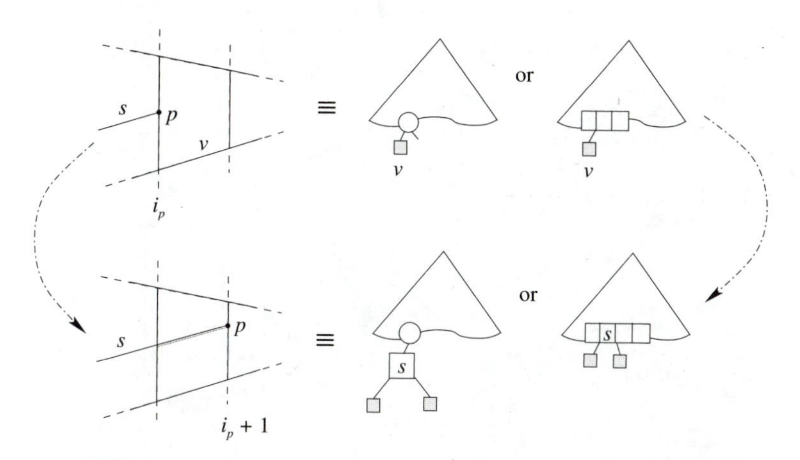


**Figure 8.5.** This is the second case of leaf operations when an event happens. Again, the leaf $v$ is replaced by a small subtree. However, here the exact operation depends on the type of parent node of $v$. If the parent is a vertical node, then the new subtree is straightforward. If the parent is a multi-way (horizontal) node, then we must insert a new splitting line into that parent.

split through segment $s$ (either by a new node, or by an existing one) might have been performed higher up in the tree, if we had constructed the BSP from scratch. To restore this property, we push the new horizontal split up as far as possible. Note that it might end up in a multi-way node that already exists.

Let $f' = s \cap \Delta(v')$, where $f'$ is the new EF, and $v'$ is the new horizontal node. Let $v''$ be the left sibling of $v'$ (if any), and $f'' = s \cap \Delta(v'')$ (if any). Now, if $v''$ is a horizontal node, too, and $s$ is being stored there, too, then $s$ is a long segment with respect to $\Delta(v') \cup \Delta(v'')$. Therefore, the horizontal split with $s$ could (and should) have been done higher up in the tree.

In that case, we perform the operation as depicted in Figure 8.6. If the parent of the new horizontal node $\mu'$ is a horizontal node itself, then we do not create the stand-alone node $\mu'$. Instead, we insert the fragment $f' \cup f''$ there.

Then we repeat this process with $v'$ = that node which now stores the fragment of $s$. So, in a nutshell, we push the initially new fragment up through the tree, thereby merging it with other fragments of the same segment $s$.

Diminishing a segment (the opposite of augmenting a segment) by an EF works analogously by traversing the BSP top-down until we reach a node that contains that EF. This happens exactly one level above a leaf, at which point we perform the opposite operations to those in Step 1.

Figure 8.7 shows a simple example of such a combinatorial restructuring (in this example, no propagation up through the tree is necessary).

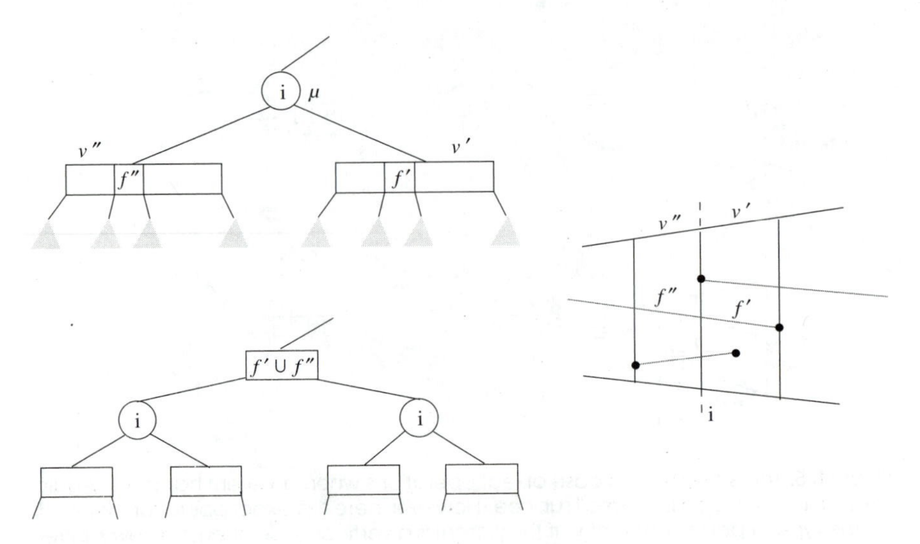


**Figure 8.6.** The properties of the new BSP are restored by pushing up the new fragment through the tree while merging it with other fragments of the same segment.

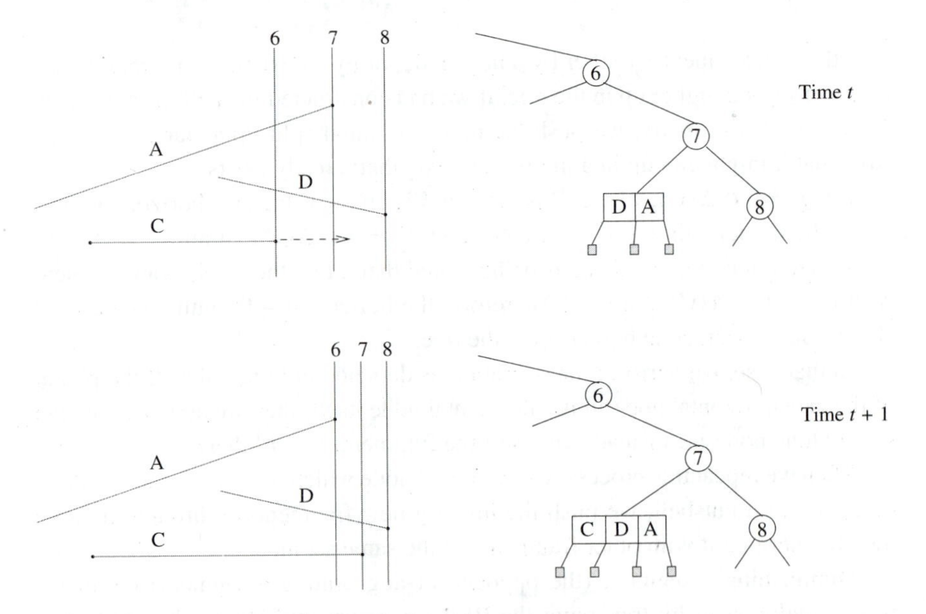


**Figure 8.7.** A simple example of a combinatorial restructuring of a kinetic BSP.

To obtain a BSP in the true sense, we have to replace the multi-way nodes by binary subtrees. We would like to refer the interested reader to [de Berg et al. 01] for the details. Overall, we can state the following lemma.

**Lemma 8.2.** *Let $S$ be a set of $n$ moving segments in the plane, which are disjoint at all times. Then there is a kinetic BSP over $S$ of size $O(n \log n)$ and worst-case depth $O(\log^2 n)$. It has only external events, and the expected response time is $O(\log n)$ (worst-case $O(\log^2 n)$).*

# 9

# Degeneracy and Robustness

A geometric data structure for a set of geometric objects must be computed by an algorithm. An algorithm that handles geometric input and produces geometric output is called a *geometric algorithm.* The decisions and the output of a geometric algorithm depend on *geometric predicate*s and on the computation of new and intermediate geometric objects. *Geometric computation* has two components: a numerical part and a combinatorial part. Numerical computation is involved in both the construction of new objects and in the evaluation of geometric predicates. The combinatorial part takes care of the representation and correctness of the combinatorial structure of (sub)results.

For example, the *Voronoi diagram* of a set $P$ of points in 2D is a subdivision of the plane into cells of nearest neighborship. That is, a *Voronoi cell* of a *Voronoi site* $p \in P$ is the set of all points that are closer to $p$ than to any other point $q \in P$; see Figure 9.1 for an example and Chapter 6 for a detailed discussion.

The combinatorial part of the Voronoi diagram is given by the graph of the subdivision. The exact coordinates of the vertices of the graph and the corresponding edges, which are also called *Voronoi vertices* and *Voronoi edges*, respectively, represent the numerical part of the problem. Note that other combinatorial representations of the Voronoi diagram are valid as well. For example, we can store a list of clockwise-ordered neighboring sites for every site $p$. This represents the structure of the Voronoi diagram without any numerical information.

Another example of a geometric problem is the computation of the *convex hull* of a set $P$ of points in 2D. The convex hull of a set $P$ is the smallest convex polygon that contains all points. We can describe the combinatorial structure of the convex hull by a circular list of successive points on the hull stemming from

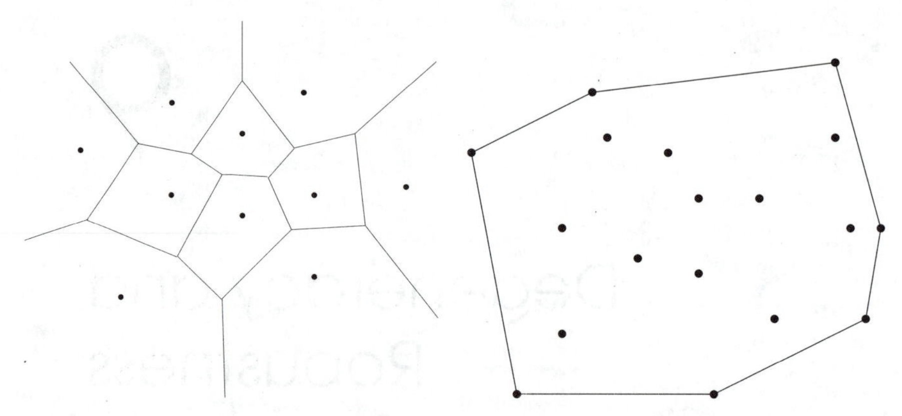

**Figure 9.1.** The Voronoi diagram for a set of sites subdivides the plane into regions of the nearest neighborship.

**Figure 9.2.** The convex hull for a set of points is the smallest convex set that contains all points.

$P$; see Figure 9.2. A numerical part is not necessary for the output; it is only necessary for the computation of this list.

There are two difficulties for the practical implementation of many geometric algorithms. On one hand, the problem of the numerical part lies in the precision of arithmetic, which influences also the combinatorial results and the flow of an algorithm. Thus, we are looking for *robustness* of the underlying arithmetic.

On the other hand, for the flow of the algorithm itself, we have to study input scenarios of geometric objects that are not independent of each other. Many geometric algorithms behave well if the given objects are assumed to be in *general position*. For example, a set of points in the plane is in general position if no three points are colinear and no four points are cocircular. Non-general positions are also denoted as *degeneracy*. The influence of and how to deal with degeneracy is the subject of Section 9.5.

There are many surveying articles considering robustness issues and degeneracy in geometric computing. We will try to give an overview and use elements mainly from [Fortune 96], [Schirra 00], [Shewchuk 97], [Yap 97b], [Sugihara 00], [Goldberg 91], [Michelucci 97], [Santisteve 99], and [Mehlhorn and Näher 94], and additionally from [Guibas et al. 89, Burnikel et al. 99, Burnikel et al. 94, Fortune and Van Wyk 93, Fortune 89, Yap 97a, Sugihara and Iri 89, Sugihara et al. 00].

First, we discuss two simple examples in 2D and 3D and point out the influence of impreciseness in Section 9.1. As a side effect, we will introduce a data structure that efficiently represents the combinatorial structure of graphs in different dimensions. Later, we will turn over to the problem of degeneracy in the combinatorial flow of an algorithm; see Section 9.5.

# 9.1. Examples of Instability in Geometric Algorithms

## 9.1.1. Intersections of Line Segments

First, we consider the *arrangement* $A$ of a set of segments in the plane; see Figure 9.3. The set of segments represents a subdivision of the plane into cells, and we need to find a well-suited representation of the subdivision, i.e., a data structure that represents the combinatorial structure of the arrangement. The combinatorial structure of the arrangement consists of one-dimensional vertices, edges, and 2D cells. We may assume that we want to have access to all cells of the arrangement systematically. A single cell is given by the ordered sequence of the surrounding segments and the intersections in between. Thus, the whole arrangement can be represented by a planar graph $G$. Logically, a graph $G = (V, E)$ is given by, a set of vertices $V$ and a set of edges $E$. Physically, we will try to build a *doubly connected edge list* (DCEL). The efficient DCEL data structure, introduced in [Guibas and Stolfi 85], includes the combinatorial information of $G$ and does not waste space for it. In fact, it is the most space-efficient way of storing a subdivision.

The DCEL of a planar graph gives access to the following information for every edge $e$ of $G$:

- a reference to the source and target vertices $v_1$ and $v_2$ of $e$,
- a reference to the left and right face $F_1$ and $F_2$ adjacent to $e$,
- a reference to the *previous* incoming edge $p_{v_1}$ (respectively $p_{v_2}$) at $v_1$ (respectively $v_2$),

| Edge | Vertex | | Faces | | Incoming Edges | | | |
|---|---|---|---|---|---|---|---|---|
| | Start | End | Left | Right | Next Start | Prev. End | Next End | Prev. End |
| $e$ | $v_1$ | $v_2$ | $F_1$ | $F_2$ | $n_{v_1}$ | $p_{v_1}$ | $n_{v_2}$ | $p_{v_2}$ |

**Table 9.1.** References for a single edge $e$ in the DCEL of a graph in 2D.

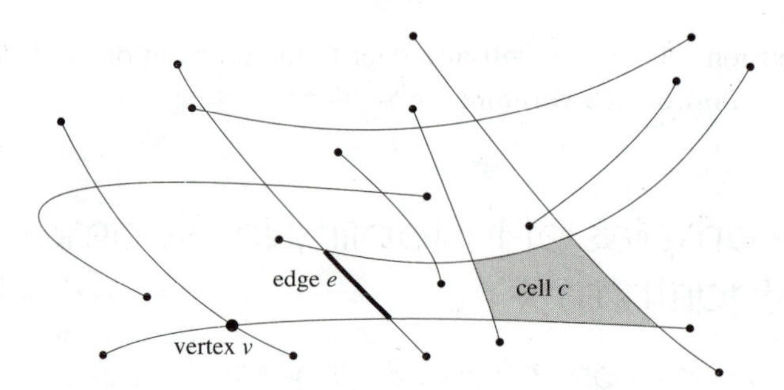


**Figure 9.3.** The arrangement of a set of segments represents a subdivision of the plane into cells represented by vertices and edges.

- a reference to the *next* incoming edge $n_{v_1}$ (respectively $n_{v_2}$) at $v_1$ (respectively $v_2$).

Additionally, the DCEL contains:

- a reference to a starting edge for every face $F$,
- a reference to the coordinates for every vertex $v$.

We further assume that we have access to the vertices, edges, and faces by their names. The terms *previous* and *next* refer to an ordering of the incoming edges in a clockwise or counterclockwise manner. The information of the reversed edge $\overline{e} = (v_2, v_1)$ is implicitly contained. Figure 9.4 shows the information of the DCEL for a single edge $e$; see also Figure 9.8 for an example of the DCEL in 3D. With the help of the DCEL, we can perform a simple walk-through in the whole graph in time linear in the number of edges.

Any algorithm that builds the DCEL of $A$ has to compute the intersections of the segments and has to order the segments by angle if three segments have a

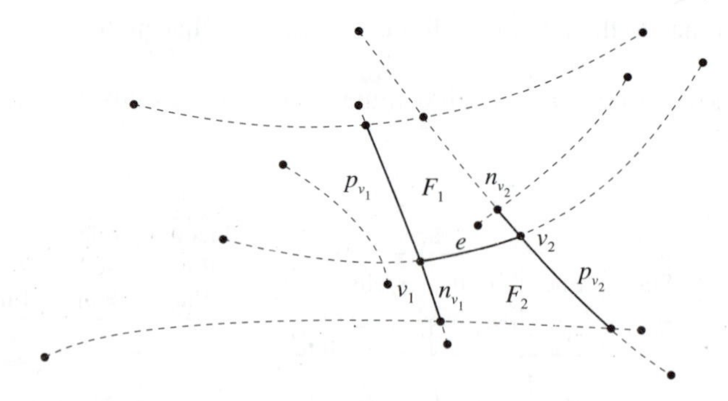


**Figure 9.4.** An edge of the arrangement and its information in the DCEL.

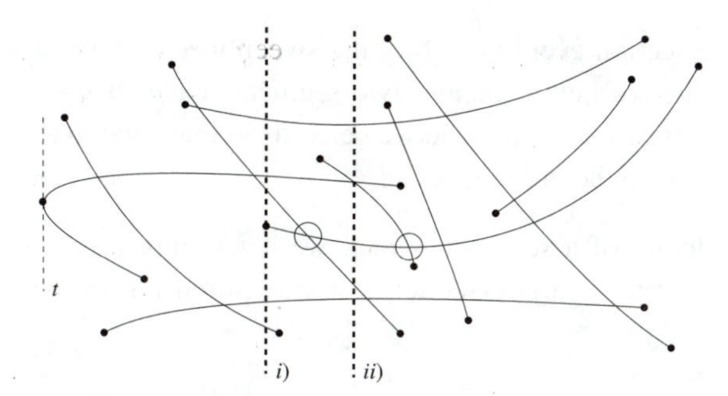


**Figure 9.5.** The sweep status structure contains a sorted list of the segments met by the sweepline. At position i), a new segment has to be inserted and the intersections with its neighbors will create future events. At position ii), such a new event changes the order of the segments, intersections are computed, and further future events may be created.

common intersection. This means that geometric predicates have to be evaluated precisely and intersections have to be computed exactly.

The subproblem of computing all intersections can be efficiently solved with the famous *sweep* paradigm; see [Bentley and Ottmann 79].

Let us consider a sketch of this idea. We assume that the segments are $x$-monotone and that intersections can be computed efficiently. If a segment is not $x$-monotone, we may split it into several segments if a vertical tangent occurs. For example, at the the vertical tangent $t$ in Figure 9.5, we can split a non-monotone segment into two monotone segments. First, we sort the endpoints of the segments by $x$-coordinates and insert them into an *event list*. Now we will visit all *events* successively by a vertical *sweepline* that moves from right to left. During the sweep, we will maintain the important information along the sweepline in a *sweep status structure*. In case of the arrangement, the sweep status structure contains a sorted list of segments that intersect the sweepline. The list is sorted by the $y$-coordinates of the intersections. The idea is that just before two line segments intersect, they must have been neighbors in the sweep status structure. See Figure 9.5 for an example of the main steps in the sweep algorithm. The following three main events must be handled:

- If a start point of a segment is met, we insert the new segment into the sorted list of segments. The new segment may have intersections with the neighboring segments in the segment list. We compute all *next* intersections with respect to the $x$-coordinates and insert them by $x$ order into the event list. Note that at most two intersections with neighboring segments are inserted in the event list;

- If an intersection event is met by the sweepline, we have to rearrange the order of the segments. Again, two segments have changed their order in the sweep status structure, and we perform an intersection test with the new neighbors as in the previous case;
- If an endpoint of a segment is met, we delete this segment in the sweep status structure. An intersection test is required for the new neighboring segments.

As an invariant, the sweep status structure always contains the sorted list of segments intersected by the sweepline. The neighboring segments are tested for intersections. Obviously, the invariant belongs to the combinatorial part of the algorithm.

During the sweep, we will collect all intersections, which are part of the information needed for the DCEL. Let us assume for a moment that we have to deal with line segments represented by their endpoints; that is, let $l_i = (p_i, q_i)$, where $p_i = (p_{ix}, p_{iy})$ and $q_i = (q_{ix}, q_{iy})$. If an intersection $(x, y)$ of $l_1$ and $l_2$ exists, it can be computed adequately by

$$x = \frac{\begin{vmatrix} b_1 & c_1 \\ b_2 & c_2 \end{vmatrix}}{\begin{vmatrix} a_1 & b_1 \\ a_2 & b_2 \end{vmatrix}} \qquad y = -\frac{\begin{vmatrix} a_1 & c_1 \\ a_2 & c_2 \end{vmatrix}}{\begin{vmatrix} a_1 & b_1 \\ a_2 & b_2 \end{vmatrix}},$$

where $a_i = (q_{iy} - p_{iy})$, $b_i = (p_{ix} - q_{ix})$, and $c_i = (p_{iy} q_{ix} - p_{ix} q_{iy})$ for $i = 1, 2$; see also Section 9.4.2. This is the numerical part of the algorithm.

If the $x$-coordinates of the intersections are not computed exactly, two actually neighboring segments will eventually not appear as neighbors during the sweep. The event list may have an incorrect order. In this inconsistent situation, we have lost information for the DCEL due to impreciseness. The invariant is no longer valid, and the result will be incomplete. Note that for the flow of the algorithm, it suffices to compare the $x$-coordinates of intersections or endpoints of line segments, rather than computing the intersection coordinates. However, comparisons may be erroneous as well.

Ramshaw's braided lines example [Nievergelt and Hinrichs 93, Nievergelt and Schorn 88] shows that naively applied floating-point arithmetic cannot guarantee preciseness for intersections. Unfortunately, the lines $l_1 : 4.3 \times x/8.3$ and $l_2 : 1.4 \times x/2.7$ seem to have several intersections if floating-point arithmetic with base 10, precision 2, and rounding to nearest is used; see Figure 9.6. For the details of floating-point arithmetic; see Section 9.3.1.

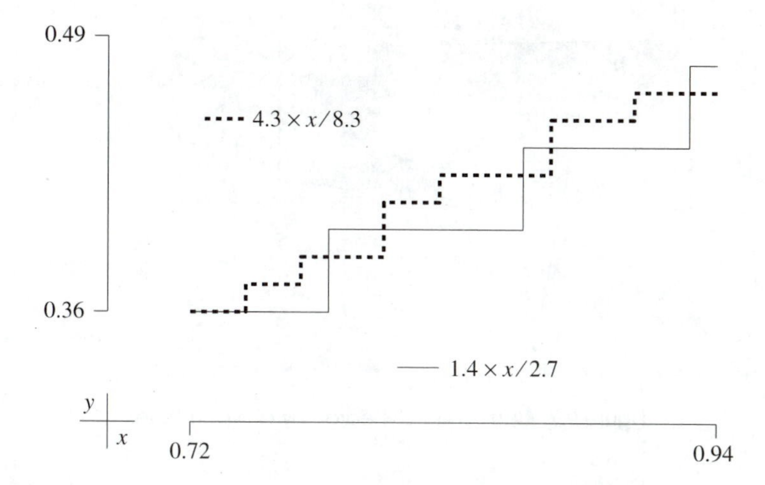


**Figure 9.6.** The lines $l_1 : 4.3 \times x/8.3$ and $l_2 : 1.4 \times x/2.7$ appear to have several intersections using floating-point evaluation with base 10, precision 2, and round to nearest.

The running time of the presented algorithm is given by $O((n + k) \log n)$, where $k$ denotes the number of intersections.

## 9.1.2. Cutting a Polyhedron by a Hyperplane

Let us consider a simple 3D example where we do not need to compute intermediate results, which seems to be the problem of the previous section. We want to find the intersecting edges of a convex polyhedron with a hyperplane in 3D; see Figure 9.7. Let us assume that the polyhedron is given by a graph $G = (V, E)$. The underlying data structure should be a DCEL in 3D. In analogy to the 2D case, the DCEL contains the following information for every edge $e$ of every tetrahedron $T$:

- a reference to the source and target vertices $v_1$ and $v_2$ of $e$,
- a reference to the left and right face $F_1$ and $F_2$ adjacent to $e$,
- a reference to the next edge $n_{v_2}$ with respect to the left face $F_1$,
- a reference to the previous edge $p_{v_1}$ with respect to the right face $F_2$.

The orientations *left* and *right* correspond to the orientation of the edge $e$. We consider the edge from the outer side of $T$. We can move one adjacent face into the plane of the other without an intersection of the faces. Now with respect to

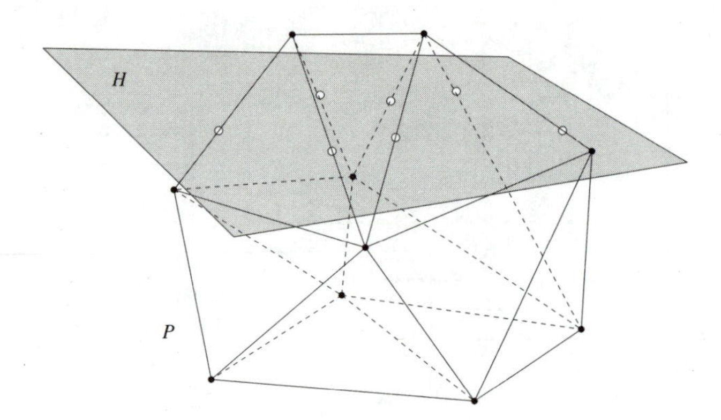


**Figure 9.7.** Cutting a polyhedron by a hyperplane.

the orientation of the edge, one face lies to the left of $e$ and the other edge lies to the right of $e$. These orientations are uniquely determined. Every edge has an orientation from the start point to the endpoint.

Additionally, we keep track of the following information:

- a reference to the normal for every face $F$,
- a reference to a starting edge for every face $F$,
- a reference to the coordinates for every vertex $v$.

We can assume that access to the vertices, edges, and faces is given by their names. Figure 9.8 together with Table 9.2 show the DCEL of a single tetrahedron. Note that for every edge $e_i = (v_{i_1}, v_{i_2})$, there also has to be an entry $\overline{e_i} = (v_{i_2}, v_{i_1})$, and a link between $e_i$ and $\overline{e_i}$ is stored.

A general graph in 3D may represent a set of polyhedra. For example, the Voronoi diagram in 3D represents a subdivision of the plane into convex polyhe-

| Edge | Vertex | | Faces | | Edges | |
|---|---|---|---|---|---|---|
| | Start | End | Left | Right | Next | Prev. |
| $e_1$ | $v_1$ | $v_2$ | $F_1$ | $F_2$ | $e_5$ | $e_6$ |
| $e_2$ | $v_2$ | $v_3$ | $F_4$ | $F_2$ | $e_3$ | $e_1$ |
| $e_3$ | $v_3$ | $v_4$ | $F_4$ | $F_3$ | $e_5$ | $e_6$ |
| $e_4$ | $v_1$ | $v_4$ | $F_3$ | $F_1$ | $e_3$ | $e_1$ |
| $e_5$ | $v_2$ | $v_4$ | $F_1$ | $F_4$ | $e_4$ | $e_2$ |
| $e_6$ | $v_1$ | $v_3$ | $F_2$ | $F_3$ | $e_2$ | $e_4$ |

**Table 9.2.** References in the DCEL of a polyhedron.

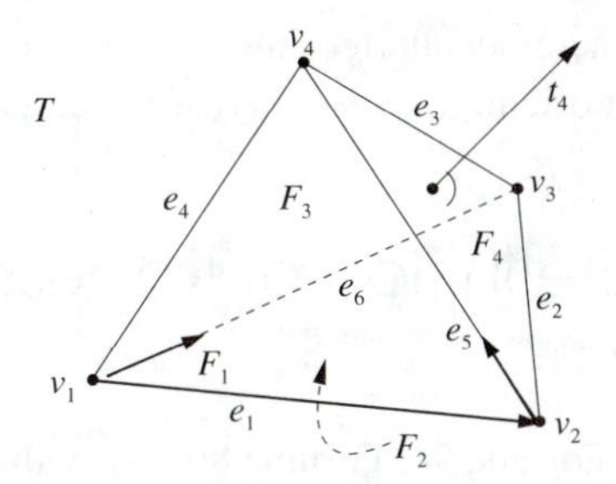


**Figure 9.8.** The DCEL information of a tetrahedron $T$. The faces $F_1$ and $F_2$ of the oriented edge $e_1$ have next edge $e_5$ and previous edge $e_6$.

dra; see Chapter 6. In this case, every polyhedron is represented by the corresponding information. Additionally, for every edge $e$, there is a reference to the next polyhedron of $e$, sorted in clockwise or counterclockwise order. Altogether, we can easily trace the graph, visiting all edges and vertices.

For a simple geometric algorithm, we need to know whether a vertex of the polyhedron $P$ lies above or below the hyperplane $H$. There is a simple and efficient predicate for this test. We have to compute the sign of the determinant

$$\text{Det}(a, b, c, v) = \begin{vmatrix} a_x & a_y & a_z & 1 \\ b_x & b_y & b_z & 1 \\ c_x & c_y & c_z & 1 \\ v_x & v_y & v_z & 1 \end{vmatrix},$$

where $v = (v_x, v_y, v_z)$ denotes a vertex of $P$ and the vectors $\vec{a} = (a_x, a_y, a_z)$, $\vec{b} = (b_x, b_y, b_z)$, and $\vec{c} = (c_x, c_y, c_z)$ span the hyperplane $H$. If and only if $\text{Det}(a, b, c, v) > 0$ holds, vertex $v$ is below the *oriented* plane $H$ formed by $a$, $b$, and $c$. The plane is meant to be oriented as follows. If we *see* the points $a$, $b$, and $c$ in a counterclockwise order, the plane is seen from *above*. In Section 9.4 we will give a formal motivation and a generalized form of this so-called *orientation test*.

Now, the problem arises that the sign of the determinant $\text{Det}(a, b, c, v)$ cannot be evaluated precisely in all cases. For example, a face of $P$ may be almost parallel and very close to the hyperplane $H$. In this case, roundoff errors may produce incorrect results, and any algorithm will fail to compute the correct result. On the other hand, it might even happen that a vertex of $P$ lies exactly on $H$. Any algorithm has to make a decision for a *degenerate* case.

Note that a simple incremental algorithm correctly computes the convex hull in 3D by using a correct orientation test only; see [Sugihara 94].

## 9.2. Formal Definition of Robustness and Stability

We follow the lines of [Fortune 96, Fortune 89] and will give a formal definition of robustness. In a mathematical sense, a *geometric problem* is a mapping from an input set of geometric objects to an output set of geometric objects.

Fortune [Fortune 96, Fortune 89] suggests to define a geometric problem by a function

$$P : R^{\omega} \rightarrow C,$$

where $R^{\omega} = \cup_n \mathbb{R}^n$ and $C$ is discrete, i.e., for each $n$, the set $\{P(x)|x \in \mathbb{R}^n\}$ is finite. For example, the 2D convex hull problem maps $\mathbb{R}^n$ for even $n$ onto cyclically ordered subsets $\{1, 2, \ldots, \frac{n}{2}\}$ indicating the indices of the counterclockwise sequence of the points on the convex hull. By representing the DCEL in an appropriate manner, the arrangement problem in Section 9.1.1 can be easily transformed into this framework.

To be more precise and to be consistent with the formal definition of degeneracy in Section 9.5.1, we use the following concept.

**Definition 9.1. (Geometric problem.)** A geometric problem is a function

$$P : X \rightarrow Y.$$

The input space $X = \mathbb{R}^{nd}$ has the standard Euclidean topology. The output space $Y$ is the product $C \times \mathbb{R}^m$ of a finite space $C$ with discrete topology and the Euclidean space $\mathbb{R}^m$.

Here $m$, $n$, and $d$ are integers, and $C$ is a discrete space that represents the combinatorial part of the output; for instance a planar graph or an ordered list of vertices. The parameter $d$ represents the dimension of the input space, and $n$ represents the number of input points. This formalism is equivalent to the definitions in Section 9.5.1.

For example, the 2D convex hull problem maps $\mathbb{R}^{2n}$ onto cyclically ordered subsets $\{1, 2, \ldots, n\}$ indicating the indices of the counterclockwise sequence of the points on the convex hull. The set of all ordered subsets of $\{1, 2, \ldots, n\}$ represents the discrete set $C$.

The arrangement problem in Section 9.1.1 can be easily transformed into this framework. $C$ represents the planar graph of the arrangement; and the intersections are represented within $\mathbb{R}^m$.

**Definition 9.2. (Selective geometric problem.)** A geometric problem is called *selective* if the output $P(x) \in C$ selects only elements out of the input set $x$, for example, the convex hull problem is selective. A *constructive* geometric problem represents new objects in the output $P(x)$.

For example, the DCEL of the line segment arrangement of Section 9.1.1 consists of vertices stemming from the intersections of two line segments. In this sense, the corresponding algorithm is constructive.

For $x \in X$, $\mathcal{A}(x) \in Y$ represents the outcome of a geometric algorithm $\mathcal{A}$ designed to solve $P$. An algorithm $\mathcal{A}$ *computes* $P$ *exactly* for $x \in X$, if $\mathcal{A}(x) = P(x)$.

**Definition 9.3. (Robust geometric algorithm.)** A geometric algorithm $\mathcal{A}$ is called *robust*, if for all $x \in X$, there is $x' \in X$ such that $\mathcal{A}(x) = P(x')$; i.e., the output of the algorithm gives at least a correct result for a correct input $x'$. The algorithm $\mathcal{A}$ is called *stable*, if for all $x \in X$, there is $x' \in X$ *near to* $x$ such that $\mathcal{A}(x) = P(x')$.

Robustness guarantees that the output of the algorithm is correct for some perturbation of the input. The algorithm is stable if the perturbation is small. Note that we cannot compare two outputs $A(x)$ and $A(x')$ in the same way. In many applications, the output $A(x)$ of an algorithm is not continuous in $x$. For example, the resulting graph of the arrangement problem may change fundamentally, although the input is only slightly perturbed.

A stable algorithm is accompanied with a measure of the perturbation bound.

**Definition 9.4. (Relative error.)** Algorithm $\mathcal{A}$ for problem $P$ has *relative error* $f(n, \epsilon)$, if for all $x \in X$ with $|x| \in O(n)$, there is an $x' \in X$ *near to* $x$ so that $\mathcal{A}(x) = P(x')$ and

$$\frac{|x - x'|}{|x|} \leq f(n, \epsilon).$$

Here, $\epsilon$ refers to the accuracy of the given number representation and $| \sim |$ means the maximum norm. For a stable algorithm, $f(n, \epsilon)$ should be as small as a function in $n$ and $\epsilon$. A very stable algorithm should be independent from $n$ and should guarantee $f(n, \epsilon) \in O(\epsilon)$.

We will give a simple motivation for this concept. Let us assume that an algorithm $\mathcal{A}$ internally handles polynomials with degree at most $k$, i.e., every computation can be performed by evaluation of multi-variable polynomials of degree $\leq d$. For example, $p(x, y) = x^3y^5 + x^4y^2 + 7$ is a multi-variate polynomial of degree 8. Let us further assume that a machine error $\epsilon$, see Section 9.3.1 for details, is given. This means that a number can be represented in the computer by approximately $k = -\log \epsilon$ bits. Rounding all numbers approximately to $k/d$ bits allows exact computations. This is a perturbation of $O(\epsilon^{-d})$. Altogether, in the given situation, a perturbation bound of $f(n, \epsilon) \in O(\epsilon^{-d})$ should be achievable. This is denoted also as a perturbation bound *benchmark*; see [Fortune 89].

## 9.3. Geometric Computing and Arithmetic

Geometric algorithms are designed under the assumption of the *Real Random Access Machine* (Real RAM) computer model, which includes exact arithmetic on real numbers and allows a computer-independent analysis of complexity and running time in the famous $O$-notation.

For the Real RAM [Preparata and Shamos 90], we assume that each storage location is capable of holding a single real number and that the following operations are available at unit costs:

- standard arithmetic operations $(+, -, \times, /)$,
- comparisons $(<, \leq, =, \neq, \geq, >)$,
- extended arithmetic operations: $k$th root, trigonometric functions, exponential functions, and logarithm functions (if necessary).

Unfortunately, we cannot represent everyday decimal numbers in the binary system of computers. For example, the simple decimal number 0.1 has no exact representation in the binary system. We can give only a non-ending approximation by 0.0001100110011 ..., which is a short representation for $1 \cdot 2^{-4} + 1 \cdot 2^{-5} + 1 \cdot 2^{-8} + 1 \cdot 2^{-9} + 1 \cdot 2^{-12} + 1 \cdot 2^{-13} + \cdots$. We conclude that exact arithmetic cannot be achieved for all decimal numbers directly.

Altogether, the Real RAM assumption does not always hold. Fast standard floating-point arithmetic running on modern microcomputers is accompanied with roundoff errors.

There are two main ways to overcome the problem: either we use fast standard floating-point arithmetic and try to live with the intrinsic errors or we have to implement the Real RAM by developing exact arithmetic. Sometimes, the solution is somewhere in between. As long as there are no severe problems, we can live with the roundoff errors of fast floating-point operations. If some serious problems arise, we will switch over to time-consuming exact operations.

The adaptive approach is also the basis for the paradigm called *exact geometric computation* (EGC); for example, considered in [Yap 97a] and [Burnikel et al. 99]. For EGC, the emphasis is on the geometric correctness of the output, rather than the exactness of the arithmetic. One tries to ensure that at least the decisions and the flow of the algorithm are correct and consistent. As long as no severe inconsistency appears, there is no need to perform cost-consuming exact calculations. EGC often is accompanied by an EGC library that guarantees exact comparisons; see Section 9.7 for a detailed discussion of existing packages.

First, in Section 9.3.1, we will have a look at the main properties of standard floating-point arithmetic available on almost every microcomputer. We will consider the cost impact of exact arithmetic in Section 9.3.2. Imprecise and adaptive arithmetic variants are presented in Section 9.3.3, and finally, the principles of EGC are shown in Section 9.3.4.

## 9.3.1. Floating-Point Arithmetic

Floating-point arithmetic for microcomputers is defined in the IEEE standards 754 and 854; see [Goldberg 91] and [IEEE 85].

For obvious reasons, hardware-supported arithmetic is restricted to binary operations. Therefore, the IEEE 754 standard is defined for the radix $\beta = 2$. Fortunately, the standard includes rules for conversion to *human* decimal arithmetic. Therefore, with a little loss of precision, we can work with decimal arithmetic in the same manner. For example, programming languages make use of the conversion standard and give access to floating-point operations with radix $\beta = 10$, which depend on hardware-supported fast floating-point operations with radix $\beta = 2$.

The standard IEEE 854 allows either radix $\beta = 2$ or radix $\beta = 10$. In contrast to IEEE 754, it does not specify how numbers are encoded into bits. Additionally, the precisions *single* and *double* are not specified by particular values in IEEE 854. They have constraints for allowable values instead. With respect to encoding into bits and conversions to decimals, repeating the features of the standard IEEE 754 will give us some more insight.

The main advantage of standardization is that floating-point arithmetic is supported by almost every type of hardware in the same way. Thus, software implementations based on the standard are portable to almost every platform. With respect to the underlying arithmetic, floating-point operations should not produce different outputs on different platforms.

The floating-point arithmetic of a programming language should guarantee the same results on every platform and should be based on the standard. So, in some sense, a programming language analogously builds a standard of its own, following the rules of arithmetic and rounding as required in IEEE 754. Therefore, it is well justified to have a closer look at IEEE 754.

**The main features of IEEE 754.** We will give a short overview of the features and go into detail for some of them afterwards.

**Definition 9.5. (Floating-point representation.)** A floating-point representation consists of a *base* $\beta$, a *precision* $p$, and an *exponent range* $[e_{min}, e_{max}] = E$. A floating-point number is represented by

$$\pm d_0.d_1\, d_2 \ldots d_{p-1} \cdot \beta^e,$$

where $e \in E$ and $0 \leq d_i < \beta$. It represents the number

$$\pm\left(\sum_{i=0}^{p-1} d_i \beta^{-i}\right) \cdot \beta^e .$$

The representation is called *normalized* if $d_0 \neq 0$.

For example, the rational number $\frac{1}{2}$ can be represented for $p = 4$ and $\beta = 10$ by 0.500 or in a normalized form by $5.000 \times 10^{-1}$.

A direct bit encoding for floating-point numbers is available only for $\beta = 2$. Unfortunately, not all numbers can be represented exactly by floating-point numbers with $\beta = 2$. For example, the rational $\frac{1}{10}$ has the endless representation $1.100110011\ldots \times 2^{-4}$.

For arbitrary $\beta$, we have $\beta^p$ possible significands and $e_{max} - e_{min} + 1$ possible exponents. Taking one bit for the sign into account, we need

$$\log_2(e_{max} - e_{min} + 1) + \log_2(\beta^p) + 1$$

bits for the representation of an arbitrary floating-point number. The precise encoding is specified in IEEE 754 for $\beta = 2$ and not specified for $\beta \neq 2$.

The IEEE 754 standard includes the following.

- Representation of floating-point numbers: see Definition 9.5.

- Precision formats: the precision formats are single, single extended, double, and double extended. For example, single precision makes use of 32 bits with $p = 24$, $e_{max} = +127$, and $e_{min} = -126$.

- Format layout: the system layout of the precision formats is completely specified; see Section 9.3.1 for a detailed example of the single format.

- Rounding: roundoff and roundoff errors for arbitrary floating-point arithmetic cannot be avoided, but the rounding rules should create reproducible results; see Section 9.3.1 for a detailed discussion of roundoff errors and rounding with respect to arithmetic operations.

- Basic arithmetic operations: the standard includes rules for standard arithmetic. Addition, subtraction, multiplication, division, square root, and remainder have to be evaluated exactly first and rounded to nearest afterwards.

- Conversions: conversions between floating-point and integer or decimal numbers have to be rounded and converted back exactly. For example, computation of $\frac{1}{3} \times 3$ should give the result 1 even though $\frac{1}{3}$ must be rounded. Further discussions on conversion are given in Section 9.3.1.

- Comparisons: comparisons between numbers with the predicates $<$, $\leq$, or $=$ have to be evaluated exactly also for different precisions.

- Special values: some special numbers NaN, $\infty$, and $-\infty$ have an own fixed representation. Arithmetic with these numbers is standardized.

In the following, we will present the main ideas of the standard and its hardware realization.

**Single format.** The standard makes use of a format width of 32 bits for single precision $p = 24$ with $e_{max} = +127$ and $e_{min} = -126$. The first bit represents the sign, the next 8 bits represent the exponent, and the last 23 bits represent the significand. We will make use of the normalized form. Thus, 23 bits are sufficient to represent 24 significands if zero is a special value.

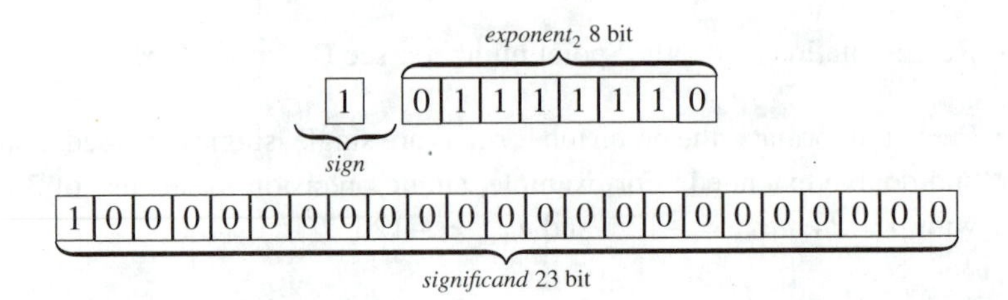


The exponent field should represent both negative and positive exponents. Therefore we use a bias, which is 127 here. An exponent of zero means that $(0 - \text{bias}) = -127$ is the exponent, whereas an exponent of 255 results in the maximal exponent $(255 - \text{bias}) = 128$. Altogether, the given example represents the number

$$(-1)^1 \times (1.10\ldots0) \times 2^{(126-127)} = -1.1 \times 2^{-1} = -0.75\,.$$

Exponents of -127 (all exponents bits are zero) or +128 (all exponents bits are one) are reserved for special numbers; see the section on special values.

Roundoff errors. Roundoff errors can be expressed in an absolute and in a relative sense. In the following, a number $z$ is referred to be *exact* if it is represented exactly with infinite precision with respect to the given radix $\beta$. First, we specify how an exact value is represented as a floating-point value rounded by the round-to-nearest rule.

**Definition 9.6. (Round to nearest.)** An exact value $z$ is rounded to the nearest floating-point number $z'$ if $z'$ is representable within the given precision $p$ and $z'$ minimizes $|z - z'|$ under all representable floating-point numbers.

If two representable floating-point numbers $z'$ and $z''$ exist with $|z - z'| = |z - z''|$, we make use of the so-called tie-breaking *round-to-even* rule, which takes the floating-point number with a biggest absolute value. Alternatively, we can apply a *round-to-zero* tie-breaking rule, which takes the floating-point number with the smallest absolute value.

For example, for $p = 3$ and basis $\beta = 10$, the exact value $z = 0.3158$ is rounded to $z' = 0.316$. The floating-point values $7.35 \times 10^2$ and $7.36 \times 10^2$ have the same absolute distance to $z = 735.5$. In this case, *round-to-even* will prefer $7.36 \times 10^2$. The most natural way of measuring rounding errors is in units in last place (ulps).

**Definition 9.7. (Units in last place.)** The term units in the last place, *ulps* for short, refers to the absolute error of a floating-point expression $d_0.d_1\, d_2 \ldots d_{p-1} \cdot \beta^e$ with respect to the exact value $z$. That is, the absolute error is within

$$|d_0.d_1\, d_2 \ldots d_{p-1} - z/\beta^e|\beta^{p-1}$$

units in the last place.

For example, let precision $p = 3$ and basis $\beta = 10$. If $3.12 \times 10^{-2}$ appears to be the result of a floating-point operation of the exact value $z = 0.0314$, we have an absolute error of $|3.12 - 3.14|10^2 = 2$ units in the last place. Similarly, if the exact value is $z = 0.0312159$, the absolute error is within 0.159 ulps. Let us consider an example with an exact value represented by infinite precision with the same basis. Let $3.34 \times 10^{-1}$ approximate the fractional expression $\frac{1}{3}$, which, in turn, has an exact infinite representation $0.33333333\ldots$ with $\beta = 10$. The absolute error is given by $= 0.666\ldots$ ulps.

Note that rounding to the nearest possible floating-point value always has an error within $\leq \frac{1}{2}$ ulps. When analyzing the rounding error caused by various formulas, a relative error is a better measure. During the evaluation of arithmetic, subresults are successively converted by the round-to-nearest rule.

**Definition 9.8. (Relative error.)** The *relative error* of a given floating-point expression $d_0.d_1\, d_2 \ldots d_{p-1} \cdot \beta^e$ with respect to the exact value $z$ is represented by

$$\left| \frac{d_0.d_1\, d_2 \ldots d_{p-1} \cdot \beta^e - z}{z} \right|.$$

For example, the relative error when approximating $z = 0.0312159$ by $3.12 \times 10^{-2}$ is $|0.0000159/0.0312159| \approx 0.000005$.

Let us consider the relative error of rounding to the nearest possible floating-point value. We can compute the relative error that corresponds to $\frac{1}{2}$ ulp. When an exact number is approximated by the closest possible floating-point value $d_0.d_1\, d_2 \ldots d_{p-1} \cdot \beta^e$, the absolute error can be as large as

$$\overbrace{0.00\ldots0}^{p}\frac{\beta}{2} \cdot \beta^e,$$

which is $\frac{\beta}{2}\beta^{-p} \cdot \beta^e$. The numbers with values between $\beta^e$ and $\beta^{e+1}$ have the same absolute error of $\frac{\beta}{2}\beta^{-p} \cdot \beta^e$; therefore, the relative error ranges between $\frac{1}{2}\beta^{-p}$ and $\frac{\beta}{2}\beta^{-p}$, which gives

$$\frac{1}{2}\beta^{-p} \leq \frac{1}{2}\text{ ulp} \leq \frac{\beta}{2}\beta^{-p};$$

that is, the relative error corresponding to $\frac{1}{2}$ ulps can vary by a factor of $\beta$. Now $\epsilon = \frac{\beta}{2}\beta^{-p}$ is the largest relative error possible when exact numbers are rounded to the nearest floating-point number, $\epsilon$ is denoted also as the *machine epsilon.*

Arithmetic operations. The standard includes rules for standard arithmetic operations such as addition, subtraction, multiplication, division, square root, and remainder. The main requirement is that the result of the imprecise arithmetic operation $x$ op $y$ of two floating-point values of precision $p$ must be equal to the exact result of $x$ op $y$, rounded to the nearest floating-point value within the given precision.

We will try to demonstrate how we can fulfill the requirement efficiently. For example, subtraction can be performed adequately with a fixed number of additional digits; similar results hold for other operations.

First, we will extract the problems. For example, for $p = 3$, $\beta = 10$, $x = 3.15 \times 10^6$, and $y = 1.25 \times 10^{-3}$, we have to calculate

$$\begin{aligned} x &= 3.15 \times 10^6, \\ y &= 0.000000000125 \times 10^6, \\ x - y &= 3.149999999875 \times 10^6, \end{aligned}$$

which is rounded to $3.15 \times 10^6$. Obviously, we have used many unnecessary digits here. Efficient floating-point arithmetic hardware tries to avoid the use of many digits. Let us assume that, for simplicity, the machine shifts the smaller operand to the left and discards *all* extra digits. This will give

$$\begin{aligned} x &= 3.15 \times 10^6, \\ y &= 0.00 \times 10^6, \\ x - y &= 3.15 \times 10^6; \end{aligned}$$

i.e., exactly the same result. Unfortunately, this does not work in the following example. For $p = 4$ and with $x = 10.01$ and $y = 9.992$, we achieve

$$\begin{aligned} x &= 1.001 \times 10^1, \\ y &= 0.999|2 \times 10^1, \\ x - y &= 0.002 \times 10^1, \end{aligned}$$

whereas the correct answer is 0.018. The computed answer differs by 200 ulps and is wrong in every digit. The next lemma shows how bad the relative error can be for the simple *shift-and-discard* rule.

**Lemma 9.9.** *For floating-point numbers with precision p and basis $\beta$ computing subtraction using p digits, the relative error can be as large as $\beta - 1$.*

*Proof:* We can achieve a relative error of $\beta - 1$, if we choose $x = 1.00\ldots0$ and $y = .\underbrace{dd\ldots d}_{p}$ with $d = (\beta - 1)$. The exact value is $x - y = \beta^{-p}$, whereas using $p$ digits and shift-and-discard gives

$$\begin{aligned} x &= 1.00\ldots00, \\ y &= 0.dd\ldots dd|d, \\ x - y &= 0.00\ldots01, \end{aligned}$$

which equals $\beta^{-p+1}$. Thus, the absolute error is $|\beta^{-p} - \beta^{-p+1}| = |\beta^{-p}(1-\beta)|$, and the relative error is $\left|\frac{\beta^{-p}(\beta-1)}{\beta^{-p}}\right| = \beta - 1$. □

Efficient hardware solution makes use of a restricted number of additional *guard digits*. As an example, we will show that just one extra digit achieves relatively precise results for subtraction. In general, three extra bits are sufficient for fulfilling the needs of the standard for all operations; see [Goldberg 91].

With one extra digit in the given example, we obtain

$$\begin{aligned} x &= 1.001|0 \times 10^1, \\ y &= 0.999|2 \times 10^1, \\ x - y &= 0.001\,8 \times 10^1, \end{aligned}$$

which gives the exact result $1.8 \times 10^{-2}$. More generally, the following theorem holds for a single guard digit.

**Theorem 9.10.** *For floating-point numbers with precision p and basis $\beta$ computing subtraction using $p + 1$ digits (one guard digit), the relative error is smaller than $2\epsilon$, where $\epsilon$ denotes the machine epsilon $\frac{\beta}{2}\beta^{-p}$.*

*Proof:* Let $x > y$. We scale $x$ and $y$ so that $x = x_0.x_1x_2\ldots x_{p-1} \times \beta^0$ and $y = 0.0\ldots0y_{k+1}\ldots y_{k+p} \times \beta^0$. Let $\bar{y}$ denote the part of $y$ truncated to $p + 1$ digits. Thus, we have

$$\begin{array}{ccccccccccc|cccc} x & = & x_0 & . & x_1 & \ldots & x_k & x_{k+1} & \ldots & x_{p-1} & & & & & \\ y & = & 0 & . & 0 & \ldots & 0 & y_{k+1} & \ldots & y_{p-1} & & y_p & y_{p+1} & \ldots & y_{k+p} \\ \bar{y} & = & 0 & . & 0 & \ldots & 0 & y_{k+1} & \ldots & y_{p-1} & & y_p & & & \end{array}$$

With an extra guard digit, $x - y$ is calculated by $x - \bar{y}$ and rounded to be a floating-point number in the given precision. Thus, the result equals $x - \bar{y} + \delta$ where $\delta \leq \frac{\beta}{2}\beta^{-p}$.

The relative error is given by

$$\frac{|x - \bar{y} + \delta - x + y|}{|x - y|} = \frac{|y - \bar{y} + \delta|}{|x - y|}.$$

Additionally, we have

$$\begin{array}{rcccccccccccccc} y - \bar{y} & = & & 0 & . & 0 & \dots & 0 & y_{k+1} & \dots & y_p & y_{p+1} & \dots & y_{k+p} \\ & & - & 0 & . & 0 & \dots & 0 & y_{k+1} & \dots & y_p & & & \\ \hline & = & & 0 & . & 0 & \dots & 0 & 0 & \dots & 0 & y_{p+1} & \dots & y_{k+p} \\ & < & & 0 & . & 0 & \dots & 0 & 0 & \dots & 0 & (\beta - 1) & \dots & (\beta - 1) \end{array},$$

which gives

$$y - \bar{y} < (\beta - 1)\left(\beta^{-(p+1)} + \cdots + \beta^{-(p+k)}\right).$$

For $k = 0, -1$, we are already finished, since $x - y$ is either exact or, due to the guard digit, rounded to the nearest floating-point value with a rounding error at most $\epsilon$. So, in the following, let $k > 0$. If $|x - y| \geq 1$, we conclude

$$\begin{aligned} \frac{|x - \bar{y} + \delta - x + y|}{|x - y|} = \frac{|y - \bar{y} + \delta|}{|x - y|} &\leq \frac{|x - \bar{y} + \delta - x + y|}{|x - y|} = \frac{|y - \bar{y} + \delta|}{1} \\ &\leq \beta^{-p}\left((\beta - 1)(\beta^{-1} + \cdots + \beta^{-k}) + \frac{\beta}{2}\right) \\ &< \beta^{-p}\left(1 + \frac{\beta}{2}\right). \end{aligned}$$

If $|x - \bar{y}| < 1$, we conclude $\delta = 0$ because the result of $x - \bar{y}$ has no more than $p$ digits. The smallest value $x - y$ can achieve is

$$1.0 - 0.\overbrace{0\dots0}^{k}\overbrace{(\beta - 1)\dots(\beta - 1)}^{p} > (\beta - 1)\left(\beta^{-1} + \dots + \beta^{-k}\right).$$

In this case, the relative error is bounded by

$$\begin{aligned} \frac{|x - \bar{y} + \delta - x + y|}{|x - y|} &= \frac{|y - \bar{y}|}{|x - y|} \\ &< \frac{(\beta - 1)\beta^{-p}(\beta^{-1} + \cdots + \beta^{-k})}{(\beta - 1)\left(\beta^{-1} + \dots + \beta^{-k}\right)} \\ &= \beta^{-p} \\ &< \beta^{-p}\left(1 + \frac{\beta}{2}\right). \end{aligned} \tag{9.1}$$

If $|x - \bar{y}| \geq 1$ and $|x - y| < 1$, we conclude $|x - \bar{y}| = 1$ and $\delta = 0$. In this case, Equation (9.1) applies again. For $\beta = 2$, the bound $\beta^{-p}\left(1 + \frac{\beta}{2}\right)$ exactly achieves $2\epsilon$. □

Note that a relative error of less than $2\epsilon$ already gives a good approximation but may be not precise enough for the standard, because $\frac{1}{2}$ ulp $< 2\epsilon$.

In the following, we will denote the inexact versions of the operations of $+$, $-$, $\times$, and $/$ by $\oplus$, $\ominus$, $\otimes$, and $\oslash$. For convenience, we sometimes denote $x \times y$ by $xy$ and $x/y$ by $\frac{x}{y}$, as usual.

The standard requires that for an arithmetic operation of two floating-point values in the given precision, the result will have a relative error smaller than the machine epsilon $\epsilon$.

Therefore, for two floating-point values $x$ and $y$, we have

$$\begin{aligned} x \oplus y &= (x + y)(1 + \delta_1), \\ x \ominus y &= (x - y)(1 + \delta_2), \\ x \otimes y &= xy(1 + \delta_3), \\ x \oslash y &= \frac{x}{y}(1 + \delta_4), \end{aligned} \tag{9.2}$$

where $|\delta_i| \leq \epsilon$ holds for $i = 1, 2, 3, 4$.

**Conversions from and to decimal numbers.** So far, you have seen that the decimal world and the binary world of the computer are not equivalent. Therefore, the standard provides for conversions from binary to decimal and back to binary, or from decimal to binary and back to decimal, respectively.

For example, the data type `float` of the Java programming language represents floating-point numbers with basis $\beta = 10$ in a range from $\pm 3.4 \times 10^{38}$ with precision $p = 6$. This refers to the single format in IEEE 754. The maximal exponent in IEEE 754 is 128, and we have $2^{128} = 3.40282\ldots \times 10^{38}$. The restriction of the precision to 6 refers to conversions from decimal to binary and back to decimal. It can be shown that if a decimal number $x$ with precision 6 is transformed to the nearest binary floating-point $x'$ with precision 24, it is possible to transform $x'$ uniquely back to $x$. This cannot be guaranteed for decimal numbers with precision $\geq 7$; see Theorem 9.12.

We exemplify the problems of a decimal-binary-decimal conversion cycle with rounding to nearest. Let us assume that binary and decimal are given by a similar finite precision. If decimal $x$ is converted for the first time to the nearest binary $x'$, we cannot guarantee that the nearest decimal of $x'$ results in the original value $x$; see Figure 9.9.

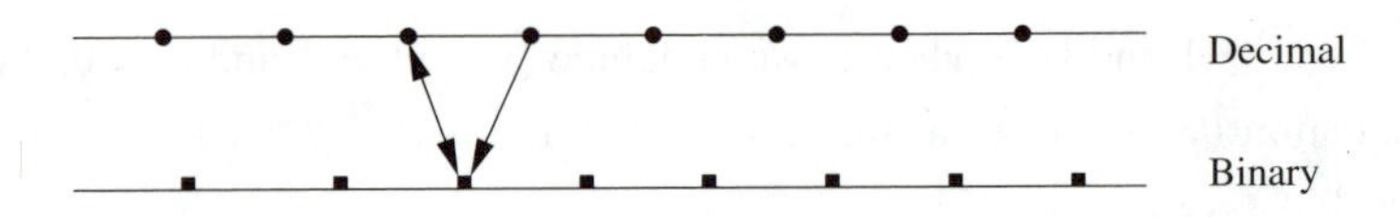


**Figure 9.9.** Conversion cycle by rounding to nearest with a first-time error.

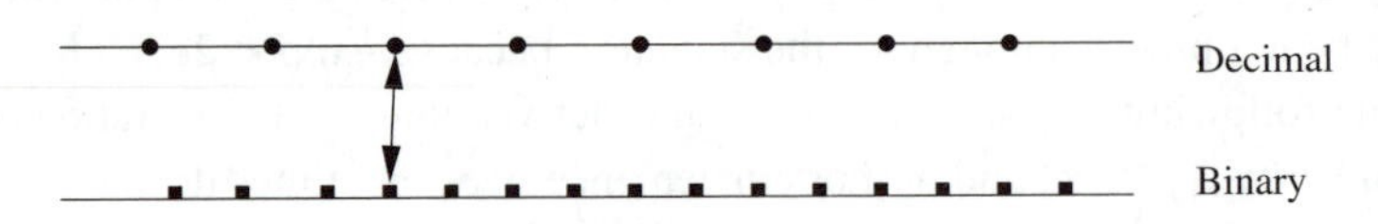


**Figure 9.10.** Conversion cycle by rounding to nearest with the correct result.

If the binary format has enough extra precision, rounding back to the nearest decimal floating-point value results in the original value; see Figure 9.10. In all cases, the error cannot increase; that is, repeated conversions after the first time give the same binary and decimal values.

We can calculate the corresponding sufficient precisions precisely and we will give a proof for the binary-decimal-binary conversion cycle.

**Theorem 9.11.** *Let $x$ be a binary floating-point number with precision 24. If $x$ is converted to the nearest decimal floating-point number $x'$ with precision 8, we cannot convert $x'$ back to the original $x$ by rounding to the nearest binary with precision 24. A precision of nine digits for the decimal is sufficient.*

*Proof:* We consider the half-open interval $[1000, 1024) = [10^3, 2^{10})$. Binary numbers with precision 24 inside this interval obviously have 10 bits to the left and 14 bits to the right of the binary point. In the given interval, we have 24 different integer values from 1000 to 1023. Therefore, there are $24 \times 2^{14} = 393216$ different binary numbers in the interval. Decimal numbers with precision 8 inside this interval obviously have four bits to the left and four bits to the right of the decimal point. Analogously, there are only $24 \times 10^4$ different decimal numbers in the same interval. So eight digits are not enough to represent each single-precision binary by a corresponding decimal.

To show that precision nine is sufficient, we will show that the spacing between binary numbers is always greater than the spacing between decimal numbers. Consider the interval $[10^n, 10^{n+1}]$. In this interval, the spacing between two decimal numbers with precision 9 is $10^{(n+1)-9}$. Let $m$ be the smallest integer so that $10^n < 2^m$. The spacing of all binary numbers with precision 24 in $[10^n, 2^m]$ is $2^{(m-24)}$. The spacing gets larger from $2^m$ to $10^{(n+1)}$. Altogether, we have $10^{(n+1)-9} < 2^{(m-24)}$, and the spacing of the binary numbers is always larger. □

For the decimal-binary-decimal conversion cycle, a similar result can be proven.

**Theorem 9.12.** *Let $x$ be a decimal floating-point number with precision 7. If $x$ is converted to the nearest binary floating-point number $x'$ with precision 24, we cannot convert $x'$ back to the original $x$ by rounding to the nearest binary value with precision 7. Precision* 6 *for the decimal value is sufficient.*

Special values. IEEE 754 reserves exponent fields values of all 0s and all 1s to denote special values. We will exemplify the notions in the single format.

*Zero* cannot be represented directly because of the leading 1 in the normalized form. Therefore, zero is denoted with an exponent field of zero and a significand field of zero. Note that 0 and -0 are different values.

1 | 0 0 0 0 0 0 0 0
0 0 0 0 0 0 0 0 0 0 0 0 0 0 0 0 0 0 0 0 0 0 0

*Infinity* values $+\infty$ and $-\infty$ are denoted with an exponent of all 1s and a significand of all 0s. The sign bit distinguishes between $+\infty$ and $-\infty$.

1 | 1 1 1 1 1 1 1 1
0 0 0 0 0 0 0 0 0 0 0 0 0 0 0 0 0 0 0 0 0 0 0

After an overflow, the system will be able to continue to evaluate expression by using arithmetic with $+\infty$ or $-\infty$.

*NaN* denotes illegal results, and they are represented by a bit pattern with an exponent of all 1s and a non-zero significand.

1 | 1 1 1 1 1 1 1 1
1 0 0 0 0 0 0 0 1 0 0 1 0 0 0 0 0 0 0 0 0 0 0

*Arithmetic with special values* allows operations to proceed after expressions run out of range. Arithmetic with a NaN subexpression always executes to NaN. Other operations can be found in the following table.

| Operation | Result |
|---|---|
| $n/\pm 0$ | NaN |
| $\infty - \infty$ | NaN |
| $\pm\infty/\pm\infty$ | NaN |
| $\pm\infty \times 0$ | NaN |
| $\infty + \infty$ | $\infty$ |
| $\pm\infty \times \pm\infty$ | $\pm\infty$ |
| $n/\pm\infty$ | 0 |

## 9.3.2. Exact Arithmetic

We have shown that fast floating-point arithmetic comes along with unavoidable roundoff errors. Therefore, sometimes using exact number packages is recommended. This so-called *exact arithmetic* approach refers to the fact that whenever a geometric algorithm starts, the corresponding objects have to be represented in a finite precision for the computer. In this sense, the input objects are always discrete. Therefore, one can guarantee the exact judgment of predicates by using a constant precision, though possibly much higher than the precision of the input.

A simple example is given in [Sugihara 00]. Let us assume that for the polyhedron problem of Section 9.1.2, we have points $p_i = (x_i, y_i, z_i) \in I^3$ with integer coordinates only. Now there is a fixed large positive integer K with

$$|x_i|, |y_i|, |z_i| \leq K$$

for all $p_i$, and we are looking for the sign of

$$\mathrm{Det}(p_k, p_l, p_m, p_n) = \begin{vmatrix} x_k & y_k & z_k & 1 \\ x_l & y_l & z_l & 1 \\ x_m & y_m & z_m & 1 \\ x_n & y_n & z_n & 1 \end{vmatrix} .$$

The Hadamard inequality [Gradshteyn and Ryzhik 00] states that the absolute value of the determinant of a matrix is bounded by the product of the length of the column vectors. More precisely, for every $n \times n$ $A = (a_{ij})$ with $a_{ij} \in \mathbb{R}$ and $|A| \neq 0$, we have $|A|^2 \leq \prod_{i=1}^{n}(\sum_{j=1}^{n} a_{ij}^2)$. This means that

$$|\,\mathrm{Det}(p_k, p_l, p_m, p_n)| \leq$$
$$\sqrt{1^2 + 1^2 + 1^2 + 1^2} \times \left(\sqrt{K^2 + K^2 + K^2 + K^2}\right)^3 = 16K^3 .$$

Hence, it suffices to use multiple-precision integers that are long enough to represent $16K^3$.

As another example, we can consider the sweep algorithm for the arrangement in Section 9.1.1. The sweep is consistent as long as the comparison of the $x$-coordinates of intersections of line segments or endpoints of segments is correct. The $x$-coordinates of line segment intersections are given by values $\frac{a}{b}$ where $a$ and $b$ denote $2 \times 2$ determinants with entries stemming from the subtraction of endpoint coordinates; see Section 9.4.2. More precisely, $a$ is a multi-variate polynomial of degree 3 and $b$ is a multi-variate polynomial of degree 2.

Therefore, we have to compute the sign of $\frac{a_1}{b_1} - \frac{a_2}{b_2}$ exactly, which is equivalent to the sign of $a_1 b_2 - a_2 b_1$. The last expression can be represented by a fifth degree multi-variate polynomial in the coordinates of the endpoints of line segments, see Section 9.4.2. Therefore, we have to adjust the precision of the arithmetic so that we can calculate five successive multiplications and some additions exactly. If the coordinates are given with precision $k$, it can be shown that it suffices to perform all computations with precision $6k$, due to the polynomial complexity of the sign test.

Unfortunately, multiple-precision computation may produce unnecessarily high computational costs. Additionally, the problem of degeneracy is not solved.

We summarize some useful exact number packages and analyze the performance impact.

**Integer arithmetic.** The idea of a multiple-precision integer system is very simple. An arbitrarily big, but finite, integer value can be stored exactly in the binary system of a computer if we allow enough space for its representation. Roughly speaking, we will use a software solution for the representation of numbers.

Big-integer, or more generally, big-number packages are part of almost every programming language (for example, Java, C++, Object Pascal, and Fortran) and are implemented for every computer algebra system (for example, Maple, Mathematica, and Scratchpad). Additionally, independent software packages are given (for example, BigNum, GMP, LiDIA, Real/exp, and PARI), and special computational geometry libraries (for example, LN, LEDA, LEA, CoreLibrary, and GeomLib) provide access to arbitrary precision numbers; more details are given in Section 9.7. The packages provide for number representation and standard arithmetic.

We consider the computational impact and the storage impact of big-integer arithmetic. They are closely related. If we need more bits for the representation of the inputs of an arithmetic operation, the operation costs increase.

Let us first consider the computational impact of big-integer arithmetic. Obviously, the evaluation of determinants is a central problem of computational geometry; see Section 9.1.2 and Section 9.4. A $d \times d$ determinant with integer entries can be naively computed by Gaussian elimination; see [Schwarz 89]. We assume that every input value can be represented by at most $b$ bits. Using floating-point arithmetic, we can eliminate the matrix entries with $O(d^3)$ $b$-bit hardware-supported operations. For exact integer arithmetic, we need to avoid divisions. Therefore, during the elimination process, the bit complexity of the entries may grow steadily by a factor of 2. The number of arbitrary precision operations is still bounded by $O(d^3)$, but the number of $b$-bit operations within these $O(d^3)$ operations can be exponential in $d$, because of increasing bit length. Altogether, the overhead of operations of the integer package can be very large. More precisely, we compare a running time of $\sum_{i=1}^{n}(n-i)^2 \in O(n^3)$ floating-point operations to an exponential running time of $\sum_{i=1}^{n}(n-i)^2 2^{i-1}$ big-integer operations.

On the other hand, for the number of additional bits needed for the evaluation of a multi-variate polynomial $p$ of degree $d$. Keep in mind that we can represent a $d \times d$ determinant by such a polynomial. Let $m$ be the number of monomials in $p$ and let the coefficients of the monomial be bounded by a constant. The monomials of a multi-variate polynomial are the summands of the polynomial. For example, $x^3y^5$ and $x^4y^2$ and 7 are the monomials of $p(x, y) = x^3y^5 + x^4y^2 + 7$. It might happen that we need $bd$ bits for the representation of a monomial, taking $d$ multiplications of $b$ bit values into account. Adding two monomials of complexity $bd$ may increase the complexity by (at most) 1, adding four monomials of complexity $bd$ may increase the complexity by (at most) 2, and so on. Altogether, the bit complexity may rise to $bd + \log m + O(1)$, whereas floating-point arithmetic is restricted to complexity $b$.

**Rational arithmetic.** A rational number can be represented by two integers. If we allow the arithmetic operations $\times$, $+$, $-$, and $/$, we can simply define rational arithmetic by integer arithmetic, using the well-known rules

$$\begin{aligned}\frac{p}{q} + \frac{r}{s} &:= \frac{ps + qr}{qs} \\ \frac{p}{q} \times \frac{r}{s} &:= \frac{pr}{qs}.\end{aligned}$$

Rational computing is unavoidable in the presence of integers. For example, the intersection of two lines each represented by integers may be rational; see Section 9.4.2. For obvious reasons, big-rational values normally are integrated into big-integer packages. Furthermore, big-number packages normally implement big-float values also; see, for example, BigNum, GNU MP, and LiDIA. As in the previous section, we should have a closer look at the computational cost and the bit complexity of rationals.

A rational $\frac{p}{q}$ is represented by two integers. The bit complexity is given by the sum of the bit complexities of $p$ and $q$. Unfortunately, a single addition may double the bit complexity. The addition of two $b$ bit integer values results in a $(b + 1)$ bit integer, whereas the bit complexity of $\frac{p}{q} + \frac{r}{s} = \frac{ps+qr}{qs}$ is the sum of the bit complexities of $\frac{p}{q}$ and $\frac{r}{s}$. The bit complexity for evaluating a multi-variate polynomial of degree $d$ with $m$ monomials may rises to $2mbd + O(1)$.

Let us consider the intersection example in Section 9.3.2. We have to compute the sign of an expression $a_1b_2 - a_2b_1$ which is a multi-variate polynomial of degree 5, here with rational variables. If the corresponding integers are expressed by $b$ bits, the number of sufficient bits rise to $2 \cdot 5 \cdot b$, which gives a factor of 10.

The computational cost impact of multi-precision rationals is similar to the cost of multi-precision integers. As already shown, in cascaded computations, the bit length of numbers may increase quickly. Therefore, the worst-case complexity of $b$ bit operations can be exponential in size, whereas $b$ bit floating-point operations will be bounded by a polynomial.

It was shown in [Yu 92] that exact rational arithmetic for modeling polyhedrals in 3D is intractable. The naive use of exact rational arithmetic for computing 2D Delaunay triangulations can be 10,000 times slower than a floating-point implementation; see [Karasick et al. 89].

**Algebraic numbers.** A powerful subclass of real and complex numbers is given by algebraic numbers. An algebraic number is defined to be the root of a polynomial with integer coefficients. The polynomial represents the corresponding number.

For example, the polynomial $p(x) = x - m$ shows that every integer $m$ is algebraic. By $p(x) = nx - m$, we can represent all rationals $\frac{m}{n}$. Additionally, we may have non-rationals. For example, $\sqrt{2}$ is the root of $p(x) = x^2 - 2$. Algebraic numbers can be complex; for example, $i\sqrt{2}$ is a root of $p(x) = x^2 + 2$. Since $\pi$ and $e$ are not algebraic, algebraic numbers define a proper subset of complex numbers.

The sum, difference, product, and quotient of two algebraic numbers are algebraic, and the algebraic numbers therefore form a so-called *field*. In abstract algebra, a field is an algebraic structure in which the operations of addition, subtraction, multiplication, and division (except division by zero) may be performed, and the associative, commutative, and distributive rules that are familiar from the arithmetic of ordinary numbers hold.

For practical reasons, we are interested only in real numbers and mainly interested in the *sign* of an algebraic number.

**Definition 9.13. (Algebraic number.)** An algebraic number $x \in \mathbb{R}$ is the root of an integer polynomial. It can be uniquely represented by the corresponding integer polynomial $p(x)$ and an isolating interval $[a, b]$ with $a, b \in \mathbb{Q}$ that identifies $x$ uniquely.

This means that the isolated interval representations $(x^2 - 2, [1, 4])$ and $(x^2 - 2, [0, 6])$ represent the same algebraic number. This is not a problem here. Let us briefly explain how arithmetic with algebraic numbers works.

Let $\alpha$ and $\beta$ be algebraic, $\alpha$ be a root of $p(x) = \sum_{i=0}^{n} a_i x^i$, and $\beta$ be a root of $p(x) = \sum_{i=0}^{n} b_i x^i$. It is easy to show that the following holds:

- $-\alpha$ is a root of $\sum_{i=0}^{n}(-1)^i a_i x^i$;
- $\frac{1}{\alpha}$ is root of $x^n \sum_{i=0}^{n} a_i x^{-i}$;
- $\alpha + \beta$ is root of $\text{Res}_y(p(y), q(x-y))$;
- $\alpha\beta$ is root of $\text{Res}_y\left(p(y), y^m q\left(\frac{x}{y}\right)\right)$

The resultant $\text{Res}_x(p(x), q(x))$ of two polynomials $p(x) = \sum_{i=0}^{n} a_i x^i$ and $p(x) = \sum_{i=0}^{n} b_i x^i$ in $x$ is defined by the following determinant of an $(n+m) \times (n+m)$ matrix. We have $m$ rows with entries $a_0, a_1, \ldots, a_n$ and $n$ rows with entries $b_0, b_1, \ldots, b_m$.

$$\text{Res}_x(p(x), q(x)) := \begin{vmatrix} a_0 & a_1 & \ldots & a_n & 0 & \ldots & 0 & 0 \\ 0 & a_0 & a_1 & \ldots & a_n & \ldots & 0 & 0 \\ \vdots & \ddots & \ddots & \vdots & \vdots & \vdots & \ddots & \ddots \\ 0 & \ldots & 0 & a_0 & a_1 & \ldots & a_n & 0 \\ 0 & \ldots & 0 & 0 & a_0 & a_1 & \ldots & a_n \\ b_0 & b_1 & \ldots & b_{m-1} & b_m & 0 & \ldots & 0 \\ 0 & b_0 & b_1 & \ldots & b_{m-1} & b_m & \ldots & 0 \\ \vdots & \ddots & \ddots & \vdots & \vdots & \ddots & \ddots & \vdots \\ 0 & \ldots & 0 & b_0 & \ldots & b_{m-1} & b_m & 0 \\ 0 & \ldots & 0 & 0 & b_0 & \ldots & b_{m-1} & b_m \end{vmatrix}.$$

The corresponding matrix is denoted also as the Sylvester matrix. It is easy to show that, if $\text{Res}_x(p(x), q(x)) = 0$ holds, then $p$ and $q$ have a common root. Analogously, the resultant can be defined by

$$\text{Res}_x(p(x), q(x)) := \prod_{i=1}^{n}\prod_{j=1}^{m}(\beta_j - \alpha_i),$$

where $\alpha_i$, for $i = 1, \ldots n$ are the roots of $p(x)$ and $\beta_i$, for $i = 1, \ldots m$ are the roots of $q(x)$; for more details, see [Davenport et al. 88].

With this definition, it is easy to see that

$$\text{Res}_y(p(y), q(x-y)) = \prod_{i=1}^{n}\prod_{j=1}^{m}((x - \beta_j) - \alpha_i)$$

and

$$\text{Res}_y\left(p(y), y^m q\left(\frac{x}{y}\right)\right) = \prod_{i=1}^{n}\prod_{j=1}^{m}\left(\frac{x}{\beta_i} - \alpha_i\right)$$

have roots $\alpha_i + \beta_j$ and $\alpha_i\beta_j$, respectively. Altogether, algebraic numbers are closed under the operations $+$, $-$, $\times$, and $/$.

For a simple example, we choose $p(x) = a_0 + a_1x$ and $q(x) = b_0 + b_1x$. Now, we have $\alpha_1 = \frac{-a_0}{a_1}$ and $\beta_1 = \frac{-b_0}{b_1}$, and

$$\begin{aligned} h(x) = \text{Res}_y(p(y), q(x-y)) &= \text{Res}_y(a_0 + a_1y, b_0 + b_1(x-y)) \\ &= \text{Res}_y(a_0 + a_1y, b_0 + b_1x - b_1y) \\ &= \frac{b_0 + b_1x}{b_1} + \frac{a_0}{a_1} \end{aligned}$$

has root $x = -\frac{a_0}{a_1} - \frac{b_0}{b_1} = \alpha_1 + \beta_1$.

Algebraic numbers have the following interesting closure property.

**Theorem 9.14.** *The root of a polynomial with algebraic coefficients is algebraic.*

There are several ways of computing with algebraic numbers. With respect to geometric computing, we are mainly interested in the sign of real algebraic expressions and discuss two numerical approaches in Section 9.3.4. Several packages for algebraic computing are available; see Section 9.7 for a detailed discussion of existing packages.

### 9.3.3. Robust and Efficient Arithmetic

In this section, we will present some techniques for using fast arithmetic while trying to avoid serious errors. We will start with the popular but naive $\epsilon$-tweaking method and then move on to more sophisticated techniques. The main idea is the use of adaptive executions; i.e., cost-consumptive computations are performed only in critical cases.

**Epsilon comparison.** Epsilon comparison is the *standard* model of almost every programmer. Every floating-point result whose absolute value is smaller than a (probably adjustable) threshold traditionally denoted by $\epsilon$ is considered to be zero. This works fine in many situations, but can never guarantee robustness. Epsilon comparison is not recommended in general. Epsilon comparison is known also as *epsilon tweaking* or *epsilon heuristic*. A value for epsilon is typically found by trial and error.

Formally, epsilon tweaking defines the following relation: $a \sim b :\Leftrightarrow |a-b| < \epsilon$. Unfortunately, no $\epsilon$ can guarantee that $\sim$ is an equivalence relation. From

$|a - b| < \epsilon$ and $|b - c| < \epsilon$, we cannot conclude that $|a - c| < \epsilon$ holds. So transitivity does not hold. Therefore, epsilon comparison can fail and can lead to inconsistencies and errors.

To make epsilon comparison a bit more robust, one should apply the following simple practical rules; see also [Michelucci 97, Michelucci 96] and [Santisteve 99].

- It is useful to define several thresholds: one for lengths, another for areas, another for angle, and so on. Adjust them carefully to the problem situation.
- Try to test special situations separately by simple comparisons. For example, for line segment intersection, we may check in advance, by comparison of the coordinates, that an intersection cannot occur.
- Try to avoid unnecessary computations. For example, distances between points can be compared without the use of a erroneous square root operation.
- Use exact data representation for comparisons. For example, an intersection may be represented by the numerical value and the corresponding segments. Two intersections should be compared with the original data.
- Never use different representations of the same expression. This might introduce inconsistencies. Somehow, this contradicts the next item.
- Rearrange predicates and expressions to avoid severe cancellation of digits. Use case distinctions for the best choice; this is the theme of Section 9.4.

**Robust adaptive floating-point expressions.** We want to make use of fast floating-point operations with fixed-precision $p$. We can assume that the input data has a fixed precision $\leq p$. We follow the ideas and arguments in [Shewchuk 97]. Floating-point expressions with limited precision are expanded so that the error can be recalculated if necessary. In the following, we assume that $\beta$ and $p$ are given.

**Definition 9.15. (Expansion.)** Let $x_i$ for $i = 1, \ldots n$ be floating-point numbers with respect to $\beta$ and $p$. Then

$$x = \sum_{i=1}^{n} x_i$$

is called an expansion. $x_i$ are called components.

For example, let $p = 6$ and $\beta = 2$. Then $-11 + 1100 + -111000$ is an expansion with three components. For our purpose, we consider so-called non-overlapping expansions.

**Definition 9.16. (Non-overlapping floating-point numbers.)** Two floating-point numbers $x$ and $y$ are non-overlapping iff the smallest significant bit of $x$ is bigger than the biggest significant bit of $y$ or vice versa.

For example, $x = -110000$ and $y = 110$ are non-overlapping. It is useful to consider non-overlapping expansions $x = \sum_{i=1}^{n} x_i$ ordered by the absolute value of the subexpressions $x_i$; that is, $|x_i| > |x_{i+1}|$ for $i = 1, \ldots, (n-1)$. In this case, the sign of $x$ is determined by the sign of $x_1$.

**Definition 9.17. (Normalized expansion.)** A normalized expansion $x = \sum_{i=1}^{n} x_i$ has pairwise non-overlapping components $x_i$ and additionally $|x_i| > |x_{i+1}|$ for $i = 1, \ldots, (n-1)$ holds.

As you will see, sometimes some of the inner components may become zero. In this case, we speak also of a normalized expansion iff the condition $|x_i| > |x_{i+1}|$ for $i = 1, \ldots, (n-1)$ holds for all non-zero components.

The main idea is that we can expand floating-point arithmetic to floating-point arithmetic with expansions, thus being able to reconstruct rounding errors if necessary. In other words, instead of using more digits, we make use of more terms. In many cases, only a few terms are relevant. We try to present the main ideas of the expansion approach; more details can be found in [Shewchuk 97]. Arithmetic with expansions is based on the following observations.

According to the standard IEEE 754 for floating-point arithmetic (see Section 9.3.1, for example) for addition, the following holds:

$$a \oplus b = (a+b)(1+\delta) = (a+b) + \mathrm{err}(a \oplus b),$$

where $\delta$ and $\mathrm{err}(a \oplus b)$ can be positive or negative. In any case, we have $|\mathrm{err}(a \oplus b)| \leq \frac{1}{2}\,\mathrm{ulp}(a \oplus b)$, since the standard guarantees exact arithmetic combined with the round-to-nearest rule and the round-to-even tie-breaking rule. Unfortunately, floating-point arithmetic breaks standard arithmetic rules such as associativity. For example, for $p = 4$ and $\beta = 2$, we have $(1000 \oplus 0.011) \oplus 0.011 = 1000$ but $1000 \oplus (0.011 \oplus 0.011) = 1001$.

We can make use of the ulp notation as a function of an incorrect result. It represents the magnitude of the smallest bit in the given precision. For example, for $p = 4$, we have $\mathrm{ulp}(-1001) = 1$ and $\mathrm{ulp}(10) = 0.01$. It seems to be natural to represent $a + b$ by a non-overlapping expansion $x + y$ with $x = (a \oplus b)$ and $y = -\mathrm{err}(a \oplus b)$, provided that the error is representable within the given precision. We can construct non-overlapping expansions by addition using Algorithm 9.1.

FastTwoSum$(a, b)$ $(|a| \geq |b|)$

$x := a \oplus b$
$b_{\text{virtual}} := x \ominus a$
$y := b \ominus b_{\text{virtual}}$
**return** $(x, y)$

**Algorithm 9.1.** FastTwoSum easily computes a non-overlapping expansion.

In the following, we assume $\beta = 2$ and $p$ is fixed. Some of the proofs will make use of $\beta = 2$, which means that we cannot change the base. Fortunately, this is not a restriction. Either we make use of the conversion properties of the IEEE standard (see Section 9.3.1) or we must convert the input data of our problem by hand.

**Theorem 9.18.** *Let $a, b$ be floating-point numbers with $|a| \geq |b|$. The algorithm* FastTwoSum *computes a non-overlapping expansion $x+y = a+b$, where $x$ represents the approximation $a \oplus b$ and $y$ represents the roundoff error* $-\operatorname{err}(a \oplus b)$.

Before we prove Theorem 9.18, we consider the following example for $p = 4$ and $\beta = 2$ with $a = 111100$ and $b = 1001$. We have

| | | |
|---|---|---|
| $a$ | 111100 | |
| $b$ | 1001 | |
| $a + b$ | 1000101 | |
| | | $\downarrow$ round-to-nearest |
| $x = a \oplus b$ | 1001000 | |
| $a$ | 111100 | |
| $b_{\text{virtual}} = x \ominus a$ | 1100 | |
| $y = b \ominus b_{\text{virtual}}$ | - 0011 | |

which results in a non-overlapping representation $1001000 + -11 = 1000101 = a + b$. Note that FastTwoSum uses a constant number of $p$-bit floating-point operations.

First, we prove some helpful lemmata.

**Lemma 9.19.** *Let $a$ and $b$ be floating-point numbers.*

1. *If $|a - b| \leq |b|$ and $|a - b| \leq |a|$, then $a \ominus b = a - b$.*
2. *If $b \in \left[\frac{a}{2}, 2a\right]$, then $a \ominus b = a - b$.*

*Proof:* For the first part, we assume without loss of generality that $|a| \geq |b|$. The leading bit of $a - b$ is no bigger in magnitude than the leading bit of $b$; the smallest bit of $a - b$ is no smaller than the smallest bit of $b$. Altogether, $a - b$ can be expressed with $p$ bits. For the second part, without loss of generality, we assume $|a| \geq |b|$. The other case is symmetric using $-b \ominus -a$. We conclude that $b \in \left[\frac{a}{2}, a\right]$. And we have $\text{sign}(a) = \text{sign}(b)$. Therefore, we have $|a - b| \leq |b| \leq |a|$ and we apply the first part of the lemma. □

Now we can prove that the following subtraction is exact.

**Lemma 9.20.** *Let* $|a| \geq |b|$ *and* $x = a+b+\text{err}(a\oplus b)$ *then* $b_{\text{virtual}} := x \ominus a = x - a$.

*Proof:* Case 1. If $\text{sign}(a) = \text{sign}(b)$ or if $|b| < \left|\frac{a}{2}\right|$ holds, we conclude that $x \in \left[\frac{a}{2}, 2a\right]$, and we apply Lemma 9.19 to $-(a \ominus x)$.

Case 2. If $\text{sign}(a) \neq \text{sign}(b)$ and $|b| \geq \left|\frac{a}{2}\right|$, we have $b \in \left[-\frac{a}{2}, -a\right]$. Now $x$ is computed exactly because $x = a \oplus b = a \ominus -b$ equals $a - -b = a + b$ by Lemma 9.19. Thus, $b_{\text{virtual}} := x \ominus a = (a + b) \ominus a = b$. □

**Lemma 9.21.** *The roundoff error* $\text{err}(a \oplus b)$ *of* $a + b$ *can be expressed with* $p$ *bits.*

*Proof:* Without loss of generality, we assume $|a| \geq |b|$, and we have $a \oplus b$ is the floating-point number with $p$ bits closest to $a + b$. But $a$ is a $p$-bit floating-point number. The distance between $a \oplus b$ and $a + b$ cannot be greater then $|b|$, altogether $|\,\text{err}(a \oplus b)| \leq |b| \leq |a|$, and the error is representable with $p$ bits. □

Now we can prove Theorem 9.18.

*Proof:* The first assignment gives $x = a + b + \text{err}(a \oplus b)$. For the second assignment, we conclude $b_{\text{virtual}} = x \ominus a = x - a$ from Lemma 9.20. The third assignment is computed exactly also, because $b \ominus b_{\text{virtual}} = b + a - x = -\,\text{err}(a \oplus b)$ can be represented by $p$ bits; see Lemma 9.21. In either case, we have $y = -\,\text{err}(a \oplus b)$ and $x = a + b + \text{err}(a \oplus b)$. Exact rounding guarantees that $|y| \leq \frac{1}{2}\,\text{ulp}(x)$; therefore, $x$ and $y$ are non-overlapping. □

The algorithm FastTwoSum stems from Dekker [Dekker 71]. A more convenient algorithm, TwoSum by Knuth [Knuth 81], does not require $|a| \geq |b|$ and avoids cost-consuming comparisons.

Without a proof, TwoSum analogously computes a normalized expansion $x + y$, where $y$ represents the error term. Additionally, there are analogous results for constructing expansions for subtraction and products of two floating-point numbers. We do not go into detail and refer to [Shewchuk 97].

Up to now, we have constructed a single expansion by addition. Obviously. we have to provide for arithmetic between expansions. That is, for two normalized

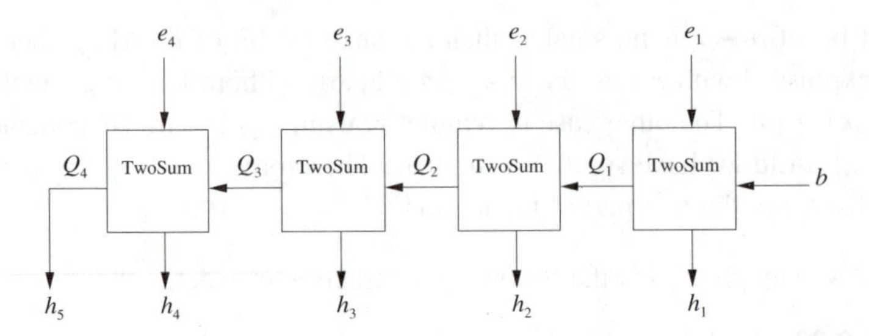


**Figure 9.11.** The scheme of the procedure GrowExpansion. (Redrawn and adapted courtesy of Jonathan Richard Shewchuk (Shewchuk 97).)

expansions $e$ and $f$ with $m$ and $n$ components, respectively, we want to compute a normalized expansion $h$ with $e + f = h$. First, we show that we can extend a normalized expansion by addition of a single $p$-bit number. The computation scheme is shown in Figure 9.11. It might happen that some components become zero, but non-zero components are still ordered by magnitude. If FastTwoSum should be applied, we have to make a case distinction in Algorithm 9.3.

**Theorem 9.22.** *Let $e = \sum_{i=1}^{m} e_i$ be a normalized expansion with $m$ components and let $b$ be a $p \geq 3$ bit floating-point number. The algorithm* GrowExpansion *computes a normalized (up to zero components) expansion*

$$h = \sum_{i=1}^{m+1} h_i = e + b$$

*with $O(m)$ $p$-bit floating-point operations.*

---

TwoSum$(a, b)$

---

$x := a \oplus b$
$b_{\text{virtual}} := x \ominus a$
$a_{\text{virtual}} := x \ominus b_{\text{virtual}}$
$b_{\text{roundoff}} := b \ominus b_{\text{virtual}}$
$a_{\text{roundoff}} := a \ominus a_{\text{virtual}}$
$y := a_{\text{roundoff}} \oplus b_{\text{roundoff}}$
**return** $(x, y)$

---

**Algorithm 9.2.** TwoSum computes a non-overlapping expansion $x + y = a + b$.

GrowExpansion($\sum_{i=1}^{m} e_i, b$)

```
Q_0 := b
for i := 1 TO m do
        TwoSum(Q_{i-1}, e_i)
end for
h_{m+1} := Q_m
return h = (h_1, h_2, ..., h_{m+1})
```

**Algorithm 9.3.** GrowExpansion computes a non-overlapping expansion $h_1 + h_2 + \cdots h_{m+1} = e + b$.

*Proof:* First, we show by induction that the invariant

$$Q_i + \sum_{j=1}^{i} h_j = b + \sum_{j=1}^{i} e_j$$

holds after the $i$th iteration of the `for` statement. Obviously, the invariant holds for $i = 0$ with $Q_0 = b$. Now assume that $Q_{i-1} + \sum_{j=1}^{i-1} h_j = b + \sum_{j=1}^{i-1} e_j$ holds, and we compute $Q_{i-1} + e_i = Q_i + b_i$ in line 3. Thus, we have $Q_{i-1} + e_i + \sum_{j=1}^{i-1} h_j = Q_i + \sum_{j=1}^{i} h_j = b + \sum_{j=1}^{i} e_j$, and the induction step is shown.

At the end, we obtain $\sum_{j=1}^{m+1} h_j = b + \sum_{j=1}^{m} e_j$.

It remains to show that the result is normalized; that is, ordered by magnitude and non-overlapping. For all $i$, the output of TwoSum has the property that $h_i$ and $Q_i$ do not overlap. From the proof of Lemma 9.21, we know that $|h_i| = |\operatorname{err}(Q_{i-1} \oplus e_i)| \leq |e_i|$. Additionally, $e$ is a non-overlapping expansion whose non-zero components are arranged in increasing order. Therefore, $h_i$ cannot overlap $e_{i+1}, e_{i+2}, \ldots$. The next component of $h$ is constructed by summing $Q_i$ with component $e_{i+1}$, the corresponding error term $h_{i+1}$ is either zero or must be bigger and non-overlapping with respect to $h_i$. Altogether, $h$ is non-overlapping and increasing (except for zero components of h). □

Let us consider a simple example for $p = 5$. Let $e = e_1 + e_2 = \texttt{111100} + \texttt{1000000}$ and $b = 1001$. We have TwoSum$(Q_0, e_1)$ = TwoSum(`1001`, `111100`) = (`1000100`, `1`) = $(Q_1, h_1)$ and TwoSum$(Q_1, e_2)$ = TwoSum(`1000100`, `1000000`) = (`10001000`, `−100`) = $(Q_2, h_2)$ and $h_3$ = `1000100`. Thus, calling GrowExpansion$(e, b)$ results in `10001000` + `−100` + `1`.

Finally, we are able to add two expansions by following the ExpansionSum procedure. The scheme of ExpansionSum is shown in Figure 9.12.

**Theorem 9.23.** *Let $e = \sum_{i=1}^{m} e_i$ be a normalized expansion with $m$ components and $f = \sum_{i=1}^{m} f_i$ be a normalized expansion with $n$ components. Let $b$ be a $p \geq 3$*

---

ExpansionSum($\sum_{i=1}^{m} e_i$, $\sum_{i=1}^{n} f_i$)

---

$h := e$
**for** $i := 1$ TO $n$ **do**
  $(h_i, h_{i+1}, \ldots, h_{i+m}) :=$ GrowExpansion($(h_i, h_{i+1}, \ldots, h_{i+m-1}), f_i$)
**end for**
**return** $h = (h_1, \ldots, h_{m+n})$

---

**Algorithm 9.4.** ExpansionSum computes a non-overlapping expansion $h_1 + \cdots + h_{m+n} = e + f$.

*bit floating-point number. The algorithm GrowExpansion computes a normalized expansion*

$$h = \sum_{i=1}^{n+m} h_i = e + f$$

*with $O(mn)$ $p$-bit floating-point operations.*

*Proof:* We give a proof sketch here. It can be shown by simple induction that $\sum_{i=1}^{n+m} h_i = \sum_{i=1}^{m} e_i + \sum_{i=1}^{n} f_i$. Following the addition scheme of Figure 9.12 one can prove successively that $\sum_{i=1}^{j-1} h_j + \sum_{i=j}^{j+m-1} h_j^{(j-1)}$ are non-overlapping and increasing for $j = 1, \ldots (n-1)$. □

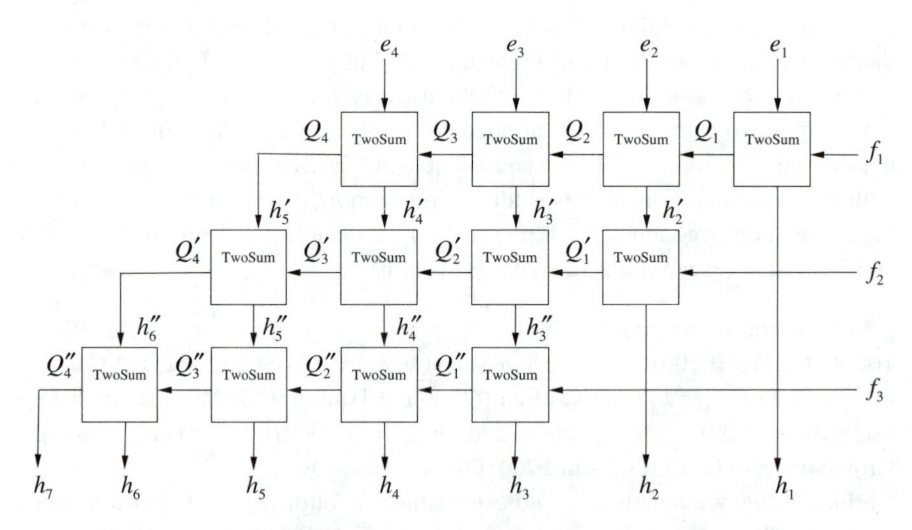


**Figure 9.12.** The addition scheme of ExpansionSum for two expansions $f = \sum_{i=1}^{3} f_i$ and $e = \sum_{i=1}^{4} e_i$. (Redrawn and adapted courtesy of Jonathan Richard Shewchuk [Shewchuk 97].)

TwoProduct$(a, b)$ ($a$ and $b$ are $p$-bit floating-point values)

$x := a \otimes b$
$(a_x, a_y) := \text{Split}\left(a, \lfloor \frac{p}{2} \rfloor\right)$
$(b_x, b_y) := \text{Split}\left(b, \lfloor \frac{p}{2} \rfloor\right)$
$\text{err}_1 := x \ominus (a_x \otimes b_x)$
$\text{err}_2 := \text{err}_1 \ominus (a_y \otimes b_x)$
$\text{err}_3 := \text{err}_2 \ominus (a_x \otimes b_y)$
$y := (a_y \otimes b_y) \ominus \text{err}_3$
**return** $(x, y)$

**Algorithm 9.5.** TwoProduct computes a non-overlapping expansion $x + y = a \times b$.

Analogous results can be proven for subtraction since we did not take the sign of the expressions into account. Therefore, there exists an algorithm TwoDiff that computes a non-overlapping expansion $x + y = a \ominus b - \text{err}(a \ominus b) = a - b$. Furthermore, there is an algorithm ExpansionDiff that computes a normalized expansion $h = \sum_{i=1}^{n+m} h_i = e - f$ for two normalized expansions $e = \sum_{i=1}^{m} e_i$ and $f = \sum_{i=1}^{n} f_i$.

For more sophisticated expressions, simple multiplications and scalar multiplication should be available. It was shown in [Shewchuk 97] that corresponding algorithms TwoProduct and ScaleExpansion exist. Technically, TwoProduct makes use of a splitting operation Split; see Algorithm 9.6. Here, a floating-point value is split into two half-precision values, and afterwards four multiplications of the corresponding fragments are performed; see [Dekker 71] and [Shewchuk 97]. In turn, ScaleExpansion makes use of TwoSum and TwoProduct.

**Theorem 9.24.** *Let $a$ and $b$ be two floating-point values with $p \geq 6$. There is an algorithm* TwoProduct *that computes a non-overlapping expansion $x + y = a \otimes b - \text{err}(a \otimes b) = a - b$ with $x = a \otimes b$ and $y = -\text{err}(a \otimes b)$.*

Split$(a, s)$ ($a$ represents a $p$-bit floating-point value, and $s$ is a splitting value with $\frac{p}{2} \leq s \leq p - 1$)

$c := (2^s + 1) \otimes a$
$big := c \ominus a$
$x := c \ominus big$
$y := a \ominus x$
**return** $(x, y)$

**Algorithm 9.6.** Split splits a floating-point value $a$ into $a = x + y$, where $x$ has $(p - s)$ bits and $y$ has $(s - 1)$ bits.

---

ScaleExpansion($\sum_{i=1}^{m} e_i, b$)

---

```
(Q_2, h_1) := TwoProduct(e_1, b)
for i := 2 TO m do
        (S_i, s_i) := TwoProduct(e_i, b)
        (Q_{2i-1}, h_{2i-2}) := TwoSum(Q_{2i-2}, s_i)
        (Q_{2i}, h_{2i-1}) := FastTwoSum(S_i, Q_{2i-2})
end for
h_{2m} := Q_{2m}
return (h_1, h_2, ..., h_{2m})
```

---

**Algorithm 9.7.** ScaleExpansion computes a non-overlapping expansion $h_1 + \cdots + h_{2m} = be$.

**Theorem 9.25.** *Let $e = \sum_{i=1}^{m} e_i$ be a normalized expansion with $p \geq 6$ and let $b$ be a floating-point value. There is an algorithm* ScaleExpansion *that computes a normalized expansion $h = \sum_{i=1}^{2m} h_i = b \times e$ with $O(m)$ $p$-bit floating-point operations.*

We omit the proofs and refer to [Shewchuk 97] or [Shewchuk 96], where some other useful operations are presented. For example, we have seen that some of the components may become zero. So, it is highly recommended that we *compress* a given expansion.

**Theorem 9.26.** *Let $e = \sum_{i=1}^{m} e_i$ be a normalized expansion with $p \geq 3$. Some of the components may be zero. There is an algorithm* Compress *that computes a normalized expansion $e = \sum_{i=1}^{n} h_i$ with non-zero components if $e \neq 0$. The largest component $h_n$ approximates $e$ with an error smaller than* ulp($h_n$).

Compress guarantees that the largest component is a good approximation for the value of the whole expression.

Now we will try to demonstrate how expansion arithmetic is used for adaptive evaluations of geometric expressions. We consider the following simple example stemming from [Shewchuk 97] and explain the main construction rules. For convenience, we make use of an expression tree. An expression tree is a binary tree that represents an arithmetic expressions. The leaves of the tree represent constants, whereas every inner node represents a geometric operation of its children. In the case of expansion arithmetic, the leaves consist of $p$-bit floating-point values.

Let us assume that we want to compute the square of the distance between two points $a = (a_x, a_y)$ and $b = (b_x, b_y)$. That is, we need to compute $(a_x - b_x)^2 + (a_y - b_y)^2$. In a first step, we construct expansions $x_1 + y_1 = (a_x - b_x)$

---

Compress($\sum_{i=1}^{m} e_i$)

---

```
Q := e_m
Bottom := m
for i := m − 1 Down TO 1 do
        (Q, q) := FastTwoSum(Q, e_i)
        if q ≠ 0 then
                g_Bottom := Q   Bottom := Bottom − 1   Q := q
        end if
end for
g_Bottom := Q   Top := 1
for i := Bottom + 1 TO m do
        (Q, q) := FastTwoSum(g_i, Q)
        if q ≠ 0 then
                h_Top := Q   Top := Top − 1
        end if
end for
h_Top := Q
n := Top
return (h_1, h_2, ..., h_n)
```

**Algorithm 9.8.** Compress computes a non-overlapping expansion $h_1 + \cdots + h_n = e$ with non-zero entries. The largest component is a good approximation of $e$.

and $x_2 + y_2 = (a_y - b_y)$ by a construction algorithm TwoSum. We conclude from Theorem 9.18 that

- $|y_1| = |\operatorname{err}(a_x \ominus b_x)| \leq \epsilon|x_1|$,
- $|y_2| = |\operatorname{err}(a_y \ominus b_y)| \leq \epsilon|x_2|$,
- $x_1 = a_x \ominus b_x$,
- $x_2 = a_y \ominus b_y$,

where $\epsilon = \frac{\beta}{2}\beta^{-p}$ represents the machine epsilon. Now $(a_x - b_x)^2 + (a_y - b_y)^2$ equals $E(x_1, x_2, y_1, y_2) = x_1^2 + x_2^2 + 2(x_1y_1 + x_2y_2) + y_1^2 + y_2^2$, and we have the following magnitudes:

- $x_1^2 + x_2^2 \in O(1)$,
- $2(x_1y_1 + x_2y_2) \in O(\epsilon)$, and
- $y_1^2 + y_2^2 \in O(\epsilon^2)$.

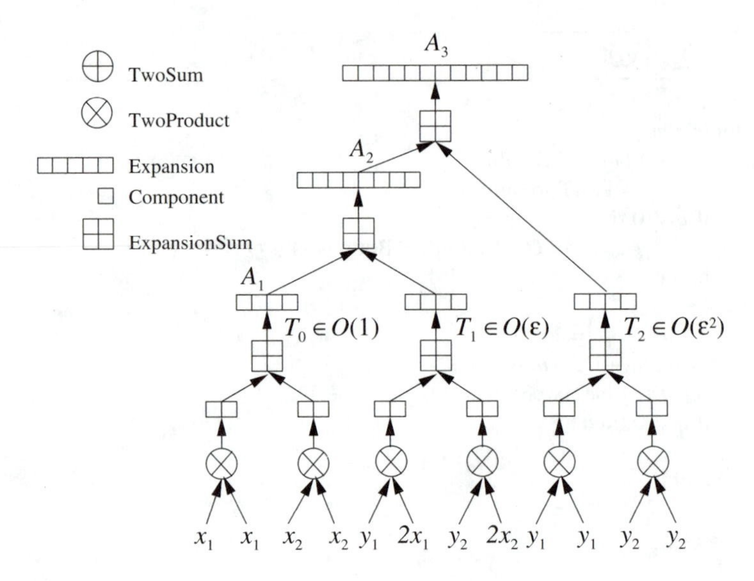


**Figure 9.13.** The simple incremental adaptive approach collects the errors by magnitude on the lowest level. (Redrawn and adapted courtesy of Jonathan Richard Shewchuk (Shewchuk 97).)

More generally, let $T_i$ be the sum of all products that contain $i$ $y$ variables. Then $T_i$ has magnitude $O(\epsilon^i)$. Now we can build different expression trees that collect $T_i$ terms in an appropriate way. A simple incremental adaptive approach is shown in Figure 9.13. The approximation $A_i$ contains all terms with size $O(\epsilon^{i-1})$ or larger. $A_3$ is the correct result. More precisely, the approximation $A_i$ is computed by the sum of the approximation $A_{i-1}$ and $T_i$. In the following, we will refer to this as the *simple incremental adaptive* method. It collects the error terms with respect to the lowest level of the expression tree, i.e., with respect to the input values. Note that we have already sorted the input terms by error magnitude.

We can repeat this principle in higher levels. For example, the result of TwoProduct$(x_1, x_1) = x + y$ consists of an error term $y$ with magnitude $O(\epsilon)$. The error terms $T_i$ can be computed adaptively. In Figure 9.14, we take incremental adaptivity to an extreme, recollecting the $O(\epsilon^i)$ error terms in all levels. This method can be denoted as the *full incremental adaptive* method. For an approximation $A_i$ with error $O(\epsilon^i)$, we need to approximate the error terms with error $O(\epsilon^i)$. Since the error term $T_k$ itself has magnitude $O(\epsilon^k)$, we have to approximate $T_k$ up to an error of $O(\epsilon^{i-k})$. In this sense, this approach is economical, especially if the needed accuracy is known in advance.

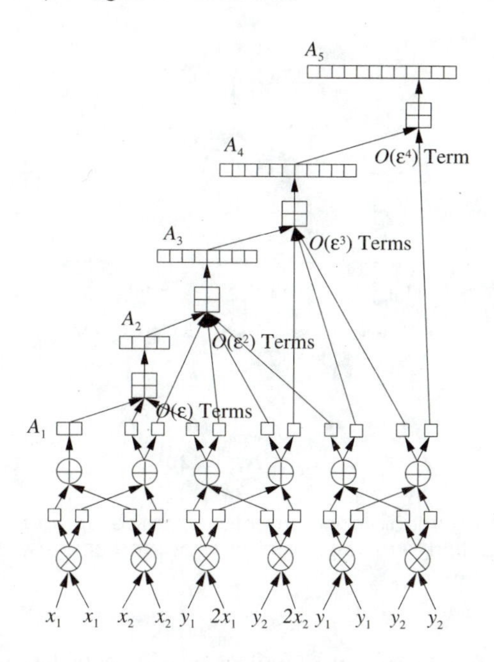


**Figure 9.14.** The full incremental adaptive approach collects the errors by magnitude on every level. (Redrawn and adapted courtesy of Jonathan Richard Shewchuk (Shewchuk 97).)

On the negative side, the full incremental adaptive approach leads to expression trees with a high complexity, since we keep track of many small error terms. Therefore, sometimes a better adaptive method with intermediate complexity could be more convenient. The *intermediate incremental adaptive* approach makes use of the approximation $A_i$ of the simple incremental method. Additionally, for the approximation of $A_{i+1}$, we compute a simple approximation $\mathrm{ct}(T_i)$ of $T_i$, thus obtaining an intermediate approximation $C_i = A_i + \mathrm{ct}(T_i)$. The method is illustrated in Figure 9.15. Sometimes, the correction term is already good enough, and the next error term is not computed.

Note that we can use the three construction schemes for the computations of the geometric predicates and expressions presented in Section 9.4. An extended example for the orientation test is given in [Shewchuk 97].

**Floating-point filters.** The previous adaptive approach makes use of floating-point arithmetic only, whereas the next two approaches come along with an exact arithmetic package. The exact package is used if a sufficient precision cannot be guaranteed by floating-point arithmetic.

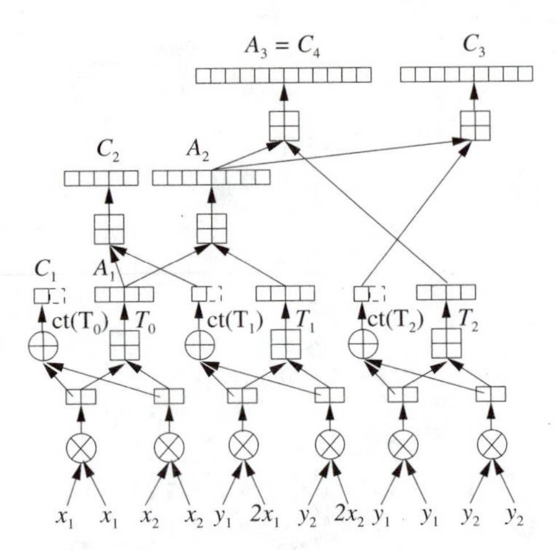


**Figure 9.15.** The intermediate incremental adaptive approach approximates the errors $T_i$ in a correction term. (Redrawn and adapted courtesy of Jonathan Richard Shewchuk (Shewchuk 97).)

We would like to reduce the number of exact calculations. In the floating-point filter approach, we replace the exact evaluation of an expression by the evaluation of guaranteed upper and lower bounds on its value. When the sign of the expression is needed and the bounding interval does not contain zero, then the sign is known. Otherwise, we have to use higher precision or exact arithmetic.

We follow the lines of [Mehlhorn and Näher 94]. Let us assume that we need to evaluate the sign of an arithmetic expression $E$. First, we will compute an approximation $\bar{E}$ using fixed-precision floating-point arithmetic. Additionally, we analyze the absolute error to obtain an error bound $\delta$ with $|E - \bar{E}| \leq \delta$. Obviously, if $\delta = 0$ or $|\bar{E}| > \delta$, the signs of $E$ and $\bar{E}$ are the same. Otherwise, if $\delta \neq 0$ and $|\bar{E}| \leq \delta$, the filter approach failed, and we have to recompute $E$ more precisely.

Let the cost for computing $E$ be denoted by Exact. Let the cost of computing $\bar{E}$ and $\delta$ be denoted by Filter. Finally, let probFilter denote the probability that the filter fails. In this case, the expected cost is Filter + probFilter × Exact. Minimizing Filter and minimizing probFilter are two contradicting goals. In an extreme case, we will have Filter = 0 and probFilter = 1 if there is no filter, and on the other side, Filter = Exact and probFilter = 0 if the exact value represents the filter.

Let us consider some examples for an inductive construction of appropriate error bounds. We assume that the error bounds $\delta_a$ and $\delta_b$ and the approximations

$\bar{a}$ and $\bar{b}$ for $a$ and $b$ have already been computed. The approximations result from floating-point operations. If we need to compute an error bound for $c = a + b$, we can make use of the following technique.

We compute $\bar{c} := \bar{a} \oplus \bar{b}$, and we have

$$\begin{aligned} |c - \bar{c}| = |a + b - (\bar{a} \oplus \bar{b})| &\leq |a + b - (\bar{a} + \bar{b})| + |\bar{a} + \bar{b} - (\bar{a} \oplus \bar{b})| \\ &\leq |a - \bar{a}| + |b - \bar{b}| + |\bar{a} + \bar{b} - (\bar{a} \oplus \bar{b})| \\ &\leq \delta_a + \delta_b + \epsilon(\bar{c})\,. \end{aligned} \tag{9.3}$$

This means that the error bound $\delta_c$ can be given by the sum of the error bounds plus $\epsilon|\bar{c}|$, where $\epsilon$ denotes the machine error, as usual. If the sign of the expression is needed, we compare the absolute value of the approximation $\bar{c}$ with the error estimate.

Similarly, we can design a floating-point filter for multiplications. For concreteness, we assume that $c = a \times b$ have to be computed under the same prerequisites. Thus, we conclude that for $\bar{c} := \bar{a} \otimes \bar{b}$, the error is bounded as follows:

$$\begin{aligned} |c - \bar{c}| &= |a \times b - \bar{a} \otimes \bar{b}| \\ &\leq |a \times b - \bar{a} \times \bar{b}| + |\bar{a} \times \bar{b} - \bar{a} \otimes \bar{b}| \\ &\leq |\bar{a} \times \bar{b} - \bar{a} \otimes \bar{b}| + |a \times b - \bar{a} \times b| + |\bar{a} \times b - \bar{a} \times \bar{b}| \\ &\leq \epsilon|\bar{c}| + |\bar{a}|\delta_b + |b|\delta_a \\ &\leq \epsilon|\bar{c}| + |\bar{a}|\delta_b + \max(|\bar{b} + \delta_b|, |\bar{b} - \delta_b|)\delta_a \\ &= \epsilon|\bar{c}| + |\bar{a}|\delta_b + (|\bar{b}| + |\delta_b|)\delta_a\,. \end{aligned} \tag{9.4}$$

Altogether, $\delta_c$ is computed by three multiplications and some additions.

The presented method is called *dynamic error analysis* because we dynamically take some additional operations (for example, computing $|\bar{b} - \delta_b|$) into account. Actually, we have to perform floating-point operations for $\delta_c$, so the situation is a bit more complicated.

Other filter methods try to use error bounds that can be almost precomputed for the complete expression structure. In this case, only the floating-point operations of the expression have to be computed. Such methods are called *static error analysis* or *semi-dynamic error analysis*, and they require a few additional operations. Static and semi-dynamic filters make use of a priori estimates of the magnitude of the corresponding results.

For example, as shown previously, the error $|c - \bar{c}|$ for $\bar{c} := \bar{a} \oplus \bar{b}$ is given by $|a - \bar{a}| + |b - \bar{b}| + |\bar{a} + \bar{b} - (\bar{a} \oplus \bar{b})|$. Let us assume that there are already inductive bounds $\text{ind}_a$, $\text{ind}_b$, $\max_a$, and $\max_b$, so that $|b - \bar{b}| \leq \text{ind}_b \max_b$ and $|a - \bar{a}| \leq \text{ind}_a \max_a$. Here, $\max_b$ is a binary estimate of the magnitude of $b$, and $\text{ind}_b$ refers to the depth of the inductive estimation. As a consequence, the error estimate for $|c - \bar{c}|$ can be given by $\text{ind}_c \max_c$ with $\max_c := \max_a + \max_b$ and $\text{ind}_c := (\frac{1}{2}\epsilon + \text{ind}_a + \text{ind}_b)$. We can represent any geometric expression $E$ by its expression tree; see Figure 9.16 or Section 9.3.3. Let us assume that we have floating-point arithmetic with $\beta = 2$ and precision $p$. For every leaf $x$ of the expression tree $E$, we initialize the inductive process by $\text{ind}_x = 2^{-(p+1)} = \frac{1}{2}\epsilon$ and an appropriate value for $\max_x$. In the static version, we may choose the maximal power of two, bounding all input values $x$. In the semi-dynamic version, we may choose to take the magnitude of every leaf $x$ into account; thus, we may set $\max_x := 2^{\lceil \log x \rceil}$. In both cases, the error estimate for $E$ is completely determined. Similar results hold for multiplications and other simple operations.

Static error analysis is used in the LN package of [Fortune and Van Wyk 93], whereas semi-dynamic error analysis is part of LEDA's floating-point operations; see [Mehlhorn and Näher 00]. Floating-point filters can be easily computed in advance; unfortunately, it might happen that the estimate is two pessimistic.

**Lazy evaluation.** Lazy evaluation makes use of a well-suited representation of arithmetic expression. The arithmetic expression is represented twice: once by its formal definition, i.e., its expression tree, and once by an approximation that is an arbitrary precision floating-point number together with an error bound. As long as the approximation is good enough, we use it. If we can no longer guarantee the sign of an expression by its approximation, we will increase the precision and use the expression tree for a better approximation.

Lazy evaluation is normally defined for integer arithmetic and rational arithmetic, and comes along with an arbitrary precision integer or rational arithmetic package. Approximations are computed by floating-point arithmetic. The floating-point approximation and its error bound can be expressed also by a floating-point interval that represents the approximation.

Let us assume that we use a package of arbitrary precision rationals. As already explained, a lazy number for the arithmetic expression $\frac{a}{b-c} - \frac{5}{17}$ is given by the expression tree of $E$ (see Figure 9.16) and an approximation interval, say $[-9.23 \times 10^{-3}, 2.94 \times 10^{2}]$. Here, $a$, $b$, and $c$ represent expression trees of lazy numbers.

Inductively, the expression tree is either an integer (or a constant) or an operator together with pointers to its arguments, i.e., references to other expression trees. Note that sometimes the expression tree may become a directed acyclic graph (DAG). For example, in Figure 9.16, assume that $b \equiv a$.

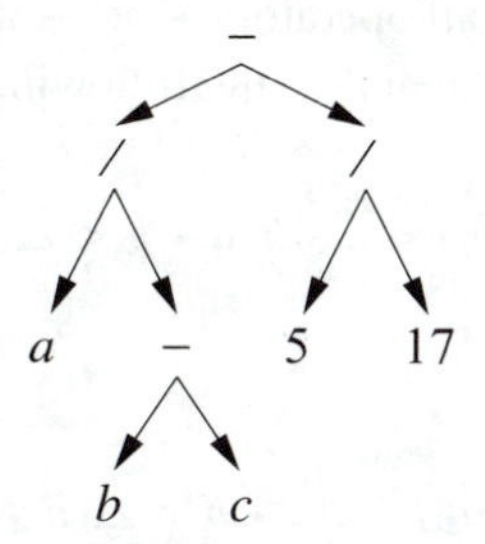


**Figure 9.16.** The expression tree of $\frac{a}{b-c} - \frac{5}{17}$.

Elementary operations for lazy numbers are performed by the following steps.

- Allocate a new tree (DAG) node for the result of the operation.
- Compute the floating-point interval with the interval arithmetic operations, see below.
- Register the name of the operation, the pointers to the operands, and the given interval at the given node.

In these steps, no exact computations will be performed. Let us assume that we have lazy numbers $a$ and $b$. In the following cases, we will have to evaluate the intervals or expressions more precisely.

- The sign of $a$ is needed and the floating-point interval contains 0.
- We have to compare $a$ and $b$ but the intervals do overlap.
- The lazy number $a$ belongs to number $b$, and $b$ has to be determined more precisely.

It might happen that a lazy number is never evaluated exactly. We still have to create the approximation intervals for lazy numbers. We assume that all lazy numbers are bounded in magnitude by the maximal and minimal representable floating-point values $M$ and $\epsilon$ on the machine. More precisely, we require $x \in ]-M, -\epsilon[\cup\{0\}\cup]+\epsilon, +M[$ for every lazy number $x$.

For a floating-point number $x$, let $\nabla(x)$ denote the next smallest representable floating-point number on the machine. *Representable* means representable within the given precision $p$, and *the next smallest* means smaller with respect to $x$. $\triangle(x)$ denotes the next biggest representable floating-point number on the machine. For example, for $p = 4$ and $x = 1$, we have $\nabla(x) = 0.9999$ and $\triangle(x) = 1.001$.

It is easy to show that for all operators $+, \times, -$, and $/$ and their corresponding floating-point versions $\oplus, \otimes, \ominus$, and $\oslash$, the following inequalities hold:

$$\nabla(a \circledast b) < a \circledast b, a \star b < \Delta(a \circledast b)$$

and

$$\nabla(a \circledast b) < a \star b < \Delta(a \circledast b)\,.$$

Therefore, we conclude that for the construction of lazy numbers, we make use of the intervals $[\nabla(a \circledast b), \Delta(a \circledast b)]$ computed by floating-point arithmetic.

Arithmetic with floating-point intervals for lazy numbers can be done as follows. Let $x$ and $y$ be lazy numbers with intervals $I_x = [a_x, b_x]$ and $I_y = [a_y, b_y]$.

- $I_{x+y} = [\nabla(a_x \oplus a_y), \Delta(b_x \oplus b_y)]$
- $I_{x \times y} = [\nabla(m), \Delta(M)]$ where $m$ $(M)$ is the minimum (maximum) of $\{a_x \otimes a_y, a_x \otimes b_y, b_x \otimes a_y, b_x \otimes b_y\}$
- $I_{-x} = [-b_x, -a_x]$
- $I_{\frac{1}{x}} = [\nabla(1.0 \oslash b_x), \Delta(1.0 \oslash a_x)]$ if $0 \notin [a_x, b_x]$

It is easy to see that the intervals may grow successively after some operations. Only precise evaluation can stop this process.

The natural way for the (interval) evaluation of a lazy expression starts at the leaves of the trees and successively propagates the results or intervals to the top (root) of the tree. During the evaluation process, some of the exact evaluated expressions may become useless. Therefore, a lazy arithmetic package should make use of a garbage collector. The lazy arithmetic package LEA [Benouamer et al. 93] is implemented in C++ and consists of three modules: an interval arithmetic module, a DAG module, and an exact rational arithmetic package.

Finally, we discuss the main advantage of lazy arithmetic versus pure floating-point filters. Let us assume that the needed precision varies a lot, i.e., sometimes low precision is enough, and in the worst case, high precision is needed. Lazy arithmetic is well suited for this situation. We can increase the precision of the calculations successively until finally the sign of the expression is determined by an interval. Floating-point filters do not support this adaptive approach.

**Separation bounds for algebraic expressions.** Another well suited adaptive arithmetic approach separates an arithmetic expression from zero. Similar to the previous sections we would like to evaluate the sign of an algebraic geometric expression given in an expression DAG. We discuss simple *separation bounds* for algebraic numbers expressed by the operations $+$, $-$, $\times$, $/$, and $\sqrt{k}$ for arbitrary integer $k$.

**Definition 9.27. (Separation bound.)** Let $E$ be an arithmetic expression with operators $+$, $-$, $\times$, $/$, and $\sqrt{k}$. The real value $\text{sep}(E)$ is denoted as a separation bound for $E$ if the following holds:

From $|E| \neq 0$, we conclude that $|E| \geq \text{sep}(E)$.

If a separation bound is given and $|E|$ equals zero, we would like to find out that $|E|$ is smaller than $\text{sep}(E)$. In this case we conclude $|E| = 0$ from the definition of the bound. Of course, we intend to minimize the computational effort for the sign evaluation and the evaluation of the separation bound.

Let us first assume that an approximation $\overline{E}$ was computed from imprecise arithmetic within an absolute error smaller than the separation bound over 2. That is, $|E - \overline{E}| \leq \frac{\text{sep}(E)}{2}$. If this is true, we can simply evaluate the sign of $E$ as follows:

- If $|\overline{E}| > \frac{\text{sep}(E)}{2}$, we conclude that $E$ and $\overline{E}$ have the same sign;

- Otherwise, $|\overline{E}| \leq \frac{\text{sep}(E)}{2}$ holds, and from $|E - \overline{E}| \leq \frac{\text{sep}(E)}{2}$, we conclude $|E| < \text{sep}(E)$. From the definition of $\text{sep}(E)$, we conclude $|E| = 0$.

We have to solve two main problems here. First, we have to find an appropriate separation bound. Second, the evaluation of an approximation $\overline{E}$ should be as cheap as possible.

The second problem is solved with an iterative approximation process. Rather than computing the approximation within an error bound of $\frac{\text{sep}(E)}{2}$, we increase the precision adaptively until finally the approximation is good enough; see [Burnikel et al. 01]. Let us assume that the separation bound was already computed successfully. The iterative sign test of an expression $E$ with $\text{sep}(E)$ works as follows.

The following is the iterative sign test with separation bound sketch:

- Initialize $\Delta$ by 1;

- Compute an approximation $\overline{E}$ of $E$ so that $|E - \overline{E}| \leq \Delta$ holds;

- If $|\overline{E}| > \Delta$ holds, the sign of $E$ and $\overline{E}$ coincidence and we are done;

- Otherwise, from $|\overline{E}| \leq \Delta$ we conclude $|E| \leq 2\Delta$. Now we check whether $2\Delta \leq \text{sep}(E)$ holds. If this is true, we conclude $E = 0$ from the definition of the separation bound. Otherwise, we decrease $\Delta$ by $\Delta := \frac{\Delta}{2}$ and repeat to compute an approximation with respect to $\Delta$.

We adaptively increase the precision. The given iterative test can be implemented in a straightforward manner using an adaptive arithmetic. For example, if we use floating-point arithmetic with arbitrary precision, we will increase the bit length by 1 in every iteration. Obviously, the maximal precision depends on the size of the separation bound and is given by $\log\left(\frac{1}{\text{sep}(E)}\right)$ because $2\Delta \leq \text{sep}(E)$ is finally fulfilled.

It remains to compute appropriate separation bounds. The computation of a separation bound of an expression $E$ can be represented in an execution table. The execuction table is successively applied to the DAG of the expression and results in a separation bound; see Theorem 9.28. For example, Table 9.3 shows the execution table for a separation bound of algebraic expressions with square roots only; see [Burnikel et al. 00]. Note that many of the standard expressions used in geometric algorithms and presented in Section 9.4.2 do not use arbitrary roots. The presented separation-bound technique is incorporated into data type `leda real` of the LEDA system; see [Mehlhorn and Näher 00]. For example, if the table is applied to the division-free expression $E = (\sqrt{2} - 1)(\sqrt{2} + 1) - 1$, then we will achieve the bounds $u(E) = 4 + 2\sqrt{2}$ and $l(E) = 1$. Applying Theorem 9.28 we can get a separation bound of

$$E \geq 3.11 \times 10^{-3}$$

for non-zero $E$. Using precision $p = 4$, we can conclude that $E$ equals zero.

Theorem 9.28 was proven in [Burnikel et al. 00].

| Expression | $u(E)$ | $l(E)$ |
|---|---|---|
| Integer n | $\|n\|$ | 1 |
| $E_1 \pm E_2$ | $u(E_1) \times l(E_2) + u(E_2) \times l(E_1)$ | $l(E_1) \times l(E_2)$ |
| $E_1 \times E_2$ | $u(E_1) \times u(E_2)$ | $l(E_1) \times l(E_2)$ |
| $E_1/E_2$ | $u(E_1) \times l(E_2)$ | $l(E_1) \times u(E_2)$ |
| $\sqrt{E}$ | $\sqrt{u(E)}$ | $\sqrt{l(E)}$ |

**Table 9.3.** The execution table for arithmetic expression with integers and square roots.

**Theorem 9.28.** *Let $E$ be an algebraic expression with integer input values and operators $+$, $-$, $/$, $\times$, and $\sqrt{}$. Let $k$ denote the number of $\sqrt{}$ operations in $E$ and let $D(E) := 2^k$. We can assume that $E$ is represented by a* DAG. *If we apply the execution table (Table 9.3), we obtain values $u(E)$ and $l(E)$ such that*

1. $|E| \leq u(E)l(E)^{D(E)^2-1}$;

2. $|E| \neq 0$ *implies* $|E| \geq \frac{u(E)}{l(E)u(E)^{D(E)^2}}$.

*If $E$ is division-free then $l(E) = 1$ and $D(E)^2$ can be replaced by $D(E)$.*

A comprehensive overview of existing separation bounds, their execution tables, and their analysis can be found in [Li 01]; see also [Burnikel et al. 00, Burnikel et al. 01, Li and Yap 01a].

### 9.3.4. Exact Geometric Computations (EGC)

EGC combines exact arithmetic or robust adaptive arithmetic with the idea that impreciseness in calculations sometimes can be neglected, whereas the combinatorial structures should always be correct. Thus, it guarantees that all standard geometric algorithms designed for the Real RAM will run as desired; i.e., the flow and the output of the algorithm are correct in a combinatorial sense.

More precisely, in the terminology of EGC intensively considered in [Yap 97a], we differ between *constructional* steps and *combinatorial* steps of an algorithm. In the former step, new elements are computed; the latter one causes the algorithm to branch. Different combinatorial steps lead to different computational paths. Each combinatorial step should depend on the evaluation of a geometric predicate. Altogether, to ensure exact combinatorics, it suffices to ensure that all predicates are evaluated exactly. For the numerical part of an algorithm, EGC requires that the result is represented exactly, although it might happen that the exact result can be approximated only within the given arithmetic. For example, we can uniquely represent an intersection of line segments by its endpoints, but the calculation of the coordinates may be erroneous.

Summarizing the EGC approach requires that

- the algorithm is driven by geometric predicates only,
- geometric expressions are represented uniquely, and
- the sign of a geometric predicate is evaluated correctly.

EGC is sort of a paradigm that can be fulfilled by many approaches. We have to ensure that the underlying arithmetic produces correct results for the sign of predicates. Therefore, exact integer, rational, or algebraic arithmetic is suitable. But also the adaptive approaches of lazy evaluation, separation bounds, floating-point filters, and adaptive floating-point expansions presented in the previous sections may lead to EGC, provided that the precision can be adapted adequately.

The idea of EGC is well supported by the CGAL library; see [Overmars 96], which is a C++ library of geometric algorithms and data structures. The corresponding algorithms are designed under the Real RAM model, and the flow of the algorithms are driven by geometric predicates. Therefore, we can combine CGAL with an arbitrary number package that fulfills the ECG paradigm. For example, we can make use of exact multi-precision rationals or floats of the C++ package GMP, or we can use adaptive arithmetic realized in the C++ library LEDA; see [Mehlhorn and Näher 00], which is based on floating-point filters, separation bounds, and algebraic numbers; see also [Burnikel et al. 95], [Karamcheti et al. 99a], [Li and Yap 01b], and [Li 01].

## 9.4. Robust Expressions and Predicates

In this section, we will present useful expressions and predicates that occur very often in geometric algorithms. There are many different ways for evaluating a geometric expression or predicate. A careful analysis often leads to relatively stable results. In Section 9.3.1, we saw that without a guard digit, the relative error obtained when subtracting nearby numbers can be very large. We should take this into account. Appropriate representations of geometric expressions are useful for almost all arithmetic strategies considered in Section 9.3.

We consider two simple examples stemming from [Shewchuk 99] and [Goldberg 91]. Both examples depend on floating-point roundoff errors (see Section 9.3.1). We will see that it is worth thinking about a robust representation of arithmetic expressions. A formula can sometimes be rearranged in order to eliminate catastrophic cancellation. The third example stemming from [Goldberg 91] shows a formal justification of the rearrangement, due to backward-error analysis. The given rearrangement ideas can be justified formally. Later in Section 9.4.2, we will summarize a list of useful robust expressions. Many expressions will be represented by the sign of determinants. The evaluation of determinants is incorporated into standard computer algebra systems. For an efficient and robust evaluation of the sign of determinants, see also [Clarkson 92], [Kaltofen and Villard 04] or [Avnaim et al. 97].

### 9.4.1. Examples of Formula Rearrangement

**The area of a simple polygon.** The area of a simple polygon in two dimensions can be computed by a simple formula. Let $p_1, p_2, p_3, \ldots, p_n$ denote the counterclockwise sequence of the vertices of a polygon $P$. The signed area of the polygon is

$$\text{Area}(P) = \frac{1}{2}\left(\sum_{i=1}^{n} p_{i_x} p_{i+1_y} - p_{i_y} p_{i+1_x}\right), \tag{9.5}$$

where $p_i = (p_{i_x}, p_{i_y})$ and $p_{n+1} := p_1$. Area($P$) is positive if $P$ is given in counterclockwise order. If $P$ is given in clockwise order, Area($P$) is negative. This can be easily shown by induction on the number of vertices.

The *robustness* of Equation (9.5) depends on the relationship of the absolute value of the coordinates to the area of the polygon. If the area is relatively small and the absolute value of the coordinates are relatively high, i.e., the distance to the origin is relatively high, then roundoff errors in floating-point operations may lead to inaccurate results. Therefore, we translate the polygon near to the origin by using the points $p_i{}' = (p_i - p_n)$. It is easy to see that

$$\text{Area}(P) = \frac{1}{2}\left(\sum_{i=1}^{n-2} p'_{i_x} p'_{i+1_y} - p'_{i_y} p'_{i+1_x}\right) \tag{9.6}$$

holds. This simply follows from the fact that the area is invariant due to a translation of the polygon. The last formula is much more robust than Equation (9.5), since we do not have severe cancellations.

For example, let us consider the rectangle $P$ given by $p_1 = (301, 300)$, $p_2 = (301, 301)$, $p_3 = (300, 301)$, and $p_4 = (300, 300)$; see Figure 9.17.

Using Equation (9.5) for $\beta = 10$ and $p = 4$, we obtain the incorrect results $301 \otimes 301 = 90600$ and $301 \otimes 300 = 90300$ due to the round-to-nearest rule. This gives

$$\begin{aligned}\text{Area}(P) &= \tfrac{1}{2}\,(90600 - 90300 + 90600 - 90300 \\ &\qquad +90000 - 90300 + 90000 - 90300) = 0,\end{aligned}$$

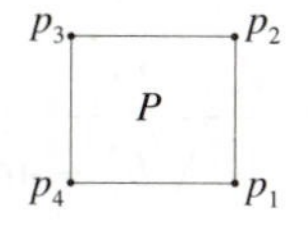


**Figure 9.17.** Computing the area of a rectangle $P$.

whereas using $p'_i = p_i - p_4$, $P'$ is given by $p'_1 = (1, 0)$, $p'_2 = (1, 1)$, $p'_3 = (0, 1)$, and $p'_4 = (0, 0)$. Now, for $P'$, the exact result is calculated by

$$\begin{aligned}\text{Area}(P) &= \tfrac{1}{2} \quad (1 - 0 + 1 - 0 \\ &\qquad\quad +0 - 0 + 0 - 0) = 1\,.\end{aligned}$$

Keep in mind that normally there is a speed trade-off we have to pay for robustness. Obviously in Equation (9.6) we need more floating-point operations than in Equation (9.5).

Translation near to the origin is a robustness paradigm, and we will use this principle especially for geometric predicates and constructors; see Section 9.4.2.

**The solutions of a quadratic equation.** The solution of a quadratic equation

$$ax^2 + bx + c = 0 \tag{9.7}$$

is given by

$$\begin{aligned}x_1 &= \frac{-b + \sqrt{b^2 - 4ac}}{2a} \\ x_2 &= \frac{-b - \sqrt{b^2 - 4ac}}{2a}.\end{aligned} \tag{9.8}$$

The expression $b^2 - 4ac$ could be the result of a floating-point subtraction of two floating-point multiplications. If $b^2$ and $4ac$ are rounded to nearest, they are accurate within $\frac{1}{2}$ ulp. Now subtraction may cause many of the accurate digits to disappear. Therefore, the result of the difference may have an error of many ulps.

For example, for $p = 3$ and $\beta = 10$, let $b = 3.34$, $a = 1.22$, and $c = 2.28$. The exact value of $b^2 - 4ac$ is 0.0292, whereas $b \otimes b = 11.2$, $4 \otimes a \otimes c = 11.1$, and $11.2 \ominus 11.1$ results exactly in 0.1. This means that the overall error is in 70.8 ulps.

If $b^2 \approx 4ac$, cancellation is not a problem in Equation (9.8), since $|b|$ will dominate the denominator. If $b^2 \gg 4ac$, then $b^2 - 4ac$ does not cause cancellations, but $\sqrt{b^2 - 4ac} \approx |b|$ is critical. One of the formulas of Equation (9.8) may produce the elimination of accurate digits. Therefore, we multiply the numerator and the denominator by $-b - \sqrt{b^2 - 4ac}$ and $-b + \sqrt{b^2 - 4ac}$, respectively, and we obtain

$$\begin{aligned}x_1 &= \frac{2c}{-b - \sqrt{b^2 - 4ac}} \\ x_2 &= \frac{2c}{-b + \sqrt{b^2 - 4ac}}.\end{aligned} \tag{9.9}$$

Let $b^2 \gg 4ac$. We use the following rule: if $b > 0$, compute $x_2$ by Equation (9.8) and $x_1$ by Equation (9.9); if $b < 0$, compute $x_1$ by Equation (9.8) and $x_2$ by Equation (9.9).

The expression $x^2 - y^2$. Although one might expect that an additional subtraction (addition) will cause a cancellation of accurate digits, we normally prefer the righthand side of

$$x^2 - y^2 = (x - y)(x + y). \tag{9.10}$$

By Theorem 9.10, the subtraction (addition) of two values without rounding errors (here $x$ and $y$) will have a small relative error. Multiplication of two quantities with a small relative error results in a product of small relative error.

If $x \gg y$ or $y \gg x$, we will prefer the left-hand side of Equation (9.10). We do not have accurate digit eliminations here. Additionally, the rounding error of the smaller value ($x^2$ or $y^2$) does not affect the final subtraction. In comparison to $(x - y)(x + y)$, we have to deal only with two rounding errors instead of three (for $(x - y)$, $(x + y)$, and for the product $(x - y)(x + y)$).

We will give a formal justification for the given result. The standard technique for a theoretical analysis is called *backward-error analysis*. We can assume that subtraction, addition, and multiplication are performed as required in the standard; see Equation (9.2).

Therefore, we have to compare the relative error of

$$\begin{aligned} (x \ominus y) \otimes (x \oplus y) &= (x - y)(1 + \delta_1) \otimes (x + y)(1 + \delta_2) \\ &= (x - y)(x + y)(1 + \delta_1)(1 + \delta_2)(1 + \delta_3) \end{aligned} \tag{9.11}$$

to the relative error of

$$\begin{aligned} (x \otimes x) \ominus (y \otimes y) &= x^2(1 + \gamma_1) \ominus y^2(1 + \gamma_2) \\ &= (x^2(1 + \gamma_1) - y^2(1 + \gamma_2))(1 + \gamma_3) \\ &= ((x^2 - y^2)(1 + \gamma_1) + (\gamma_1 - \gamma_2)y^2)(1 + \gamma_3). \end{aligned} \tag{9.12}$$

The relative error of Equation (9.11) is

$$\left| \frac{(x^2 - y^2)(1 + \delta_1)(1 + \delta_2)(1 + \delta_3) - (x^2 - y^2)}{(x^2 - y^2)} \right| = |(1+\delta_1)(1+\delta_2)(1+\delta_3)-1|,$$

which is equal to $\delta_1 + \delta_2 + \delta_3 + \delta_1\delta_2 + \delta_2\delta_3 + \delta_1\delta_3 + \delta_1\delta_2\delta_3 \leq 3(\epsilon + \epsilon^2) + \epsilon^3$.

The relative error of Equation (9.12) is

$$\left| \frac{((x^2 - y^2)(1 + \gamma_1) + (\gamma_1 - \gamma_2)y^2)(1 + \gamma_3) - (x^2 - y^2)}{(x^2 - y^2)} \right| = \left| (\gamma_1 + \gamma_3 + \gamma_1\gamma_3 + (1 + \gamma_3)\frac{(\gamma_1 - \gamma_2)y^2)}{(x^2 - y^2)} \right|,$$

which gives a problem if $x$ and $y$ are nearby. In this case, $(\gamma_1 - \gamma_2)y^2$ can be as large as $(x^2 - y^2)$, and the relative error appears to be approximately bigger than $(1 + \gamma_3)$. If $x \gg y$ holds, the term

$$(1 + \gamma_3)\frac{(\gamma_1 - \gamma_2)y^2}{(x^2 - y^2)}$$

is negligible, and we obtain a result almost independent from $\gamma_2$ and approximately smaller than $2\epsilon + \epsilon^2$. Analogously, we can argue for $y \gg x$.

### 9.4.2. A Summary of Robust Expressions

**The area of a triangle.** Assume that the length of the sides $a$, $b$, and $c$ of triangle $T$ are given. Note that in Section 9.4.1 the coordinates of the points were given. Then the area of $T$, Area($T$), may be computed by

$$\text{Area}(T) = \sqrt{s(s - a)(s - b)(s - c)}\,, \tag{9.13}$$

where

$$s = \frac{a + b + c}{2}.$$

Without loss of generality, we can assume that $a \geq b \geq c$. Otherwise, we rename the identifiers. Similar to the considerations in the previous section, Equation (9.13) has severe cancellations if $a \approx b + c$ holds, i. e., the triangle is extremely flat. In this case, we have $s \approx a$, and the term $(s - a)$ subtracts two nearby quantities while $s$ may have rounding errors. In this case, we might lose many accurate digits. We rearrange Equation (9.13) as follows

$$\text{Area}(T) = \frac{1}{4}\sqrt{(a + (b + c))(c - (a - b))\,(c + (a - b))(a + (b - c))}\,. \tag{9.14}$$

Equation (9.14) is much more accurate because it avoids severe cancellation at hand. It can be shown that the four factors are non-critical. For example, $a \ominus b$ is computed exactly with a guard digit. By backward-error analysis, the following result can be proven; see [Goldberg 91].

**Theorem 9.29.** *If the machine epsilon $\epsilon$ is not greater than* 0.005*, then Equation* (9.14) *is correct within a relative error not greater than* $11\epsilon$.

If the triangle is given by the three endpoints, we can use the oriented area formula, detailed below.

In the following, we turn over to a list of robust geometric predicates and constructors.

**Orientation test in 2D.** The orientation test is used in many geometric algorithms. Given three points $p$, $q$, and $r$ in the plane, we want to find out whether chain$(p,q,r)$ makes a left turn or right turn running from $p$ over $q$ to $r$. In the mathematical sense, a counterclockwise (left) turn is meant to be positive. One may think also of the orientation of the quadrants of a coordinate system in 2D.

The orientation test has various applications. For example, if we would like know whether two line segments $l_1 = (p_1, q_1)$ and $l_2 = (p_2, q_2)$ will have an intersection, we test whether the endpoints of $l_2$ do not lie on one side of $l_1$ and vice versa. Therefore, we have to check that chain$(p_1, q_1, p_2)$ and chain$(p_1, q_1, q_2)$, as well as chain$(p_2, q_2, p_1)$ and chain$(p_2, q_2, q_1)$, do not have the same orientation.

The semantic of the orientation test in 2D (O2D) is as follows:

$$\text{O2D}(p,q,r) = \begin{cases} > 0 & : \ \text{chain}(p,q,r) \text{ is in counterclockwise order,} \\ < 0 & : \ \text{chain}(p,q,r) \text{ is in clockwise order,} \\ = 0 & : \ p, q \text{ and } r \text{ are colinear.} \end{cases}$$

For the computation of the orientation test of $p$, $q$, and $r$, we make use of the *signed area* of the parallelogram spanned by $p$, $q$, and $r$ or equivalently determined by the vectors $\overrightarrow{(p-r)}$ and $\overrightarrow{(q-r)}$.

$$\begin{aligned} \text{O2D}(p,q,r) &= ((p_x - q_x)\left(p_y + q_y\right) + (q_x - r_x)\left(q_y + r_y\right) \\ &\quad +(r_x - p_x)\left(r_y + p_y\right)) \\ &= \begin{vmatrix} p_x & p_y & 1 \\ q_x & q_y & 1 \\ r_x & r_y & 1 \end{vmatrix} \\ &= \left(1 \cdot \begin{vmatrix} q_x & q_y \\ r_x & r_y \end{vmatrix} + (-1) \cdot \begin{vmatrix} p_x & p_y \\ r_x & r_y \end{vmatrix} + 1 \cdot \begin{vmatrix} p_x & p_y \\ q_x & q_y \end{vmatrix}\right) \\ &= \frac{1}{2}\begin{vmatrix} p_x - r_x & p_y - r_y \\ q_x - r_x & q_y - r_y \end{vmatrix}. \end{aligned}$$

The last expression is more accurate than the former ones. We simply translate all points by $-r$, so that the relation between the absolute value of the points and the correlation to each other avoids severe cancellations; see also Section 9.4.1.

We are mainly interested in the sign of O2D; therefore, O2D sometimes is defined as a Boolean predicate, i.e., $\text{O2D}(p, q, r) = \text{TRUE}$ if $\text{chain}(p, q, r)$ is in counterclockwise order and so on; see also Section 9.6 .

As a consequence for the signed area of the triangle of $p$, $q$, and $r$, we have

$$\text{SignArea}(p, q, r) = \frac{1}{2}\,\text{O2D}(p, q, r);$$

i. e., half of the area of the parallelogram.

**Area of a triangle in 3D.** For two vectors $\vec{u} = (u_x, u_y, u_z)$ and $\vec{w} = (w_x, w_y, w_z)$ in $\mathbb{R}^3$, the vector cross product is defined by

$$\vec{u} \times \vec{w} := \vec{i}\,(u_y w_z - u_z w_y) - \vec{j}\,(u_x w_z - u_z w_x) + \vec{k}\,(u_x w_y - u_y w_x),$$

which sometimes is denoted by the determinant

$$\begin{vmatrix} \vec{i} & \vec{j} & \vec{k} \\ u_x & u_y & u_z \\ w_x & w_y & w_z \end{vmatrix},$$

where $\vec{i} = (1, 0, 0)$, $\vec{j} = (0, 1, 0)$, and $\vec{k} = (0, 0, 1)$.

Let $p$, $q$, and $r$ define a triangle $d$ in 3D. The vector cross product $v = \overrightarrow{(p - r)} \times \overrightarrow{(q - r)}$ results in a vector $\vec{v}$ orthogonal to $d$, and the area of the triangle in $3D$ is given by half of the length of $\vec{v}$.

For $p = (p_x, p_y, p_z)$, let $p_{xy}$ denote $(p_x, p_y)$ in the $xy$-plane. $p_{xz}$ and $p_{yz}$ are defined analogously. Obviously, the $x$-coordinate of $v_x$ of $v$ is given by

$$\text{O2D}(p_{yz}, q_{yz}, r_{yz}) = \begin{vmatrix} p_y - r_y & p_z - r_z \\ q_y - r_y & q_z - r_z \end{vmatrix}.$$

Analogously, we have $v_y = \text{O2D}(p_{zx}, q_{zx}, r_{zx})$ and $v_z = \text{O2D}(p_{xy}, q_{xy}, r_{xy})$. Altogether, the area of the triangle $d$ in 3D is given by

$$V = \frac{\sqrt{\text{O2D}(p_{yz}, q_{yz}, r_{yz})^2 + \text{O2D}(p_{zx}, q_{zx}, r_{zx})^2 + \text{O2D}(p_{xy}, q_{xy}, r_{xy})^2}}{2}.$$

Note that the consideration of the cross product fits exactly to the 2D case. We can extend the 2D vectors $\overrightarrow{(p - r)}$ and $\overrightarrow{(q - r)}$ to the 3D setting $\overrightarrow{(p - r)} =$

$(p_x - r_x, p_y - r_y, 0)$ and $\overrightarrow{q - r} = (q_x - r_x, q_y - r_y, 0)$. The vector cross product results in

$$\overrightarrow{(p - r)} \times \overrightarrow{(q - r)} := \vec{k}((p_x - r_x)(q_y - r_y) - (p_y - r_y)(q_x - r_x)),$$

and the length of this vector equals $|\,\mathrm{O2D}(p, q, r)|$.

**Orientation test in 3D.** In 3D, a similar test can be applied. Having four points $p$, $q$, $r$, and $s$ in 3D, we want to find out whether $s$ lies below or beyond a 2D plane $E$ spanned by the $p$, $q$, and $r$. Obviously, we have to specify what below and above mean exactly, which can be done by fixing the orientation of $E$. If we *see* the points $p$, $q$, and $r$ from $s$ in a counterclockwise order, the plane is meant to be seen from *above*, which gives a negative value. The orientation test $\mathrm{O3D}(p, q, r, s)$ is positive if $s$ lies below the oriented plane.

The semantic of the orientation test in 3D (O3D) is as follows:

$$\mathrm{O3D}(p, q, r, s) = \begin{cases} > 0 & : \text{chain}(p, q, r) \text{ is seen in clockwise order from } s, \\ < 0 & : \text{chain}(p, q, r) \text{ is seen in counterclockwise order from } s, \\ = 0 & : p, q, r \text{ and } s \text{ are colinear.} \end{cases}$$

There is a nice *thumb* rule for the test. Take your left hand and try to point with slightly curled fingers to the points $p$, $q$, and $r$ in a circular and clockwise manner, starting with the leftmost finger, thus describing an oriented plane. If your (hopefully shorter) thumb points to $s$, O3D returns a positive value.

Geometrically speaking and in analogy to O2D, the expression $\mathrm{O3D}(p, q, r, s)$ is the signed volume of the parallelepiped determined by the vectors $\overrightarrow{(p - s)}$, $\overrightarrow{(q - s)}$, and $\overrightarrow{(r - s)}$. It is positive if the points are in position as described earlier; see also Figure 9.18.

Altogether, it suffices to evaluate the following determinant.

$$\mathrm{O3D}(p, q, r, s) = \begin{vmatrix} p_x & p_y & p_z & 1 \\ q_x & q_y & q_z & 1 \\ r_x & r_y & r_z & 1 \\ s_x & s_y & s_z & 1 \end{vmatrix} = \begin{vmatrix} p_x - s_x & p_y - s_y & p_z - s_z \\ q_x - s_x & q_y - s_y & q_z - s_z \\ r_x - s_x & r_y - s_y & r_z - s_z \end{vmatrix}$$

Due to [Shewchuk 99], the first expression can be evaluated with fewer operations, whereas the second one makes use of transformation and is more robust even if the points are almost coplanar.

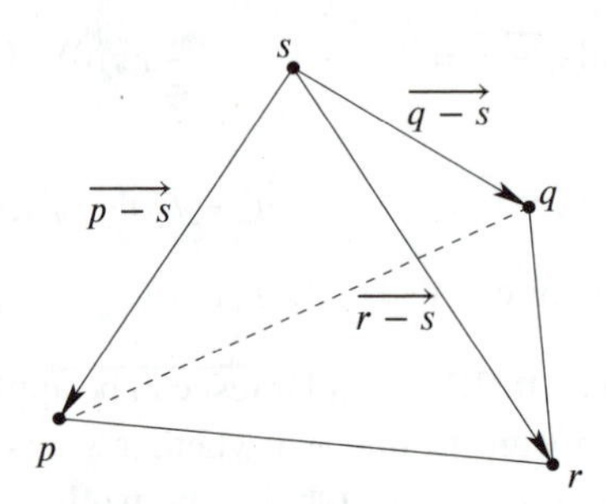


**Figure 9.18.** The chain $p$, $q$, and $r$ is clockwise if we look from the top. The vertex $s$ lies below the oriented plane and $\mathrm{O3D}(p,q,r,s)$ is positive.

$\mathrm{O3D}(q,r,s,p)$ determines the signed volume of the parallelepiped represented by four points $p$, $q$, $r$, and $s$; the signed volume of the corresponding tetrahedron (see Figure 9.18) is given by

$$\mathrm{SignArea}(p,q,r,s) = \frac{\mathrm{O3D}(p,q,r,s)}{6}.$$

We have seen that the orientation test in 3D is somehow grounded in the orientation test of 2D. A geometric interpretation can be given as follows. Let us assume that we rotate the points $p$, $q$, $r$, and $s$ to points $p'$, $q'$, $r'$, and $s'$ so that $p'$, $q'$, and $r'$ lie in a plane parallel to the $xy$-plane and $s'_z$ has the largest $z$-coordinate. In other words, the vector $(0,0,1)$ is perpendicular to the plane spanned by $p'$, $q'$, and $r'$, and $s'$ lies above the plane. Now $\mathrm{O3D}(p,q,r,s) = \mathrm{O3D}(p',q',r',s')$ is given by $-\mathrm{O2D}(p',q',r')$.

This holds in any dimension! For the orientation test in dimension $d+1$, we rotate the points $p_1, p_2, \ldots, p_{d+2}$ so that $p'_1, p'_2, \ldots, p'_{d+1}$ builds a plane perpendicular to the $d+1$-dimensional vector $(0,\ldots,0,1)$ and $p'_{d+2}$ has the largest $d+1$-coordinate. Now the sign of the orientation test of $p_1, p_2, \ldots, p_{d+2}$ in dimension $d+1$ is given by the inverse sign of the orientation test of $p'_1, p'_2, \ldots, p'_{d+1}$ in dimension $d$; see also "Generalization of the orientation and incircle tests to arbitrary dimensions" later in this section. Thus, if $p_{d+2}$ sees the positive plane from above, the orientation test returns a negative value.

Testing colinearity in 3D. O2D tests for colinearity in 2D, and O3D tests for coplanarity in 3D. So one might ask also whether three points in 3D are colinear. By ignoring the corresponding coordinates, we can project the points onto the $xy$-, $yz$- and $xz$-plane, respectively. We can choose any pair of the three planes and apply the O2D test for the corresponding points on both planes.

Let $p = (p_x, p_y, p_z)$, $q = (q_x, q_y, q_z)$, and $r = (r_x, r_y, r_z)$ denote three points in 3D. If all $z$-coordinates are the same, we can apply the O2D test for the points $p' = (p_x, p_y)$, $q' = (q_x, q_y)$, and $r' = (r_x, r_y)$. Otherwise, let $p_{xz}$ denote $(p_x, p_z)$ in the $xy$-plane; $p_{xz}$ and $p_{yz}$ are defined analogously. We apply the orientation tests $\mathrm{O2D}(p_{xz}, q_{xz}, r_{xz})$ and $\mathrm{O2D}(p_{yz}, q_{yz}, r_{yz})$. The vertices are colinear in $\mathbb{R}^3$ iff both tests give zeros.

**Incircle test in 2D.** The *incircle test* in 2D points out whether a vertex $s$ lies inside, outside, or on the circle passing through three vertices $p$, $q$, and $r$. Let $\mathrm{Circle}(p, q, r)$ denote the set of points of the circle passing through the vertices $p$, $q$, and $r$. The expression In2D has the following semantic:

$$\mathrm{In2D}(p, q, r, s) = \begin{cases} > 0 & : \quad s \text{ is inside } \mathrm{Circle}(p, q, r), \\ < 0 & : \quad s \text{ is outside } \mathrm{Circle}(p, q, r), \\ = 0 & : \quad s \text{ lies on } \mathrm{Circle}(p, q, r). \end{cases}$$

Fortunately, we are able to perform the incircle test by an orientation test in 3D. Let us assume that $\mathrm{O2D}(p, q, r)$ is positive, and let us consider the following projection. A point $a = (a_x, a_y)$ in the plane is projected onto the paraboloid $P = \{(x, y, x^2 + y^2) | (x, y) \in \mathbb{R}^2\}$ by

$$\lambda(a) = (a_x, a_y, a_x^2 + a_y^2) \in P. \tag{9.15}$$

We project all points of $\mathrm{Circle}(p, q, r)$ onto the paraboloid $P$, which gives a set $K'$ on $P$; see Figure 9.19. Now let us consider the orientations of the given planes. We assume that $\mathrm{O2D}(p, q, r)$ is positive.

The set $\mathrm{Circle}(p, q, r)$ can be expressed by $\mathrm{Circle}(p, q, r) = \{(x, y) | (x - I_x)^2 + (y - I_y)^2 = l^2\}$, where $I_x$, $I_y$, and $l$ depend on $p$, $q$, and $r$; see "Circumcenter and circumradii of triangles in 2D and 3D" later in this section. Therefore, we have

$$\begin{aligned} K' &= \{(x, y, z) | (x, y) \in \mathrm{Circle}(p, q, r), x^2 + y^2 = z\} \\ &= \{(x, y, z) | z - 2I_x x - 2I_y y + {I_x}^2 + {I_y}^2 - l^2 = 0\} \cap P, \end{aligned}$$

which shows that $K'$ is given by the intersection of $P$ with a 2D plane in 3D, uniquely determined by $\lambda(q)$, $\lambda(r)$, and $\lambda(s)$. Altogether, we have the following result.

**Theorem 9.30.** *Let $p$, $q$, $r$, and $s$ be points in $\mathbb{R}^2$. Let $E$ denote the 2D oriented plane passing through the points $\lambda(p)$, $\lambda(q)$, and $\lambda(r)$. Let $\mathrm{O2D}(p, q, r) > 0$. The point $s$ lies*

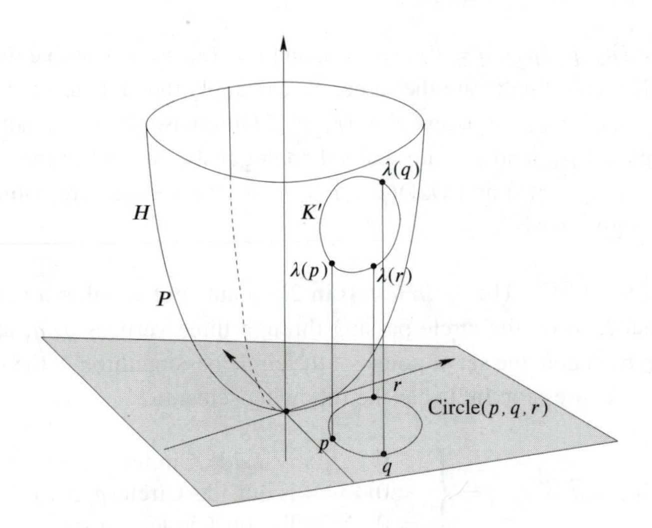


**Figure 9.19.** Projecting Circle$(p, q, r)$ onto the paraboloid results in a set of points lying in a 2D plane.

- *inside* Circle$(p, q, r)$ *if* $\lambda(s)$ *lies below* $E$ *(*In2D$(p, q, r, s) > 0$*),*
- *outside* Circle$(p, q, r)$ *if* $\lambda(s)$ *lies above* $E$ *(*In2D$(p, q, r, s) > 0$*), and*
- *on* Circle$(p, q, r)$ *if* $\lambda(s)$ *lies on* $E$ *(*In2D$(p, q, r, s) = 0$*).*

Altogether, it suffices to evaluate the following determinant.

$$\begin{aligned} \text{In2D}(p, q, r, s) &= \begin{vmatrix} p_x & p_y & p_x^2 + p_y^2 & 1 \\ q_x & q_y & q_x^2 + q_y^2 & 1 \\ r_x & r_y & r_x^2 + r_y^2 & 1 \\ s_x & s_y & s_x^2 + s_y^2 & 1 \end{vmatrix} \\ &= \begin{vmatrix} p_x - s_x & p_y - s_y & (p_x - s_x)^2 + (p_y - s_y)^2 \\ q_x - s_x & q_y - s_y & (q_x - s_x)^2 + (q_y - s_y)^2 \\ r_x - s_x & r_y - s_y & (r_x - s_x)^2 + (r_y - s_y)^2 \end{vmatrix} \end{aligned}$$

If O2D$(p, q, r)$ is negative, the sign of In2D$(p, q, r, s)$ is reversed, so that In2D$(p, q, r, s)$ is negative iff $s$ is inside Circle$(p, q, r)$ and so on.

**Insphere test in 3D.** Similar to the incircle test, we are able to perform the insphere test in 3D by an orientation test in the fourth dimension. Let $p$, $q$, $r$, $s$, and $t$ be five points in 3D. The incircle test should return whether $t$ lies inside, outside, or on the sphere passing through the four points $p$, $q$, $r$, and $s$.

Let us assume that O3D$(p, q, r, s)$ is positive. Analogously, we project the points $p, q, r$, and $s$ onto the 4D paraboloid $\{(x, y, z, w)|w = x^2+y^2+z^2, (x, y, z) \in \mathbb{R}^3\}$ and apply the orientation test in 3D to $(p_x, p_y, p_z, p_x^2+p_y^2+p_z^2)$, $(q_x, q_y, q_z, q_x^2+q_y^2+q_z^2)$, $(r_x, r_y, r_z, r_x^2+r_y^2+r_z^2)$, and $(s_x, s_y, s_z, s_x^2+s_y^2+s_z^2)$.

$$\text{In3D}(p, q, r, s, t) = \begin{vmatrix} p_x & p_y & p_z & p_x^2+p_y^2+p_z^2 & 1 \\ q_x & q_y & q_z & q_x^2+q_y^2+q_z^2 & 1 \\ r_x & r_y & r_z & r_x^2+r_y^2+r_z^2 & 1 \\ s_x & s_y & s_z & s_x^2+s_y^2+s_z^2 & 1 \\ t_x & t_y & t_z & t_x^2+t_y^2+t_z^2 & 1 \end{vmatrix}$$

$$= \begin{vmatrix} p_x-t_x & p_y-t_y & p_y-t_y & (p_x-t_x)^2+(p_y-t_y)^2+(p_z-t_z)^2 \\ q_x-t_x & q_y-t_y & q_y-t_y & (q_x-t_x)^2+(q_y-t_y)^2+(q_z-t_z)^2 \\ r_x-t_x & r_y-t_y & r_y-t_y & (r_x-t_x)^2+(r_y-t_y)^2+(r_z-t_z)^2 \\ s_x-t_x & s_y-t_y & s_y-t_y & (s_x-t_x)^2+(s_y-t_y)^2+(s_z-t_z)^2 \end{vmatrix}$$

Let Sphere$(p, q, r, s)$ denote the sphere passing through $p$, $q$, $r$, and $s$. The semantic of In3D is as follows:

$$\text{In3D}(p, q, r, s, t) = \begin{cases} > 0 & : \ t \text{ is inside Sphere}(p, q, r, s), \\ < 0 & : \ t \text{ is outside Sphere}(p, q, r, s), \\ = 0 & : \ t \text{ lies on Sphere}(p, q, r, s). \end{cases}$$

**Generalization of the orientation and incircle tests to arbitrary dimensions.** Let $p_1, p_2, \dots, p_{d+1}$ be a sequence of $d + 1$ point in $\mathbb{R}^d$, where $p_i = (p_{i_1}, p_{i_2}, \dots, p_{i_d})$. The orientation test in dimension $d$ decides on which side of the oriented hyperplane passing through $p_1, p_2, \dots, p_d$ the point $p_{d+1}$ lies. The value of the following determinant OdD is the signed volume of the simplex spanned by $p_1, p_2, \dots, p_{d+1}$, up to a constant depending on the dimension $d$.

$$\text{OdD}(p_1, p_2, \dots, p_{d+1}) = \begin{vmatrix} p_{1_1} & p_{1_2} & \cdots & p_{1_d} & 1 \\ p_{2_1} & p_{2_2} & \cdots & p_{2_d} & 1 \\ \vdots & \vdots & \vdots & \vdots & \vdots \\ p_{d+1_1} & p_{d+1_2} & \cdots & p_{d+1_d} & 1 \end{vmatrix} \tag{9.16}$$

If the orientation test is performed more than once for the same hyperplane, it is worth storing some intermediate results. We can simply expand the determinant about the last row, thus obtaining a linear combination

$$\text{OdD}(p_1, p_2, \dots, p_{d+1}) = a_1 p_{d+1_1} + a_2 p_{d+1_2} + \cdots + a_d p_{d+1_d} + a_{d+1},$$

where $a_i$ is determined by a determinant of the first $d$ rows of Equation (9.16). More precisely, $a_1, a_2, \ldots, a_{d+1}$ represent the coefficients of the hyperplane passing through $p_1, p_2, \ldots, p_d$; i.e., the hyperplane passing through $p_1, p_2, \ldots, p_d$ is given by

$$\{(x_1, x_2, \ldots, x_d) \in \mathbb{R}^d \mid a_1x_1 + a_2x_2 + \cdots + a_dx_d + a_{d+1} = 0\}.$$

Membership in a $d$-dimensional sphere can be calculated by an orientation test in dimension $d+1$. If $\mathrm{OdD}(p_1, p_2, \ldots, p_{d+1})$ is positive, we check whether $p_{d+2}$ is inside the sphere passing through $p_1, p_2, \ldots, p_{d+1}$ by evaluating the determinant

$$\mathrm{IndD}(p_1, p_2, \ldots, p_{d+2}) = \\ \begin{vmatrix} p_{1_1} & p_{1_2} & \cdots & p_{1_d} & p_{1_1}^2 + p_{1_2}^2 + \cdots p_{1_d}^2 & 1 \\ p_{2_1} & p_{2_2} & \cdots & p_{2_d} & p_{2_1}^2 + p_{2_2}^2 + \cdots p_{2_d}^2 & 1 \\ \vdots & \vdots & \vdots & \vdots & \vdots & \vdots \\ p_{d+2_1} & p_{d+2_2} & \cdots & p_{d+2_d} & p_{d+2_1}^2 + p_{d+2_2}^2 + \cdots p_{d+2_d}^2 & 1 \end{vmatrix}. \quad (9.17)$$

Similar to the preceding predicates for Equation (9.16) and Equation (9.17), there is an obvious rearrangement that makes use of simple transformations.

The correctness of the insphere test in dimension $d$ can be proven with the same arguments as in "Insphere test in 3D" earlier in this section. It can be shown that the projections of $d$ points $p_i = (p_{i_1}, p_{i_2}, \ldots, p_{i_d})$ onto the $(d+1)$-dimensional hyperparaboloid $H = \{(x_1, x_2, \ldots, x_d, x_1^2 + \cdots x_d^2) | (x_1, x_2, \ldots, x_d) \in \mathbb{R}^d\}$ lie in a common $d$-dimensional hyperplane. A query point $p_{d+2}$ is inside the $d$-dimensional hypersphere passing through $p_1, \ldots, p_{d+1}$ iff the corresponding projection of $p_{d+2}$ onto $H$ lies below the hyperplane of

$$\{(p_{i_1}, p_{i_2}, \ldots, p_{i_d}, p_{i_1}^2 + p_{i_2}^2 + \cdots + p_{i_d}^2) | i = 1, \ldots, d+1\},$$

provided that $p_1, \ldots, p_{d+1}$ are oriented adequately.

Up to now, we have considered geometric predicates. Next, we turn to constructors. We will compute intersection results in various aspects. Although we can make use of robust transformations at the end of the calculations, we must transform the results back to the correct positions, which might cause cancellations (see also Section 9.4.1).

#### Intersection of lines in 2D and 3D.

The intersection of two lines $l_1$ and $l_2$ depends on the representation of $l_1$ and $l_2$. If $l_i = \{(x, y) \in \mathbb{R}^2 | a_ix + b_iy + c_i = 0\}$, then the intersection $(I_x, I_y)$ is given by

$$I_x = \frac{\begin{vmatrix} b_1 & c_1 \\ b_2 & c_2 \end{vmatrix}}{\begin{vmatrix} a_1 & b_1 \\ a_2 & b_2 \end{vmatrix}} \quad I_y = -\frac{\begin{vmatrix} a_1 & c_1 \\ a_2 & c_2 \end{vmatrix}}{\begin{vmatrix} a_1 & b_1 \\ a_2 & b_2 \end{vmatrix}}.$$

If $l_1$ is given by a pair of points $(p, q)$ and $l_2$ is given by a pair of points $(r, s)$, we can calculate $a_1 = (q_y - p_y)$, $a_2 = (s_y - r_y)$, $b_1 = (p_x - q_x)$, $b_2 = (r_x - s_x)$, $c_1 = (p_y q_x - q_y p_x)$, and $c_2 = (r_y s_x - s_y r_x)$, which gives

$$I_x = \frac{\begin{vmatrix} p_x - q_x & p_x q_y - q_x p_y \\ r_x - s_x & r_x s_y - r_y s_x \end{vmatrix}}{\begin{vmatrix} q_y - p_y & p_x - q_x \\ s_y - r_y & r_x - s_x \end{vmatrix}} \quad I_y = -\frac{\begin{vmatrix} q_y - p_y & p_y q_x - q_y p_x \\ s_y - r_y & r_y s_x - s_y r_x \end{vmatrix}}{\begin{vmatrix} q_y - p_y & q_x - p_x \\ s_y - r_y & r_x - s_x \end{vmatrix}}.$$

Equivalently, we can express the intersection point by the linear combination $(I_x, I_y) = p + \gamma(q - p)$, where

$$\gamma = \frac{\begin{vmatrix} r_x - p_x & r_y - p_y \\ s_x - p_x & s_y - p_y \end{vmatrix}}{\begin{vmatrix} q_y - p_y & q_x - p_x \\ r_y - s_y & r_x - s_x \end{vmatrix}} \tag{9.18}$$

Note that the unique denominator

$$\begin{vmatrix} a_1 & b_1 \\ a_2 & b_2 \end{vmatrix}$$

is zero iff $l_1$ and $l_2$ are parallel. Parallelism can be tested by evaluating a single determinant. The unique denominator is not equivalent to an orientation test in 2D but, for example, the numerator of Equation (9.18) is given by O2D$(r, s, p)$, which makes use of a transformation by $p$.

For $(I_x, I_y) = p - \gamma(q - p)$, sometimes it might be helpful to rearrange $p$, $q$, $r$, and $s$. The denominator will change only its sign, but we may have less severe cancellations if $(I_x, I_y)$ is close to $p$; see also Section 9.4.1. Therefore, if $(I_x, I_y)$ appears to be close to $q$, $r$, or $s$, we will swap the closest of the points to $p$.

For two lines in 3D, we can compute the intersection of the lines $l_1$ passing through $p$ and $q$ and $l_2$ passing through $r$ and $s$ by

$$(I_x, I_y, I_z) = q + \frac{(p - s) \times (r - s) \cdot (r - s) \times (q - s)}{|(p - s) \times (r - s)|^2}(p - s).$$

Again, we can improve preciseness if $(I_x, I_y, I_z)$ is close to $q$, and eventually we swap $p$ and $q$, which has no influence on the denominator.

If there is no intersection, $(I_x, I_y, I_z)$ is a point on $l_1$ nearest to $l_2$, and the distance between $l_1$ and $l_2$ is given by

$$\frac{|\,\mathrm{O3D}(p, q, r, s)|}{|(p - s) \times (r - s)|}.$$

**Intersection of a line and a plane in 3D.** Similar to the considerations in the previous section, we can express the intersection of a line passing through $p$ and $q$ and a plane passing through $r$, $s$, and $t$ by $(I_x, I_y, I_z) = p - \gamma(q - p)$. where

$$\gamma = \frac{\begin{vmatrix} p_x - r_x & p_y - r_y & p_z - r_z \\ s_x - r_x & s_y - r_y & s_z - r_z \\ t_x - r_x & t_y - r_y & t_z - r_z \end{vmatrix}}{\begin{vmatrix} p_x - q_x & p_y - q_y & p_z - q_z \\ s_x - r_x & s_y - r_y & s_z - r_z \\ t_x - r_x & t_y - r_y & t_z - r_z \end{vmatrix}}.$$

As in the previous cases, it should be more robust to swap $q$ and $p$ if $q$ is closer to $(I_x, I_y, I_z)$. We do not need to recompute the denominator in this case.

The denominator is zero iff the line and the plane run parallel. In this case, we have to check whether the line lies totally inside the plane by checking whether $p$ is a linear combination of $r$, $s$, and $t$. This can be done by the Gaussian elimination technique of a system of linear equations; for a robust consideration of this technique, see for example [Schwarz 89].

**Circumcenter and circumradii of triangles in 2D and 3D.** For example, the circumcenter of a circle passing through three points $p$, $q$, and $r$ in the plane is relevant for the computation of a Voronoi diagram in 2D. If we consider the triangle spanned by $p$, $q$, and $r$, the circumcenter $(I_x, I_y)$ is given by the intersection of the three perpendicular bisectors of the sides of the triangle where

$$I_x = r_x + \frac{\begin{vmatrix} (p_x - r_x)^2 + (p_y - r_y)^2 & p_y - r_y \\ (q_x - r_x)^2 + (q_y - r_y)^2 & q_y - r_y \end{vmatrix}}{2\begin{vmatrix} p_x - r_x & p_y - r_y \\ q_x - r_x & q_y - r_y \end{vmatrix}}$$

and

$$I_y = r_y + \frac{\begin{vmatrix} p_x - r_x & (p_x - r_x)^2 + (p_y - r_y)^2 \\ q_x - r_x & (q_x - r_x)^2 + (q_y - r_y)^2 \end{vmatrix}}{2\begin{vmatrix} p_x - r_x & p_y - r_y \\ q_x - r_x & q_y - r_y \end{vmatrix}}.$$

The denominator represents O2D$(p, q, r)$, and the coordinates are unstable if the three points are almost colinear. The corresponding radius of the circle can be efficiently computed by

$$l = \frac{|(q-r)||(r-p)||(p-q)|}{2\begin{vmatrix} p_x - r_x & p_y - r_y \\ q_x - r_x & q_y - r_y \end{vmatrix}}.$$

Altogether, we have Circle$(q, r, s) = \{(x, y) \in \mathbb{R}^2 | (x - I_x)^2 + (y - I_y)^2 = l^2\}$. The center of the triangle in 3D can be computed by

$$(I_x, I_y) = r + \frac{\left[|p-r|^2(q-r) - |q-r|^2(p-r)\right] \times \left[(p-r) \times (q-r)\right]}{2|(p-r) \times (q-r)|^2}.$$

**Circumcenter and circumradii of a tetrahedron.** The center $(I_x, I_y)$ of the circumsphere of the tetrahedron expressed by four points $p$, $q$, $r$, and $s$ in 3D is given by

$$(I_x, I_y) = s + \frac{|p-s|^2(q-s) \times (r-s) + |q-s|^2(r-s) \times (p-s) + |r-s|^2(p-s) \times (q-s)}{2\begin{vmatrix} p_x - s_x & p_y - s_y & p_z - s_z \\ q_x - s_x & q_y - s_y & q_z - s_z \\ r_x - s_x & r_y - s_y & r_z - s_z \end{vmatrix}}.$$

The corresponding radius $l$ reads

$$\frac{\left| |p-s|^2(q-s) \times (r-s) + |q-s|^2(r-s) \times (p-s) + |r-s|^2(p-s) \times (q-s) \right|}{2\begin{vmatrix} p_x - s_x & p_y - s_y & p_z - s_z \\ q_x - s_x & q_y - s_y & q_z - s_z \\ r_x - s_x & r_y - s_y & r_z - s_z \end{vmatrix}}.$$

The denominator is equivalent to O3D$(p, q, r, s)$, and it is unstable if the four points are almost coplanar.

### 9.4.3. Efficient Evaluation of a Determinant

In the preceding section, we considered many predicates and constructors based on evaluation of determinants. The efficient evaluation of determinants is well supported by geometric libraries that support the exact geometric computation paradigm introduced in [Yap 93]; see also Section 9.3.4. For example, the LEDA library (http://www.algorithmic-solutions.com), the Core Library (http://www.cs.

nyu.edu/exact/core/), and the Real/Expr library (http://www.cs.nyu.edu/exact/realexpr/Det.html) support efficient determinant evaluations. On the other hand, robust and fast determinant calculations are also implemented in common computer algebra systems and can be supported by the use of exact number packages. See Section 9.7 for a detailed overview of existing packages. On the theoretical side, for an efficient and robust evaluation of the sign of determinants, see [Clarkson 92], [Kaltofen and Villard 04], or [Avnaim et al. 97].

## 9.5. Degeneracy

In the preceding section, we already noticed some of the influences of degeneracy. For example, if three points are almost colinear or if four points appear to be almost coplanar, the computation of the coordinates of circumcenters and circumspheres may become unstable. These cases are based on the fact that roundoff errors prevent us from exact calculations. In the following, we will assume that we are able to calculate geometry-based expressions exactly. In this case, we are able to *detect* degeneracy and only the algorithmic problem of dealing with degenerate cases remains. We will focus on the algorithm-dependent degeneracy.

Many geometric algorithms are designed and analyzed under the assumption that degenerate cases do not occur. The main reason is that degeneracy may evoke case distinctions that let the solution appear inelegant, and sometimes completely destroys the complexity analysis. Additionally, degenerate cases are not useful for explaining the intrinsic idea of a geometric algorithm.

### 9.5.1. Formal Definition of Degeneracy

We repeat the notations from [Emiris and Canny 92]. Formally, we consider a geometric problem $P$ as a mapping between topological spaces $X$ and $Y$. The input space $X = \mathbb{R}^{nd}$ has the standard Euclidean topology. The output space $Y$ is the product $C \times \mathbb{R}^m$ of a finite space $C$ with discrete topology and the Euclidean space $\mathbb{R}^m$. Here $m$, $n$, and $d$ are non-negative integers, and $C$ is a discrete space that represents the combinatorial part of the output, such as a planar graph or an ordered list of vertices. The parameter $d$ represents the dimension of the input space, and $n$ represents the number of input points. Note that this formalism is equivalent to the definitions in Section 9.2.

**Definition 9.31. (Degenerate problem.)** A problem instance $x \in \mathbb{R}^{nd}$ of a geometric problem $P$ is said to be *degenerate* iff $P$ has a discontinuity at $x$.

For example, let $d = 2$ and consider $n$ points in the plane. Let $P(x)$ denote the Voronoi diagram of the input points (see Chapter 6); that is, a straight line planar graph, represented in $C$, and the coordinates of its vertices represented in $\mathbb{R}^m$. There is a discontinuity at $x$ if there are four points on a circle that represent a common vertex.

Under the EGC paradigm (see Section 9.3.4), any geometric algorithm that solves a geometric problem should be driven by geometric predicates only. That is, every branch of an algorithm depends on the sign of a predicate. Therefore, we can consider a geometric algorithm $A$ as a simple decision tree, where the decision nodes test the sign of a (usually low degree) polynomial of the input variables.

**Definition 9.32. (Degenerate algorithm.)** An instance $x$ of a geometric algorithm $A$, where $A$ solves problem $P$, is called *degenerate* if the computation of $A$ on input $x$ contains a predicate test with output zero.

Typical cases for degeneracy are three points on a line, four points on a circle, or two points with the same $x$-coordinate. Degenerate cases may lead to difficult case analysis in a corresponding algorithm, so they should be avoided. In the following, the input is said to be in *general position* if degeneracy does not occur. Degeneracy is closely related to corresponding predicates; for example, the case of three points on a line corresponds to O2D, and the case of four points on a circle corresponds to In2D.

The following methods increase the computational complexity of a corresponding algorithm. We will make use of the Real RAM model described in the beginning of Section 9.3. Only the four basic operations $\{+, -, \times, /\}$ are allowed between real numbers. We consider the maximal computation cost of the longest computation path in the decision tree of a geometric algorithm. Two different complexity measures will be discussed. In the *algebraic model*, we count unit cost for every vertex in the computation path along the decision tree. In the *bit model*, we also keep track of the bit size of the corresponding operands, which is more realistic. For integers of size $O(b)$, addition and subtraction need $O(b)$, whereas multiplication and division require $M(b)$ bit operations, where $M(b) := O(b \log b \log\log b)$ is an upper bound on the bit complexity of any operation in the RAM; see [Aho et al. 74]. The total bit complexity is given by the maximum of the sum of the bit operations of any computation path.

### 9.5.2. Symbolic Perturbation

In this section, we will discuss the generic technique of a symbolic perturbation. The input is perturbed symbolically so that a general position is enforced. Thus, for a probably degenerate input $x$, we obtain a non-degenerate input $x(\epsilon)$ and compute $P(x(\epsilon))$ with algorithm $A$. In a post-processing step, we obtain the answer $P(x)$ from $P(x(\epsilon))$.

There are several perturbation methods discussed in the literature. The idea of a perturbation of the input values goes back to symmetry breaking rules in linear programming; see [Dantzig 63]. The right-hand side of the $i$th constraint

$$\sum_{j=1}^{n} a_{ij} x_j + x_{n+i} = b_i$$

is perturbed by an infinitesimal quantity that depends on the index $i$ of the constraint and becomes

$$\sum_{j=1}^{n} a_{ij} x_j + x_{n+i} := b_i(\epsilon) = b_i + \epsilon^i,$$

where $x_1, x_2, \ldots, x_n$ are the original variables, $x_{n+i}$ are the slack variables, and the values $a_{ij}$ and $b_i$ are constants. The perturbation method avoids infinite cycles in a simplex algorithm.

Let us assume that $x_{ij}$ are the input coordinates of our problem. For example, we have $n$ points $x_i = (x_{i1}, x_{i2}, \ldots, x_{id})$ for $i = 1, \ldots, n$ in the $d$-dimensional Euclidean space.

**Definition 9.33. (Valid perturbation.)** Per [Emiris and Canny 92], a perturbation $x(\epsilon)$ for an input instance $x$ is called *valid* iff

- $x(\epsilon)$ is in general position,
- $x(\epsilon)$ is arbitrarily close to $x$ such that
- whenever $x$ is non-degenerate, $x$ and $x(\epsilon)$ evoke the same computation path in the corresponding geometric algorithm.

A perturbation is called *symbolic* if the perturbation values are never evaluated exactly.

Note that general position corresponds to a (set of) geometric predicate(s). Symbolic perturbation schemes are never evaluated exactly, but we have to rewrite the corresponding predicates. Depending on the perturbation scheme, this is more or less efficient. The symbolic variable $\epsilon$ will never be evaluated exactly. The focus is on the proof of the existence of a small perturbation $\epsilon$ that fulfills the validity conditions.

Edelsbrunner and Mücke [Edelsbrunner and Mücke 90] presented the perturbation scheme (denoted also as Simulation of Simplicity)

$$x_{ij}(\epsilon) := x_{ij} + \epsilon^{2i\delta+j}, \tag{9.19}$$

where $\delta > d$. The main drawback of the presented scheme is that deciding the sign of a determinant of a $d \times d$ matrix requires $\Omega(2^d)$ additional effort in the algebraic model.

Yap [Yap 87b] suggested a more general technique for polynomial test predicates $F$ of degree $k$. The input $Y = (y_1, y_2, \ldots, y_k)$ is perturbed by $y_i + \epsilon_i$ with

$$1 \gg \epsilon_1 \gg \epsilon_2 \gg \cdots \gg \epsilon_k,$$

where $\gg$ means *is significantly greater than*. The symbolic evaluation of $F(X+\epsilon)$ then makes use of the Taylor expansion for $F(Y)$, thus taking partial derivatives of the multi-variate polynomial $F$ into account. For example, for a one-variable function $F$, we have

$$F(Y+\epsilon) = \sum_{i=0}^{\infty} \frac{F^i(Y)\epsilon^i}{i!},$$

and the sign of $F(Y)$ is given by the first non-zero $\frac{F^i(Y)}{i!}$. The Taylor expansion is finite for a multi-variate polynomial $F$. Unfortunately, we must compute all partial derivatives in the worst case, which may lead to $\Omega(d^n)$ additional operations in the algebraic model.

We present a simple and efficient perturbation scheme suggested in [Emiris et al. 97] and [Emiris and Canny 92] for some widespread predicates, such as ordering, O2D, In2D, and its generalizations to arbitrary dimensions. Note that consistency within a geometric algorithm requires that exactly one scheme should be applied to all occurring predicates.

**A simple perturbation scheme for orderings.** The sweepline algorithm for computing the arrangement of a set of line segments (see Section 9.1.1), implicitly makes use of the fact that the $x$-coordinates of all segments are disjunct. This means that the event structure has a perfect ordering in the very beginning. The non-uniqueness of $x$-coordinates is a useful prerequisite for many sweepline algorithms.

In general, for a set of $n$ points $x_i = (x_{i1}, x_{i2}, \dots, x_{id})$, we require that the $k$th coordinates are disjunct. This means that $x_{i_1k} \neq x_{i_2k}$ holds for all $i_1, i_2 \in \{1, \dots, n\}$ with $i_1 \neq i_2$.

[Emiris et al. 97] suggested the following symbolic perturbation scheme

$$x_{ij}(\epsilon) := x_{ij} + \epsilon \cdot i^j . \tag{9.20}$$

You can see that $x_{i_1k} + \epsilon \cdot {i_1}^k = x_{i_2k} + \epsilon \cdot {i_2}^k$ is equivalent to $i_1 = i_2$ if $x_{i_1k} = x_{i_2k}$ holds. Therefore, the scheme in Equation (9.20) perturbs the input to a non-degenerate one with respect to the ordering predicate. The extended ordering predicate of two coordinates $x_{i_1k}$ and $x_{i_2k}$ can now be implemented as shown in Algorithm 9.9.

In Algorithm 9.9, the first assignment represents the *original* comparison. Therefore, the extended ordering predicate does not exceed the running time of the original predicate in both computation models. In the bit model, both assignments in Algorithm 9.9 are computed in $O(\log n)$. We conclude that there is a positive $\epsilon$ that fulfills validity.

**Theorem 9.34.** *The symbolic perturbation scheme, Equation* (9.20), *is valid for the ordering predicate, and the corresponding implementation is optimal in the algebraic model and in the bit model.*

**A simple perturbation scheme for the orientation test.** We concentrate on the orientation test in 2D. The following arguments hold in every dimension $d$. The O2D checks whether or not three points are colinear; see "Orientation test in 2D" earlier in this section. For convenience, we repeat the formulation of the predicate.

Let $x_1$, $x_2$, and $x_3$ be points in the plane. If $\mathrm{O2D}(x_1, x_2, x_3)$ is positive, $\mathrm{chain}(x_1, x_2, x_3)$ is in counterclockwise order. If $\mathrm{O2D}(x_1, x_2, x_3)$ is negative, $\mathrm{chain}(x_1, x_2, x_3)$ is in clockwise order. If $\mathrm{O2D}(x_1, x_2, x_3) = 0$, $x_1$, $x_2$, and $x_3$ are colinear.

Ordering($x_{i_1k}, x_{i_2k}$)

> $s := x_{i_1k} - x_{i_2k}$
> **if** $s = 0$ **then**
>     $s := i_1 - i_2$
> **end if**
> **return** sign($s$)

**Algorithm 9.9.** The ordering predicate for a simple perturbation scheme.

We will see that the scheme shown in Equation (9.20) is valid also for the orientation test in 2D. However, it is less efficient than the scheme

$$x_{ij}(\epsilon) := x_{ij} + \epsilon \left(i^j \mod q\right) \tag{9.21}$$

suggested by [Emiris et al. 97] and [Emiris and Canny 92], where $q$ is the smallest prime with $q > n$. The modulo $q$ rule in Equation (9.21) reduces the bit complexity in the computation of a special determinant, as you will see below.

It is easy to see that the scheme in Equation (9.21) is not valid for the ordering predicate. Assuming that $n = 10$ and $q = 11$, we have $8^2 \equiv 9 \mod 11$ and $3^2 \equiv 9 \mod 11$. Thus, for $x_{82} = x_{32}$, we obtain $x_{82}(\epsilon) = x_{32}(\epsilon)$. Equation (9.21) is not valid with respect to the ordering predicate, since $x(\epsilon)$ is not in general position. In the following, we will show that the scheme is valid for the orientation test predicate.

Let $n$ and $q$ be given and $d = 2$. That is, we have $n$ 2D points $x_1, x_2, \ldots, x_n$. The O2D for three input points $x_{i_1} = (x_{i_1 1}, x_{i_1 2})$, $x_{i_2} = (x_{i_2 1}, x_{i_2 2})$, and $x_{i_3} = (x_{i_3 1}, x_{i_3 2})$ is given by the determinant

$$\mathrm{O2D}(x_{i_1}, x_{i_2}, x_{i_3}) = \begin{vmatrix} x_{i_1 1} & x_{i_1 2} & 1 \\ x_{i_2 1} & x_{i_2 2} & 1 \\ x_{i_3 1} & x_{i_3 2} & 1 \end{vmatrix};$$

(see Section 9.4.2). In turn, the predicate O2D for the perturbation scheme, Equation (9.21), is given by

$$\mathrm{O2D}(x_{i_1}(\epsilon), x_{i_2}(\epsilon), x_{i_3}(\epsilon)) = \begin{vmatrix} x_{i_1 1} + \epsilon\left({i_1}^1 \mod q\right) & x_{i_1 2} + \epsilon\left({i_1}^2 \mod q\right) & 1 \\ x_{i_2 1} + \epsilon\left({i_2}^1 \mod q\right) & x_{i_2 2} + \epsilon\left({i_2}^2 \mod q\right) & 1 \\ x_{i_3 1} + \epsilon\left({i_3}^1 \mod q\right) & x_{i_3 2} + \epsilon\left({i_3}^2 \mod q\right) & 1 \end{vmatrix}.$$

The determinant is linear in every column. That is, for column vectors $a^1$, $a^2, \ldots, a^n, b$, and a constant $c$, we have

$$\begin{aligned} \mathrm{Det}(a^1, \ldots, a^{j-1}, a^j + b, a^{j+1}, \ldots, a^n) &= \mathrm{Det}(a^1, \ldots, a^{j-1}, a^j, a^{j+1}, \ldots, a^n) \\ &+ \mathrm{Det}(a^1, \ldots, a^{j-1}, b, a^{j+1}, \ldots, a^n), \\ \mathrm{Det}(a^1, \ldots, a^{j-1}, c \cdot a^j, a^{j+1}, \ldots, a^n) &= c\,\mathrm{Det}(a^1, \ldots, a^{j-1}, a^j, a^{j+1}, \ldots, a^n). \end{aligned}$$

Therefore, we conclude that

$$\mathrm{O2D}(x_{i_1}(\epsilon), x_{i_2}(\epsilon), x_{i_3}(\epsilon)) = \mathrm{O2D}(x_{i_1}, x_{i_2}, x_{i_3}) + c_1 \cdot \epsilon + \epsilon^2 \left( \begin{vmatrix} ({i_1}^1 \mod q) & ({i_1}^2 \mod q) & 1 \\ ({i_2}^1 \mod q) & ({i_2}^2 \mod q) & 1 \\ ({i_3}^1 \mod q) & ({i_3}^2 \mod q) & 1 \end{vmatrix} \right).$$

Here, $c_1$ is a constant depending on $n$ and the input values. In general, the matrix

$$\mathrm{Vand}(i_1, i_2, \ldots, i_{d+1}) := \begin{pmatrix} {i_1}^1 & {i_1}^2 & \cdots & {i_1}^d & 1 \\ {i_2}^1 & {i_2}^2 & \cdots & {i_2}^d & 1 \\ \vdots & \vdots & \vdots & \vdots & \vdots \\ {i_{d+1}}^1 & {i_{d+1}}^2 & \cdots & {i_{d+1}}^d & 1 \end{pmatrix}$$

is called the Vandermonde matrix. The determinant of the Vandermonde matrix can be computed by

$$\begin{vmatrix} {i_1}^1 & {i_1}^2 & \cdots & {i_1}^d & 1 \\ {i_2}^1 & {i_2}^2 & \cdots & {i_2}^d & 1 \\ \vdots & \vdots & \vdots & \vdots & \vdots \\ {i_{d+1}}^1 & {i_{d+1}}^2 & \cdots & {i_{d+1}}^d & 1 \end{vmatrix} = (-1)^d \prod_{k>l\geq 1}^{d+1} (i_k - i_l)\,.$$

Let $\overline{\mathrm{Vand}}$ denote the Vandermonde matrix with mod $q$ entries, that is:

$$\overline{\mathrm{Vand}}(i_1, i_2, \ldots, i_{d+1}) := \begin{pmatrix} {i_1}^1 \mod q & {i_1}^2 \mod q & \cdots & {i_1}^d \mod q & 1 \\ {i_2}^1 \mod q & {i_2}^2 \mod q & \cdots & {i_2}^d \mod q & 1 \\ \vdots & \vdots & \vdots & \vdots & \vdots \\ {i_{d+1}}^1 \mod q & {i_{d+1}}^2 \mod q & \cdots & {i_{d+1}}^d \mod q & 1 \end{pmatrix}.$$

The determinant of the manipulated Vandermonde matrix $\overline{\mathrm{Vand}}$ can be computed by

$$\begin{vmatrix} {i_1}^1 \mod q & {i_1}^2 \mod q & \cdots & {i_1}^d \mod q & 1 \\ {i_2}^1 \mod q & {i_2}^2 \mod q & \cdots & {i_2}^d \mod q & 1 \\ \vdots & \vdots & \vdots & \vdots & \vdots \\ {i_{d+1}}^1 \mod q & {i_{d+1}}^2 \mod q & \cdots & {i_{d+1}}^d \mod q & 1 \end{vmatrix} \equiv$$

$$(-1)^d \prod_{k>l\geq 1}^{d+1} (i_k - i_l) \mod q\,.$$

Now assume that $\mathrm{O2D}(x_{i_1}, x_{i_2}, x_{i_3}) = 0$ holds. Note that $q$ is prime. Then $\mathrm{O2D}(x_{i_1}(\epsilon), x_{i_2}(\epsilon), x_{i_3}(\epsilon))$ is zero if the polynomial

$$p(X) = c_1 X + X^2(|\,\mathrm{Vand}(i_1, i_2, i_3)| \mod q)$$

is equivalent to zero, which means that all coefficients are zero, or if $p$ has root $\epsilon$, which means $p(\epsilon) = 0$. If

$$0 \not\equiv (|\,\mathrm{Vand}(i_1, i_2, i_3)| \mod q) \equiv \left( \prod_{k>l\geq 1}^{3} (i_l - i_k) \mod q \right),$$

then $p$ has non-zero coefficients and $p \not\equiv 0$. In this case, the polygon $p$ has a finite number of roots. Let $\epsilon_0$ be the smallest positive root of $p$. We can choose $\epsilon < \epsilon_0$, and $p(\epsilon)$ is non-zero. Altogether, if $\mathrm{O2D}(x_{i_1}, x_{i_2}, x_{i_3}) = 0$, then $\mathrm{O2D}(x_{i_1}(\epsilon), x_{i_2}(\epsilon), x_{i_3}(\epsilon))$ can be zero only if

$$0 \equiv \left( \prod_{k>l\geq 1}^{3} (i_l - i_k) \mod q \right)$$

holds. Here, $q$ is a prime bigger than $n$, and we conclude that

$$0 \equiv \left( \prod_{k>l\geq 1}^{3} (i_l - i_k) \mod q \right)$$

is equivalent to $i_k = i_l$ for some $k > l \geq 1$. For the orientation test of three points in 2D, we do not allow two identical input points. Altogether, the perturbation scheme in Equation (9.21) is valid for O2D. Obviously, with exactly the same arguments, the perturbation scheme in Equation (9.20) is also valid for O2D.

Note that for general dimension $d$, we analogously obtain

$$\begin{aligned}\mathrm{OdD}(x_{i_1}(\epsilon), \ldots, x_{i_{d+1}}(\epsilon)) = {} & \mathrm{OdD}(x_{i_1}, \ldots, x_{i_{d+1}}) \\ & + \sum_{i=1}^{d-1} c_i \epsilon^i \epsilon^d \left( |\overline{\mathrm{Vand}}(i_1, i_2, \ldots, i_{d+1})| \right),\end{aligned}$$

and we can follow the same arguments for orientation tests in arbitrary dimensions. That is, we do not allow two identical points in the orientation test. This means, that the determinant of the Vandermonde matrix $\mathrm{Vand}(i_1, i_2, \ldots, i_{d+1})$ (modulo $q$) does not vanish. Therefore, there is an $\epsilon$ so that $\mathrm{OdD}(x_{i_1}(\epsilon), \ldots, x_{i_{d+1}}(\epsilon))$ is never zero. Altogether, the following theorem holds.

**Theorem 9.35.** *The schemes in Equations* (9.20) *and* (9.21) *are valid for the orientation test in arbitrary dimension.*

We still have to evaluate the orientation test $\mathrm{O2D}(x_{i_1}(\epsilon), x_{i_2}(\epsilon), x_{i_3}(\epsilon))$ in the plane, respectively, $\mathrm{OdD}(x_{i_1}(\epsilon), \ldots, x_{i_{d+1}}(\epsilon))$ for general dimension $d$. If

$\mathrm{OdD}(x_{i_1}, \ldots, x_{i_{d+1}})$ equals zero, we have to compute the sign of $\sum_{i=1}^{d-1} c_i \epsilon^i + \epsilon^d\,(|\,\mathrm{Vand}(i_1, i_2, \ldots, i_{d+1})| \mod q)$

It can be shown that the perturbation scheme in Equation (9.19) results also in the evaluation of a polynomial $\sum_{i=0}^{d} c_i \epsilon^i$ with $\mathrm{OdD}(x_{i_1}, \ldots, x_{i_{d+1}}) = c_0$. [Edelsbrunner and Mücke 90] suggested to evaluate the coefficients $c_i$ successively for $i = 0, \ldots, d$. The first non-zero coefficient $c_j$ determines the sign of $\mathrm{OdD}(x_{i_1}(\epsilon), \ldots, x_{i_{d+1}}(\epsilon))$. Unfortunately, we may have to compute the full determinant of $\mathrm{OdD}(x_{i_1}(\epsilon), \ldots, x_{i_{d+1}}(\epsilon))$ in this case, which gives $\Omega(2^d)$ additional operations in the algebraic model.

The schemes in Equations (9.21) and (9.20) are much more efficient in this case. For convenience, we discuss the case $d = 2$. We use the linearity of the determinant as indicated above.

$$
\begin{aligned}
&\mathrm{O2D}(x_{i_1}(\epsilon), x_{i_2}(\epsilon), x_{i_3}(\epsilon))\\
&= \begin{vmatrix}
x_{i_1 1} + \epsilon\left(i_1{}^1 \mod q\right) & x_{i_1 2} + \epsilon\left(i_1{}^2 \mod q\right) & 1\\
x_{i_2 1} + \epsilon\left(i_2{}^1 \mod q\right) & x_{i_2 2} + \epsilon\left(i_2{}^2 \mod q\right) & 1\\
x_{i_3 1} + \epsilon\left(i_3{}^1 \mod q\right) & x_{i_3 2} + \epsilon\left(i_3{}^2 \mod q\right) & 1
\end{vmatrix}\\
&= \frac{1}{\epsilon}\begin{vmatrix}
x_{i_1 1} + \epsilon\left(i_1{}^1 \mod q\right) & x_{i_1 2} + \epsilon\left(i_1{}^2 \mod q\right) & \epsilon\\
x_{i_2 1} + \epsilon\left(i_2{}^1 \mod q\right) & x_{i_2 2} + \epsilon\left(i_2{}^2 \mod q\right) & \epsilon\\
x_{i_3 1} + \epsilon\left(i_3{}^1 \mod q\right) & x_{i_3 2} + \epsilon\left(i_3{}^2 \mod q\right) & \epsilon
\end{vmatrix}\\
&= \frac{1}{\epsilon}\left|\begin{pmatrix}
x_{i_1 1} & x_{i_1 2} & 0\\
x_{i_2 1} & x_{i_2 2} & 0\\
x_{i_3 1} & x_{i_3 2} & 0
\end{pmatrix} + \epsilon\begin{pmatrix}
\left(i_1{}^1 \mod q\right) & \left(i_1{}^2 \mod q\right) & 1\\
\left(i_2{}^1 \mod q\right) & \left(i_2{}^2 \mod q\right) & 1\\
\left(i_3{}^1 \mod q\right) & \left(i_3{}^2 \mod q\right) & 1
\end{pmatrix}\right|
\end{aligned}
$$

Now let

$$
L_3 := \begin{pmatrix}
x_{i_1 1} & x_{i_1 2} & 0\\
x_{i_2 1} & x_{i_2 2} & 0\\
x_{i_3 1} & x_{i_3 2} & 0
\end{pmatrix}
$$

and

$$
\overline{\mathrm{Vand}_3} := \begin{pmatrix}
\left(i_1{}^1 \mod q\right) & \left(i_1{}^2 \mod q\right) & 1\\
\left(i_2{}^1 \mod q\right) & \left(i_2{}^2 \mod q\right) & 1\\
\left(i_3{}^1 \mod q\right) & \left(i_3{}^2 \mod q\right) & 1
\end{pmatrix}.
$$

Note that the arguments hold in every dimension; therefore, we use the index 3 for $L$ and $\overline{\mathrm{Vand}}$. Additionally, $\overline{\mathrm{Vand}}$ refers to the application of mod $q$ for every entry in the matrix Vand. The proof works also without mod $q$ but requires additional computation time for

$$
\prod_{k>l\geq 1}^{d+1} (i_k - i_l),
$$

as we will see later.

We have to compute the sign of $\frac{1}{\epsilon}\,\mathrm{Det}\left(L_3 + \epsilon\,\overline{\mathrm{Vand}_3}\right)$.

$$\mathrm{Det}(\overline{\mathrm{Vand}_3}) \equiv \left(\prod_{k>l\geq 1}^{3} (i_l - i_k) \mod q\right)$$

is non-zero, since we do not allow two identical input points. Therefore, $\overline{\mathrm{Vand}_3}$ can be inverted, and we have

$$L_3 + \epsilon\,\overline{\mathrm{Vand}_3} = \left(-\overline{\mathrm{Vand}_3}\right)\cdot\left(\left(-\overline{\mathrm{Vand}_3}\right)^{-1}\cdot L_3 - \epsilon I_3\right),$$

where $I_3$ denotes the unit matrix. Applying the rule $\mathrm{Det}(A\cdot B) = \mathrm{Det}(A)\,\mathrm{Det}(B)$ we obtain

$$\begin{aligned}
&\mathrm{O2D}(x_{i_1}(\epsilon), x_{i_2}(\epsilon), x_{i_3}(\epsilon))\\
&= \frac{1}{\epsilon}\,\mathrm{Det}\left(L_3 + \epsilon\,\overline{\mathrm{Vand}_3}\right)\\
&= \frac{1}{\epsilon}\,\mathrm{Det}\left(-\overline{\mathrm{Vand}_3}\right)\mathrm{Det}\left(\left(-\overline{\mathrm{Vand}_3}\right)^{-1}\cdot L_3 - \epsilon I_3\right)\\
&= \frac{1}{\epsilon}(-1)^3\,\mathrm{Det}\left(\overline{\mathrm{Vand}_3}\right)\mathrm{Det}\left(\left(-\overline{\mathrm{Vand}_3}\right)^{-1}\cdot L_3 - \epsilon I_3\right).
\end{aligned}$$

Let $M := \left(-\overline{\mathrm{Vand}_3}\right)^{-1}\cdot L_3$. Since $\mathrm{Det}\left(\overline{\mathrm{Vand}_3}\right)$ is known and $\left(-\overline{\mathrm{Vand}_3}\right)^{-1}$ can be easily computed, the computation of the $M - \epsilon I_3$ does not exceed the time complexity of the original predicate. Up to this point, the computation time for the determinant $\mathrm{O2D}(x_{i_1}, x_{i_2}, x_{i_3})$ subsumes the time complexity of all matrix manipulation. It remains to compute the determinant of $M - \epsilon I_3$. This is equivalent to the problem of computing the characteristic polynomial of $M$. As a result of [Keller-Gehrig 85], characteristic polynomial computation can be done within the same time complexity as *normal* determinant evaluation of a $d \times d$ matrix. Fortunately, this holds in the bit model as well as in the algebraic model.

The modulo $q$ rule in Equation (9.21) reduces the bit complexity in the computation of the Vandermonde determinant. This is the only difference between Equations (9.21) and (9.20). Equation (9.20) incurs an extra bit complexity factor $O(d)$ but is also optimal in the algebraic model. Altogether, we summarize the results.

**Theorem 9.36.** *The perturbation schemes in Equations* (9.21) *and* (9.20) *are valid for the orientation test in arbitrary dimension. The scheme in Equation* (9.21) *does not affect the time complexity in either the bit model or the algebraic model. The scheme of Equation* (9.20) *is optimal with respect to the algebraic model and* *needs an extra bit complexity factor $O(d)$ for the evaluation of* Vand *(instead of* $\overline{\text{Vand}}$*).*

**Other predicates and limitations.** It was shown in [Emiris et al. 97] and [Emiris and Canny 92] that the scheme of Equation (9.20) can be also applied to the insphere predicate for arbitrary dimensions $d$ (see Section 9.4.2). The corresponding implementations are optimal with respect to the algebraic model and incur an extra $O(d)$ factor in the bit complexity. The implementation and the proof are analogous to the arguments in the previous section and use the evaluation of a slightly changed Vandermonde matrix.

**Theorem 9.37.** *The perturbation scheme in Equation* (9.20) *is valid for the insphere test in arbitrary dimension. The implementation does not affect the time complexity in the algebraic model, but needs an extra bit complexity factor $O(d)$.*

Let us first discuss two additional schemes. The validity of the presented schemes is related to the non-singularity of the Vandermonde matrix. One can make also use of the non-singularity of a Cauchy matrix. It can be shown that the scheme

$$x_{ij}(\epsilon) := x_{ij} + \epsilon \frac{1}{i + j - 1} \tag{9.22}$$

is valid for the orientation test in arbitrary dimensions. Alternatively, one can use the scheme

$$x_{ij}(\epsilon) := x_{ij} + \epsilon q_i^{\,j}, \tag{9.23}$$

where $q_1, q_2, \ldots, q_n$ denote the first $n$ primes.

The schemes in Equations (9.22) and (9.23) do not improve the given results, but they have some relevance. [Edelsbrunner and Mücke 90] presented a list of several predicates, and [Emiris and Canny 92] stated that for each predicate in the list, one of the schemes—Equation (9.20), (9.21), (9.22), or (9.23)—is applicable.

As pointed out in [Emiris and Canny 92], the presented schemes have some limitations if the corresponding predicates work on derived objects. For example, consider the case that the input objects are the endpoints of line segments and the intersections of line segments are derived objects.

An algorithm for the 2D ham sandwich problem in [Edelsbrunner and Waupotitsch 86] makes use of three simple predicates:

- deciding whether a point lies above or below a line segment,
- comparing the coordinates of two intersections,
- comparing the distances of two intersections from a line.

The scheme of Equation (9.20) perturbs all input points in general position with respect to the predicates considered in the previous sections. But Equation (9.20) is valid only for the first test, whereas Equation (9.23) is valid for the first and the second test but not for the third one.

On the other hand, some authors in the computational geometry community do not agree that symbolic perturbation always solves degeneracy problems efficiently. Some authors suggest that degenerate cases should be handled directly. For example, in [Burnikel et al. 94], three main problems of symbolic perturbations are pointed out.

- In some settings, the running time under the symbolic perturbation scheme exceeds the direct algorithmic approach.
- The post-processing step for retrieving $P(x)$ from $P(x(\epsilon))$ for degenerate input $x$ requires some effort.
- Handling degenerate cases directly is neither difficult nor time-consuming in many applications.

For example, the running time of the algorithm of [Bentley and Ottmann 79] presented in Section 9.1.1 cannot be improved under the presence of the symbolic perturbation. [Burnikel et al. 94] suggest handling degeneracy directly on the algorithmic side and present some examples for convex hull and line segment intersections. An example for an algorithmic-dependent solution is given in the next section.

### 9.5.3. Direct Perturbation

Symbolic perturbation is generic for every input. Sometimes, it is convenient to perturb the input directly or to handle degeneracy on the algorithmic side. We will present both techniques in a simple example shown in [Klein 05]. Planar sweepline algorithms often require that the input points have no common $x$-coordinate or $y$-coordinate, respectively. For example, the sweepline algorithm in Section 9.1.1 requires that the event structure has unique events. The sweep must take care that the main invariant (which guarantees correctness) holds. That is, the sweep status structure represents the line segments within the order along the sweepline; see Section 9.1.1. This can be easily guaranteed if common $x$-coordinates of endpoints of line segments are forbidden. For the same reason, the given algorithm has to take care that the intersection points of line segments should have distinct $x$-coordinates. We will address both problems using two different techniques.

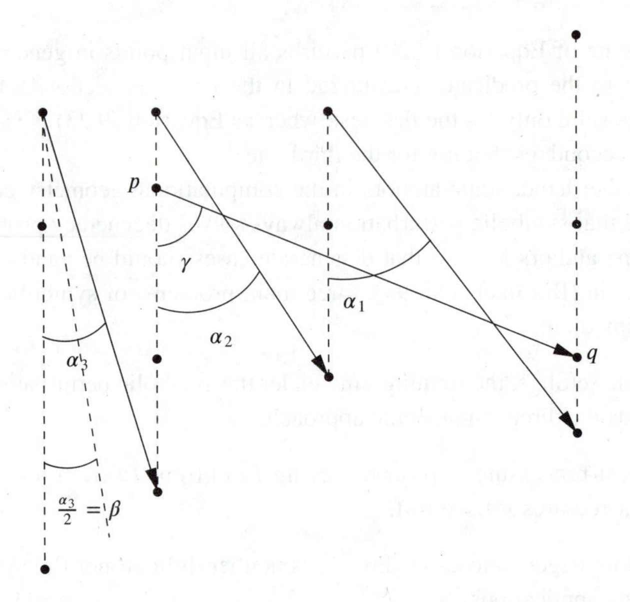


**Figure 9.20.** How to compute the rotation angle $\beta$. The angle $\gamma$ is bigger than $\alpha_1$ and $\alpha_2$ for all $p$ and $q$.

First, we can sort the input points in 2D by their $x$-coordinates and find blocks of points that run parallel to the $y$-axis; see Figure 9.20. We consider two consecutive blocks from left to right: the line passing through the highest point of the left block and the lowest point of the next block build an angle $\alpha$ with the $y$-axis. We can easily find the blocks if we sort the input points in lexicographical order.

We can rotate the coordinate system in clockwise order with angle $\beta := \frac{\alpha}{2}$, which is equivalent to a rotation of the points in counterclockwise order. That is, we apply the transformation matrix

$$\begin{pmatrix} \cos\beta & -\sin\beta \\ \sin\beta & \cos\beta \end{pmatrix}$$

to all points $(x, y)$. Obviously, this will prevent us from a degenerate situation with respect to the ordering predicate. It suffices to show that after the rotation, no two points from different blocks will have common $x$-coordinates. We consider the line $l$ passing through a point $p$ of an $y$-parallel block and an arbitrary point $q$ from a block to the right of $p$. The line $l$ builds an angle $\gamma$ bigger than $\beta$ with the (negative) $y$-axis; see Figure 9.20.

The rotation technique can be applied also to general dimensions by a transformation matrix of the form

$$\begin{pmatrix} 1 & & & & & & & & & & \\ & \ddots & & & & & & & & & \\ & & 1 & & & & & & & & \\ & & & \cos\beta & & & & -\sin\beta & & & \\ & & & & 1 & & & & & & \\ & & & & & \ddots & & & & & \\ & & & & & & 1 & & & & \\ & & & \sin\beta & & & & \cos\beta & & & \\ & & & & & & & & 1 & & \\ & & & & & & & & & \ddots & \\ & & & & & & & & & & 1 \end{pmatrix}$$

The *new* $x$ order of the input points with respect to the rotation exactly matches the lexicographical order of the points in the starting situation. Therefore, we do not have to sort the points again. Alternatively, we can use this perturbation implicitly and let the algorithm run with the lexicographical order of the endpoints.

[Gomez et al. 97] present several explicit methods for computing transformation matrices for point sets or line segment sets in 2D and 3D. They present efficient computation schemes to achieve the following non-degenerate situations:

- No two points in 2D or 3D have the same $x$-coordinate;
- No two points in 3D lie on a vertical line;
- No three points or no two line segments in 3D lie on a vertical plane.

Some attempts to compute *optimal* transformations were also made; i.e., after applying the transformation matrix, the objects are *far away* from a degenerate situation.

Additionally, in some computer graphics applications 3D objects are projected onto the 2D plane. In this case, we would like to find a non-degenerate projection. How to compute non-degenerate projections of point sets from 3D and 2D is shown, for example, in [Gomez et al. 01].

## 9.6. Imprecise Arithmetic Approach

Up to now, we tried to solve or avoid the existing problems of robustness and degeneracy for a class of geometric algorithms that are designed and analyzed in the unrealistic model of the Real RAM. Now, we turn to a completely different approach. We consider algorithms that are designed and analyzed under the presence of imprecise predicates. This means that classic results of computational geometry, such as the computation of convex hulls or Delaunay triangulations, must be reconsidered. It seems that we have to reinvent computational geometry results completely. This is the main disadvantage of the given idea.

First, we need a general framework for the use of imprecise or approximative predicates. Then we will present a classical problem solved with approximative predicates; see Section 9.6.2.

Approximative algorithms with approximative predicates in computational geometry can be found, for example, in [Fortune 92a], [Fortune 89], [Guibas et al. 93], and [Fortune and Milenkovic 91].

### 9.6.1. Epsilon Arithmetic and Approximative Predicates

Epsilon arithmetic ($\epsilon$-arithmetic) is defined in a very general sense: $\epsilon$-arithmetic is erroneous, but the errors have to be within a guaranteed error range depending on $\epsilon$. The formal definition does not specify the implementation of arithmetic operations. It is easy to see that floating-point arithmetic fulfills the definition of $\epsilon$-arithmetic.

**Definition 9.38. ($\epsilon$-arithmetic.)** $\epsilon$-arithmetic consists of at least the arithmetic operations $\oplus, \otimes, \ominus$ and $\oslash$, within a definition range $R$. The following properties hold:

- $\forall x, y \in R$ there is a $\delta < \epsilon$ such that $(x \oplus y) = (x + y)(1 + \delta)$ for $|\delta| < \epsilon$;
- Comparisons $<, \leq, =, \geq$, and $>$ are evaluated exactly.

$\epsilon$-arithmetic guarantees a relative error smaller than $\epsilon$. For example, floating-point arithmetic guarantees a relative error smaller than the corresponding machine epsilon; see Section 9.3.1.

$\epsilon$-arithmetic evokes approximative predicates; the sign of the predicate cannot be guaranteed. The *weak* predicates are defined similar to formal robustness and formal degeneracy; see Definition 9.3 and Definition 9.31.

**Definition 9.39. (Approximative predicate.)** Let $p$ be a predicate in $\mathbb{R}^{nd}$. That is, $p$ is a mapping $p : \mathbb{R}^{nd} \rightarrow B$ where $B$ represents the Boolean space {TRUE, FALSE}. $\tilde{p}$ is defined to be an approximative predicate iff

- $p(x)$ implies $\tilde{p}(x)$, and
- $\tilde{p}(x)$ implies that there is an instance $x'$ *near to* $x$ with $p(x')$.

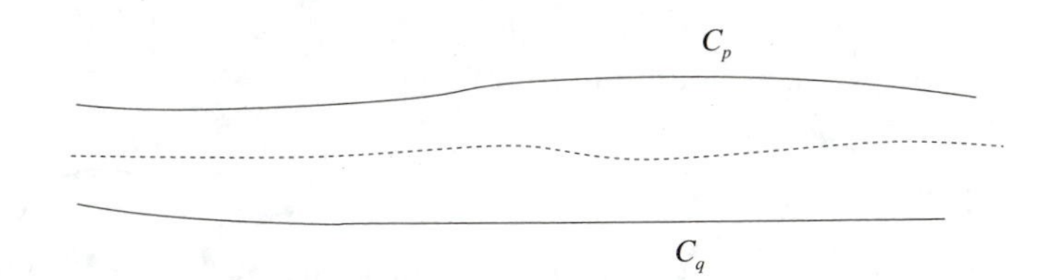


**Figure 9.21.** Two interactive approximative predicates overlap within a region.

A geometric predicate usually comes along with a disjoint *counterpart* predicate. For example, the orientation test O2D($p_1, p_2, p_3$) for three points $p_1, p_2$, and $p_3$ in 2D is positive (or TRUE) if and only if the test O2D($p_1, p_3, p_2$) is negative (or FALSE). The two predicates coincidence only for three points on a line. The line $l(p_1, p_2)$ passing through $p_1$ and $p_2$ separates the plane into two half-planes. Let $p_{1_x} < p_{2_x}$. If $p_3$ lies above $l(p_1, p_2)$, O2D($p_1, p_2, p_3$) is TRUE. If $p_3$ lies below $l(p_1, p_2)$, O2D($p_1, p_3, p_2$) is TRUE.

On the other hand, for two approximative predicates that interact in the sense indicated above, we will not be able to detect the *line* between them exactly. They will overlap for a bigger set of points. We illustrate this behavior in Figure 9.21. Assume that $\tilde{p}(x)$ is TRUE for all $x$ above the curve $C_p$ and $\tilde{q}(x)$ is TRUE for all $x$ below the curve $C_q$, and the *line* between the predicates $p$ and $q$ runs between the two curves and cannot be detected exactly. An implementation of one of the two predicates should take care that for all points $x$ within the uncertain area either $\tilde{q}(x)$ = TRUE or $\tilde{p}(x)$ = TRUE holds. Altogether, an approximative predicate has to be implemented for both predicates as follows.

**Definition 9.40. ($\epsilon$-arithmetic test.)** Let $\tilde{p}$ and $\tilde{q}$ be two approximative predicates. An *$\epsilon$-arithmetic test* for $\tilde{p}$ and $\tilde{q}$ is a Boolean program CodeP($x$) such that

- CodeP($x$) = TRUE $\Rightarrow \tilde{p}(x)$,
- CodeP($x$) = FALSE $\Rightarrow \tilde{q}(x)$.

Definition 9.40 means that the implementation of an approximative predicate takes care of its counterpart and always results in TRUE or FALSE. Note that some implementations make use of the third status UNCERTAIN, where the two approximative predicates really overlap and both predicates result in TRUE or FALSE, respectively.

In the next sections, we will present an example of an $\epsilon$-arithmetic test for approximative orientation test. An approximative incircle test predicate can be found in [Fortune 92a].

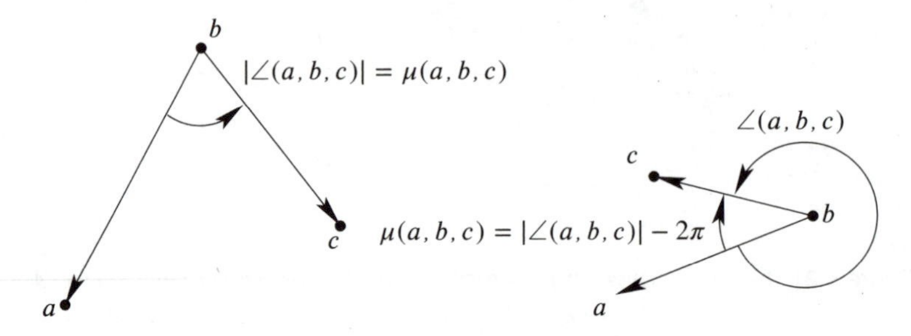


**Figure 9.22.** The angular range $\angle(a, b, c)$ and its measure $\mu(a, b, c)$.

## 9.6.2. Computing the Convex Hull

We want to design an $\epsilon$-arithmetic test for the approximative orientation test predicate and introduce some notions from [Fortune 89].

For three points $a$, $b$, and $c$ in the plane, let $\angle(a, b, c)$ denote the range between $\overrightarrow{ba}$ and $\overrightarrow{bc}$ spanned counterclockwise from $\overrightarrow{ba}$ to $\overrightarrow{bc}$; see Figure 9.22. The measure $\mu(a, b, c)$ represents the value of the complementary range of $\angle(a, b, c)$. This value can be negative or positive due to the following simple definition. If the complementary range of $\angle(a, b, c)$ has an absolute angle $\alpha$ bigger than $\frac{3\pi}{2}$, we set $\mu(a, b, c) := \alpha - 2\pi$; otherwise, $\mu(a, b, c) = \alpha > 0$; see Figure 9.22 for some examples.

**Definition 9.41. (Almost positive triangle.)** Let $\delta = D\epsilon$ for a constant $D$. The triangle $\Delta(a, b, c)$ is said to be *almost positive* if $\mu(b, a, c)$, $\mu(a, c, b)$, and $\mu(c, b, a)$ are within $[-\delta, \pi + \delta]$.

For example, the triangle $\Delta(a, b, c)$ in Figure 9.23 is almost positive if $\epsilon$ (and in turn $\delta$) is sufficiently big. If $\mu(b, a, c)$, $\mu(a, c, b)$, and $\mu(c, b, a) > 0$ hold, there is a always a real counterclockwise turn from $a$ over $b$ to $c$, or in other words, O2D$(a, b, c)$ is positive (or TRUE).

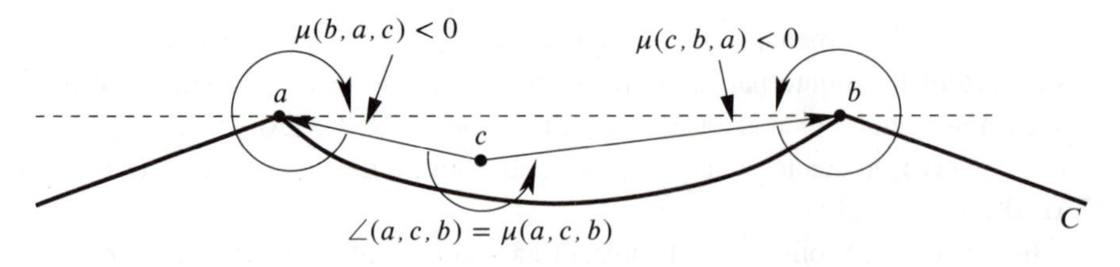


**Figure 9.23.** The triangle $\Delta(a, b, c)$ is almost positive for a fixed $\delta$. This holds for all points $c$ above the curve $C$.

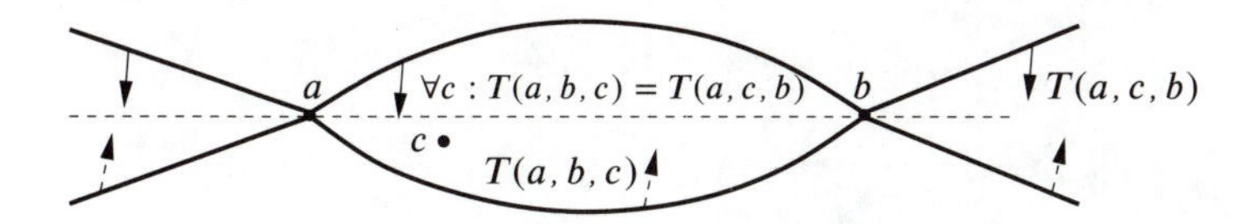


**Figure 9.24.** $\mathrm{T}(a, b, c)$ and $\mathrm{T}(a, c, b)$ overlap in the region between both curves.

If $\epsilon$ is choosen adequately, $\mu(b, a, c), \mu(a, c, b), \mu(c, b, a) \in [-\delta, \pi + \delta]$ means that there is almost a counterclockwise turn from $a$ over $b$ to $c$. For a given small $\delta$ and fixed points $a$ and $b$, the triangle $\Delta(a, b, c)$ is almost positive for all points $c$ above a continuous curve below the line passing through $a$ and $b$; see Figure 9.23. It is easy to prove that the curve starts with a line from $-\infty$ to $a$ and ends with a line from $b$ to $\infty$.

We define a simple predicate for a convex hull algorithm.

**Definition 9.42. (Approximative predicate.)** For three points $a$, $b$, and $c$ in the plane with $a_x \leq c_x \leq b_x$, let $\mathrm{T}(a, b, c)$ be an approximative predicate such that $\mathrm{T}(a, b, c) = \mathrm{TRUE}$ iff $\Delta(a, b, c)$ is almost positive.

It remains to define an $\epsilon$-arithmetic test for $\mathrm{T}(a, b, c)$ and $\mathrm{T}(a, c, b)$. The two predicates overlap between two curves, as indicated in Figure 9.24. Figure 9.24 is an instance of Figure 9.21 for the approximative orientation test.

We assume $a_x \leq c_x \leq b_x$ and $a_y \leq b_y$ for the input of the $\epsilon$-arithmetic test in Algorithm 9.10. Since comparisons can be done exactly, we can easily extend this test for a general input of $a$, $b$, and $c$, bringing the points in an appropriate position.

**Lemma 9.43.** *Let $\delta = 5\epsilon$. The procedure* TriangleTest *is an $\epsilon$-arithmetic test for $\mathrm{T}(a, b, c)$ and $\mathrm{T}(a, c, b)$.*

---

TriangleTest$(a, b, c)$ $(a_x \leq c_x \leq b_x, a_y \leq b_y)$

---

**if** $c_y \geq b_y$ **then**
  **return** TRUE
**else if** $c_y \leq a_y$ **then**
  **return** FALSE
**else if** $(a_x \ominus c_x) \otimes (b_y \ominus c_y) \ominus (a_y \ominus c_y) \otimes (b_x \ominus c_x) > 0$ **then**
  **return** TRUE
**else**
  **return** FALSE
**end if**

---

**Algorithm 9.10.** An $\epsilon$-arithmetic test for $\mathrm{T}(a, b, c)$ and $\mathrm{T}(a, c, b)$.

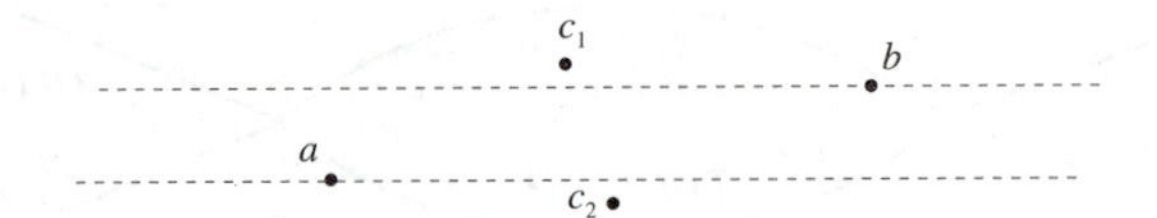


**Figure 9.25.** The point $c_1$ represents the first case in Algorithm 9.10, and $c_1$ represents the second case.

*Proof:* Figure 9.25 shows that the statement is correct for $c_y \geq b_y$ and ($c_y \leq b_y$ and $c_y \leq a_y$). It remains to show that $D := (a_x \ominus c_x) \otimes (b_y \ominus c_y) \ominus (a_y \ominus c_y) \otimes (b_x \ominus c_x) > 0$ implies $\mu(a, c, b) < \pi + \delta$. If $\mu(a, c, b) < \pi + \delta$ holds, we can conclude $\mu(b, a, c), \mu(c, b, a) \in [-\delta, \delta]$.

One can show by backward-error analysis that there is always a point $b'$ with distance $\epsilon|c - b|$ from $b$ so that $D$ achieves the correct sign for $a$, $b'$, and $c$; see Figure 9.26. Since $D > 0$ and the point $b'$ lies below the line passing through $a$ and $c$, it suffices to show that $\mu(b, c, b')$ is smaller than $\delta$. $\mu(b, c, b')$ achieves its highest value if the line passing through $c$ and $b'$ is tangent to the circle of radius $\epsilon|c - b|$ around $b$. In this case, we have $\sin(\gamma) = \epsilon$ and $\gamma \geq \mu(b, c, b')$. By simple trigonometry, we know that $\sin x < x < 5 \sin x$ for small $x$. Thus, we have $\sin(\gamma) = \epsilon = \frac{\delta}{5} < \gamma < 5\sin(\gamma) = \delta$, which finishes the proof. □

Now, we can run the following approximative upper hull algorithm on a set of $n$ points in 2D. For convenience, we show only how the left upper hull is computed. The other parts of the convex hull are computed analogously, which means that we compute a right upper hull, a right lower hull, and a left lower hull as well. First, we sort the points by increasing $x$-coordinates in $O(n \log n)$ time. Then we can compute a (upper left) $y$-monotone chain in linear time. A chain is $y$-monotone if the $y$-coordinates are increasing with respect to the $x$ order of the points. Obviously, a point that breaks the monotonicity cannot appear on the convex hull. We obtain an input as shown in Figure 9.27.

The following is the UpperLeftHull algorithm sketch:

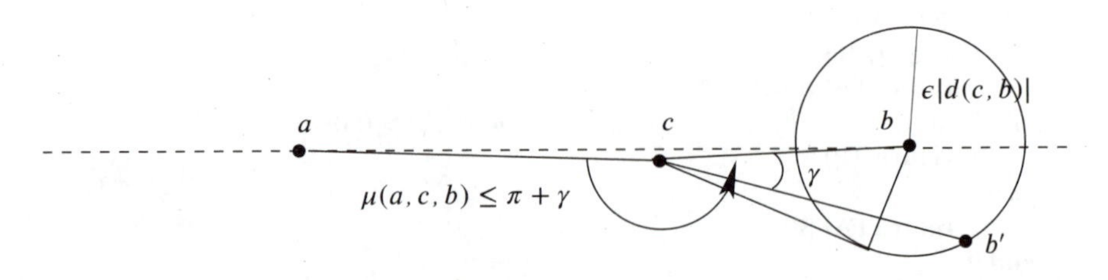


**Figure 9.26.** The angle $\gamma$ is bounded by $\gamma < 5\epsilon$.

- Insert the next point into $L$. If there are no more points, stop;
- Consider the last three elements of the current $y$-monotone point list $L$;
  - If they build an almost convex chain, repeat the process from the very beginning;
  - Otherwise, if the last three elements do not build an almost convex chain, delete the element in the middle of the three points from $L$. If there are at least three points in $L$, repeat to consider the last three elements. Otherwise, repeat from the very beginning.

The last three elements always fulfill the requirements of the TriangleTest. For example, in Figure 9.27, the algorithm may first insert $q_1$, $q_2$, $q_3$, $q_4$, and $q_5$ into $L$. Then successively the TriangleTest fails for $(q_4, q_5, q_6)$, $(q_3, q_4, q_6)$, and $(q_2, q_3, q_6)$, until the TriangleTest for $(q_1, q_2, q_6)$ yields TRUE and $q_7$ is inserted next.

**Definition 9.44. (Approximative upper hull.)** Let $q_1, q_2, \ldots, q_n$ be $n$ points in 2D sorted by $x$-coordinates. The points $q_1, q_2, \ldots, q_n$ build an *approximative upper hull* iff $\angle(q_{i-1}, q_i, q_{i+1}) \in [\pi, \pi + \delta]$.

**Lemma 9.45.** *The approximative upper hull algorithm computes an approximative upper hull in time $O(n \log n)$ for $n$ points in the plane.*

---

UpperLeftHull($q_1, q_2, \ldots, q_n$) ($q_1, q_2, \ldots, q_n$ is an $y$-monotone chain, sorted by $x$-coordinates)

---

```
L := new array
L.Insert(1, q1)
L.Insert(2, q2)
j := 2
for i := 3 TO n do
        j := j + 1
        L.Insert(j, qi)
        if j > 2 then
                while TriangleTest(L.Get(j − 2), L.Get(j − 1), L.Get(j)) = FALSE
                    do
                        L.Delete(j − 1)
                        j := j − 1
                end while
        end if
end for
```

---

**Algorithm 9.11.** Computing the upper-left hull of an $y$-monotone chain.

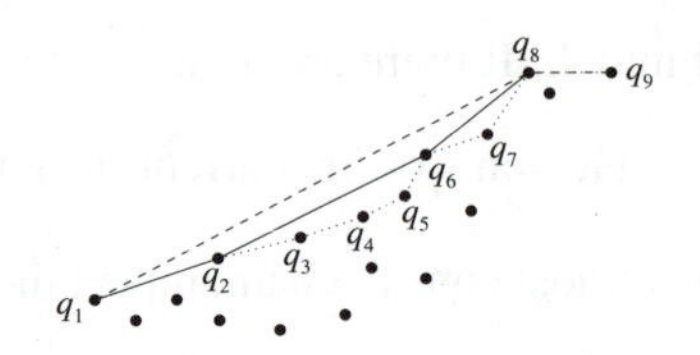


**Figure 9.27.** Points below the $y$-monotone chain $q_1, \ldots, q_9$ can never belong to the convex hull.

*Proof:* The input requirements are achieved within $O(n \log n)$ time. We successively insert the points into the current upper hull $L$. Sometimes we also delete points from $L$. We assume that $L$ is implemented like an array. These two actions determine the cost of the algorithm. Since every point is inserted or deleted only once, we have linear time complexity. □

# 9.7. Practical Recommendations and Existing Packages

So far, we have discussed several ways to address the problems of degeneracy and robustness in geometric computing. Now, we will try to categorize the given ideas and list their advantages and disadvantages. Additionally, we will give an overview of existing software packages and libraries. Practitioners may decide whether their implementation should be supported by one of the given libraries.

## 9.7.1. Imprecise and Precise Arithmetic

First, we can subdivide an algorithmic solution by imprecise or precise arithmetic usage. A precise arithmetic algorithm follows the idea of perfect arithmetic in the Real RAM model (see Section 9.3), whereas an imprecise algorithm is implemented with imprecise arithmetic, for example floating-point arithmetic (see Section 9.6).

The imprecise arithmetic approach has the following advantages:

- by construction, robustness and degeneracy problems do not exist;
- standard predicates and standard algorithms can be adapted easily.

Disadvantages of the imprecise arithmetic approach include:

- increasing time complexity;
- new inventions beyond standard predicates and algorithms necessary;

- reinvention of computational geometry;

- probably no standard libraries and software packages available.

The precise arithmetic approach has the following advantages:

- the Real RAM model allows perfect design and analysis;
- investment of more than 20 years of researcher and developer experience;
- many efficient solutions for many geometric problems;
- standard libraries and software packages available.

Disadvantages of the precise arithmetic approach include:

- the Real RAM is a theoretical model and actually not implementable;

- degeneracy and robustness must be solved additionally.

### 9.7.2. Support for EGC

Going on with Real RAM algorithms, we can further subdivide geometric algorithms by whether or not they support the EGC paradigm. The main feature in the EGC agreement is that the combinatorial computation of an algorithm is driven by geometric predicates only. EGC assumes exact predicate evaluation (due to the Real RAM), so that independently from the choice of the (exact) arithmetic, the combinatorial result is always the same. Some additional non-combinatorial results may vary with respect to the given arithmetic. See Section 9.3.4, [Yap 93], and [Yap and Dubé 95].

Non-EGC algorithms have the following advantages:

- no limitations for the design and analysis of algorithms;

- represent a bigger class of problems and algorithms.

Disadvantages of non-EGC algorithms include:

- no standardization of algorithms;

- non-flexible against the exchange of arithmetic;

- degeneracy and robustness problems must be solved internally;

- probably combinatorics and computations are not separated.

EGC algorithms have the following advantages:

- predicate-driven algorithms;
- standardization of algorithms;
- robustness and degeneracy problems are solved by definition;
- flexible arithmetic exchange;
- standardization of libraries and software packages;
- partition into the combinatorial and the computational part of the problem.

Disadvantages of EGC algorithms include:

- limited to a subclass of problems and algorithms;
- combinatorial and computational part may be separated artificially;
- shifts problems to the arithmetic packages;
- does not support direct solutions for degeneracy.

### 9.7.3. Software Packages and Libraries

Exact arithmetic for EGC can be achieved by many methods. We can distinguish between multi-precision and restricted-precision arithmetic. If the input and output are small, restricted-precision arithmetic or adaptive restricted-precision arithmetic is sufficient and fast. Otherwise, one has to choose a multi-precision package, which will increase the running time.

Computer algebra systems solve a lot of arithmetic problems efficiently. But they are designed for a more general purpose and come along with general data structures. Additionally, computer algebra systems are used in an interpreter modus and are slow against compiled solution. Finally, one has to pay for them.

General big-number packages are mainly free and they are more efficient either. Unfortunately, they are not designed solely for geometric problems. Therefore, speed-up techniques such as lazy arithmetic or floating-point filters are not implemented.

Specially designed computational geometry arithmetic packages are more efficient. The free Core Library, the free Real/Expr library, and the commercial LEDA version support the EGC paradigm by the design of geometric predicates. These three systems also support the efficient evaluation of determinants. The data type `leda real` of LEDA is made faster by lazy arithmetic, floating-point filters, and separation bounds; see Section 9.3.3. The Core Library has incorporated several separation bounds, as well [Li 01]. LiDIA and LEDA additionally contain some geometric data structures, and in LEDA, some efficient geometric algorithms are implemented. LEA is a lazy arithmetic floating-point library. All computational geometry arithmetic libraries are implemented in C++.

**Arithmetic approaches for EGC.** Restricted precision arithmetic includes:

- non-adaptive arithmetic: standard arithmetic of programming languages (Integer/Rationals);
- adaptive arithmetic: $p$-bit floating-point expansions; see Section 9.3.3;
- adaptive usage of a multi-precision arithmetic supported by many multi-precision arithmetics.

Multi-precision arithmetic includes:

- computer algebra systems
  - Maple
  - Mathematica
  - Scratchpad
  - Macsyma
- general arithmetic C++ libraries
  - BigNum [Serpette et al. 89]
  - GMP, http://www.swox.com/gmp/ [Granlund 96]
  - PARI/GP, http://pari.math.u-bordeaux.fr/ [Batut et al. 00]
  - Some other free sources for numerical computations are listed at http://cliodhna.cop.uop.edu/~hetrick/c-sources.html
- computational geometry arithmetic libraries
  - Real/Expr, http://www.cs.nyu.edu/exact/realexpr/ [Ouchi 97]
  - LEA [Benouamer et al. 93, Michelucci and Moreau 97]

  - LEDA, http://www.algorithmic-solutions.com [Mehlhorn and Näher 00]
  - LiDIA, http://www.informatik.tu-darmstadt.de/TI/LiDIA/
  - Core Library, http://www.cs.nyu.edu/exact/core/ [Karamcheti et al. 99a, Karamcheti et al. 99b]

Software packages. Finally, there are special software packages that contain data structures and algorithms. They also support the visualization and animation of the results up to some extent.

Computational geometry data structures and algorithms libraries include:

- Computational Geometry Algorithms Library (CGAL)
  - [Fabri et al. 00]
  - http://www.cgal.org/
  - implemented in C++
  - supports EGC
  - contains a lot of algorithms and data structures
  - free software
  - revision 3.1 incorporates GMP and CORE Library
  - external visualisation components can be used
  - implemented and supported by researchers and developers
  - up-to-date software
- Java Data Structures Library (JDSL)
  - [Tamassia et al. 97, Gelfand et al. 98, Goodrich and Tamassia 98, Baker et al. 99]
  - http://www.cs.brown.edu/cgc/jdsl/
  - contains GeomLib, a library for geometric applications
  - implemented in Java
  - implemented and supported by researchers and developers

  - mainly graph algorithms are implemented
  - focus is also on visualizations
  - free software
  - up-to-date software
- XYZ Geombench
  - [Nievergelt et al. 91, Schorn 90]
  - http://www.schorn.ch/geobench/XYZGeoBench.html
  - implemented in Object Pascal
  - implemented and supported by researchers and developers
  - some classical computational geometry algorithms are implemented
  - intended to support animations
  - free software
  - no longer supported
- GeoLab
  - [de Rezende and Jacometti 93a, de Rezende and Jacometti 93b]
  - http://www.cs.sunysb.edu/ algorith/implement/geolab/ implement.shtml
  - implemented in C++
  - implemented and supported by researchers and developers
  - some classic computational geometry algorithms are implemented
  - intended to support animations
  - free software
  - makes use of the XView graphics library

In summary, we highly recommend using the EGC paradigm and the CGAL library for the design of geometric algorithms. The CGAL library is well supported and contains many geometric algorithms. Practitioners can easily extend the given framework of CGAL for their own implementations. It remains to choose an appropriate supporting arithmetic package for CGAL. LEDA itself contains additional geometric data structures and algorithms, and runs well with CGAL. Unfortunately, LEDA is not available for free. Therefore, we recommend using the EGC-supporting Core Library or one of the general multi-precision arithmetic packages, for example GNU MP. Fortunately, in the latest version of CGAL, version 3.1, GNU MP and the Core Library are already incorporated.

# 10

# Dynamization of Geometric Data Structures

In this chapter, we present a generic approach for the dynamization of an *arbitrary* static geometric data structure. A simple *static* data structure may be sufficient when the set of represented geometric objects will have only few changes over time. *Static* means that you have allocated a fixed amount of space. Once created, the static structure mostly has to cope with data queries due to its geometric intention. If the set of objects varies very much over time, you need to use more complex dynamic data structures that allow efficient *insertion and deletion* of objects. It is often easy to implement a static data structure, whereas dynamic versions of the corresponding structures are more sophisticated. We present some methods to achieve dynamization by applying general transformations to the simple static data structure.

Efficient data queries are the main feature of geometric data structures. Many geometric data structures support range queries. The output is a collection of all objects within a query area $Q$. More generally, we can consider search queries. That is, we are searching for all objects that fulfill a certain property with respect to a query structure $Q$. The presented dynamization technique requires that the search query be decomposable. That is, if we split the object set into subsets, we should be able to combine the search query results for the subsets to the overall answer. For example, the nearest-neighbor search for a query point with respect to a set of given objects represents a decomposable search query. Other examples of decomposable search queries can be found in Chapter 2.

The problem of dynamization arises if object sets change over time. For example, a one-dimensional sorted array of a fixed set of objects $M$ is sufficient for simple $x$ *is element of* $M$ queries. But if the set $M$ has many changes over time, a dynamic balanced AVL tree would be more efficient. In turn, the dynamic AVL tree implementation is a bit more complicated, since rotations of the tree have to be considered for the insertion of new objects and the deletion of old objects. Additionally, the AVL-tree dynamization was *invented* for the special case of a one-dimenional search. Simple insertions and deletions into a tree can be implemented more easily.

We want to show that it is possible to dynamize a static data structure indirectly but efficiently in a simple general setting. Once this simple generic approach is implemented, it can be used for many static data structures. The necessary requirements are fulfilled for many common range query data structures. Obviously, the generic approaches cannot be optimal against a direct dynamization adapted to a single data structure, but they are easy to implement and efficient in many applications.

Generic dynamization of geometric data structures was first introduced for insertions in [Bentley 79] and [Saxe and Bentley 79]. Later on, insertions were optimally solved in [Mehlhorn and Overmars 81]. Generic dynamization for both deletion and insertion were considered in [Maurer and Ottmann 79], [van Leeuwen and Maurer 80], [van Leeuwen and Wood 80], and [Overmars and van Leeuwen 81c]. Worst-case sensitive approaches were presented in [Overmars and van Leeuwen 81a]. The main results are collected in [van Leeuwen and Overmars 81]. For details, see also [Overmars 83]. Newer results can be found in [van Kreveld 92]. A comprehensive overview in German is given in [Klein 05]. We will use the model presented there.

In Section 10.1, we start with the example of a static kd-tree implementation and discuss the methods of generic dynamization explicitly. In Section 10.2, we formalize the given problem and define some general requirements. In Section 10.3, we present the formal methods for allowing insertion and deletion in amortized efficient time. For many applications, the given dynamization technique is already efficient enough. The dynamization technique is explained in detail, and the amortized cost complexities of the new dynamic operations are shown. Similar ideas for the worst-case sensitive approach are sketched in Section 10.4.

The *effort* of the dynamization itself is amortized over time. In Section 10.5, we present examples for some other decomposable search queries and apply the presented results to their known static implementations. Many of the search structures are already implemented in the CGAL library [Overmars 96]. For practical recommendations, see also Section 9.7.

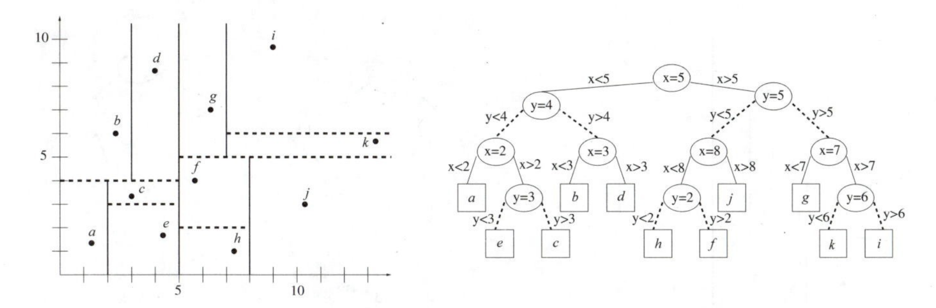


**Figure 10.1.** A single static kd-tree.

# 10.1. Example of Dynamization

We consider the balanced kd-tree introduced in Section 2.4. For a static set of $n$ points, a balanced kd-tree can be easily built by allocating space for a binary tree with $2^k$ leaves, where $2^k - 1 < n \leq 2^k$. For example, in Figure 10.1, a balanced kd-tree for a set of 11 points is given. The balanced kd-tree supports efficient range queries for rectangular ranges $Q$.

We apply a general dynamization technique. That is, we implement the dynamic insert and delete operations of the given data structure without any knowledge of its internal representation, thus avoiding complicated special rebalancing operations. Additionally, the cost of query operations should not be affected significantly. The main idea is that we transform the given problem adequately and use efficient algorithms for the full static structure.

## 10.1.1. Amortizing Kd-Tree Insert Operations Over Time

First, we will discuss the generic dynamization of insert operations and will try to demonstrate that spreading work over time is a good paradigm. Assume that a new point $l = (5.5, 8.5)$ has to be inserted into the kd-tree of Figure 10.2. We can simply insert $l$ by following the path to the corresponding rectangular region of $k$. The region is split accordingly and a new node is allocated and appended to the tree; see Figure 10.3. After some of these simple insertions, the tree is no longer balanced, but it is almost balanced. For a while, such simple insertions will not charge the query operations and insert operations significantly.

On the other hand, *many* simple insertions may unbalance the tree significantly. In this case, insert and query operations will require more effort. Therefore, after a while, we should reorganize and rebalance the tree, if we still want to perform efficient insert operations and search queries. We apply an amortized

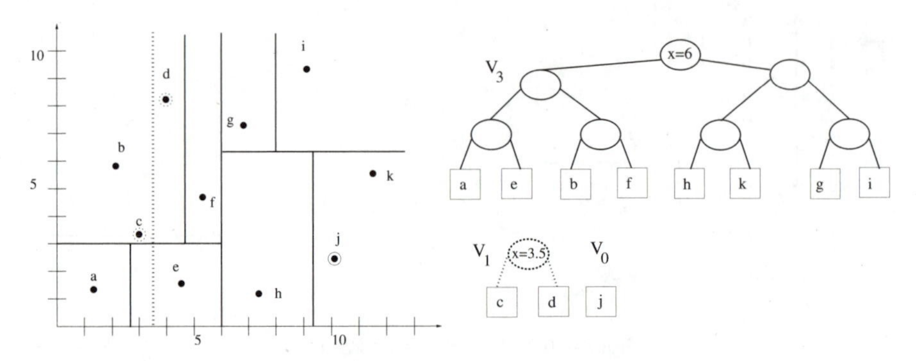


**Figure 10.2.** A set of static kd-trees.

cost model. The reorganization is costly but does not happen very often. We distribute the cost over the sequence of all operations that happens between two reorganizations. The amortized cost model is formally explained in Section 10.2.

Amortizing the cost over a sequence of operations is most promising if the cost for the rebalance operations is small. Since we are restricted to the use of the static construction algorithm of the data structure, we try to avoid reconstruction of the whole data structure. Accordingly, we transform the static structure into a set of balanced static structures and apply occasional reconstructions only to subsets of these structures. For example, in Figure 10.1, a balanced kd-tree for a set of 11 points is given, whereas in Figure 10.2, we consider a set of different static balanced kd-trees for the same set of points.

A simple insert operation may affect only small subsets. For example, if the new point $l = (5.5, 8.5)$ has to be inserted into the set of kd-trees, we would like to insert $l$ into one of the smaller structures.

In the next section, we will specify the decomposition more precisely.

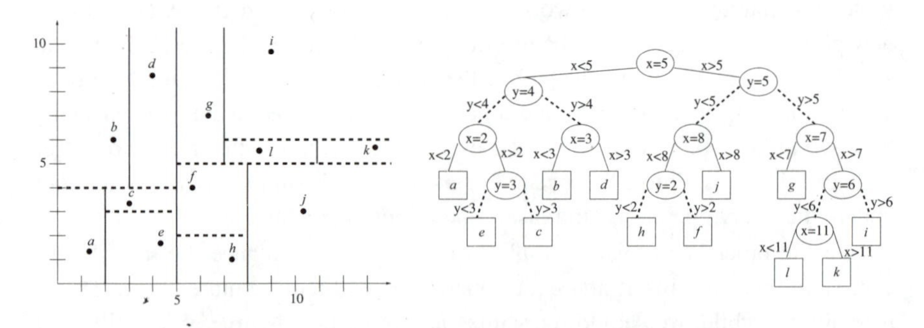


**Figure 10.3.** Inserting a single object into a static kd-tree.

## 10.1.2. A Binary Decomposition of Static Kd-Trees

Let us assume that we start with a fixed set of points. We need a general rule for subdividing a single static kd-tree into a set of static structures. Many ideas for such decomposition were presented in the beginning of the 1980s. For example, Maurer and Ottmann [Maurer and Ottmann 79] introduced an equal block method, which spread $n$ objects over $k$ structures $V_1, V_2, \ldots, V_k$ with $|V_j| \leq \frac{2n}{k}$. Thus, every $V_j$ has some room for simple insertions. This means that, in general, a simple *weak* insertion technique has to be implemented.

We will present an efficient decomposition that goes back to [Bentley 79]. As you will see, *weak* insertions are not necessary. The main idea is that we use the binary representation of the number of points. For example, the binary representation of 11 is given by 1011, which represents a linear combination

$$11 = 1 \cdot 2^3 + 0 \cdot 2^2 + 1 \cdot 2^1 + 1 \cdot 2^0$$

and results in the corresponding three substructures $|V_0| = 1$, $|V_1| = 2$, and $|V_3| = 8$; see Figure 10.2. In general, for $n$ points in the plane and the *binary representation*

$$n = a_l 2^l + a_{l-1} 2^{l-1} + \ldots + a_1 2 + a_0 \text{ mit } a_i \in \{0, 1\},$$

there is always a unique *binary decomposition* $W_n$, given as a set of kd-trees $\{V_i : a_i = 1\}$. Every $V_i$ contains exactly $2^i$ objects, and every object is contained in uniquely one $V_i$.

As previously mentioned, we would like to insert new elements in smaller structures. The main advantage is that, due to the binary decomposition, the kd-tree $V_i$ with $2^i$ elements remains valid until the number of objects of the lower-order structures $V_0, V_1, V_2, \ldots, V_{i-1}$ will exactly sum up to $2^i - 1$ elements and

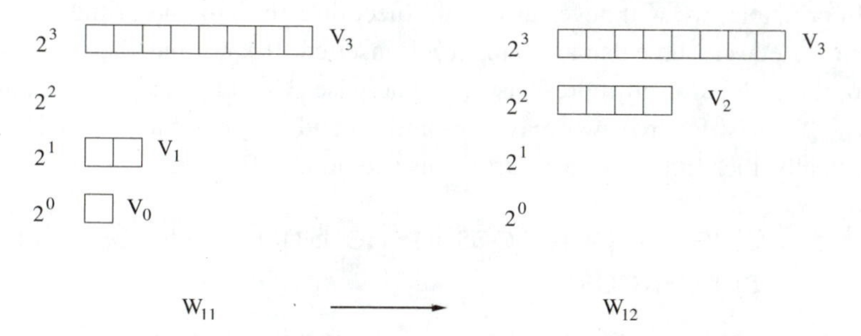


**Figure 10.4.** A change in the binary decomposition from $W_{11}$ to $W_{12}$.

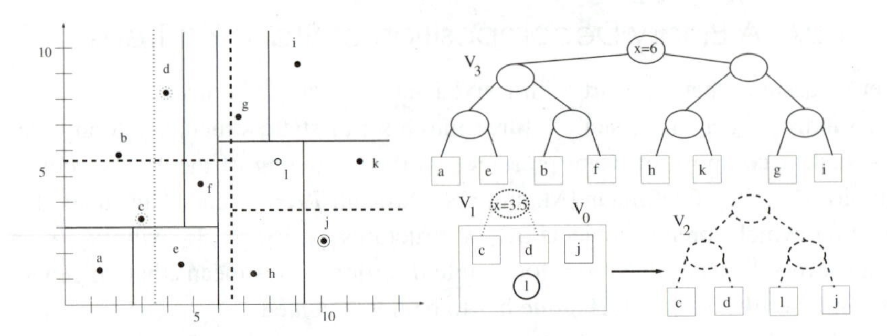


**Figure 10.5.** After insertion of $l$, we build $V_2$ from $V_1$, $V_0$, and $l$.

an additional element $p$ has to be inserted; i.e., there is room for $2^i - 1$ elements below $V_i$. If the insertion of $p$ happens, we have to construct $V_{i+1}$ out of $V_0, V_1, V_2, \ldots, V_{i-1}$ and $p$. For clarification, let us further assume that with respect to the binary decomposition, the structures $V_0, V_1, V_2, \ldots, V_{i-1}$ and $V_{i+1}$ are empty. Thus, we can insert $\sum_{j=1}^{i-1} 2^i = 2^i - 1$ elements without affecting $V_i$.

For example, if we insert $l = (5.5, 8.5)$, the number of elements below $V_2$ rise to $|V_1| + |V_0| + 1 = 2^2$ (see Figure 10.4), and we reconstruct $V_1$, $V_0$, and $l$ to $V_2$ (see Figure 10.5). Note that the binary representation of the current number of elements $12 = 1 \cdot 2^3 + 1 \cdot 2^2 + 0 \cdot 2^1 + 0 \cdot 2^0$ exactly represents the distribution of the points in the kd-tree $V_2$ and $V_3$.

According to the amortization idea just presented, a reconstruction operation for $V_i$ occurs if exactly $2^i$ insert operations are done. Additionally, the number of objects for constructing $V_{i+1}$ from $V_0, V_1, V_2, \ldots, V_{i-1}, V_i$ is also in $O(2^i)$. Thus, we can amortize the reconstruction cost for $V_{i+1}$ by $O(2^i)$ insert operations. We divide the cost for building $V_{i+1}$ by $2^i$; for details see Section 10.3.1. The binary decomposition goes back to [Bentley 79] and is denoted also as the *logarithmic method*.

In principle, we will never insert an object directly into one of the substructures $V_j$. Let us assume that an element $p$ is inserted. If $V_0$ is free, there is nothing to do; the element itself represents $V_0$. Otherwise, $V_0$ and $p$ may be combinend to $V_2$. If $V_2$ is not free, we may combine $V_0$ and $p$ and $V_2$ to $V_3$, and so on. Technically, there are only reconstructions and no insertions at all.

### 10.1.3. Query Operations in the Binary Representation of Kd-Trees

We still have to guarantee that the binary decomposition of a kd-tree supports efficient data queries. We require that the query is *decomposable*. That is, the set

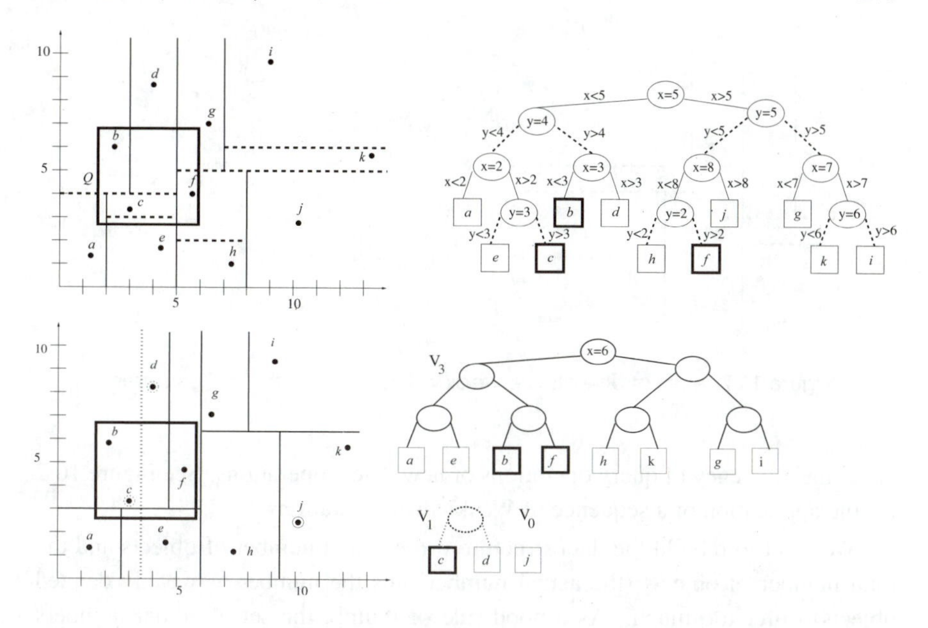


**Figure 10.6.** A kd-tree query is decomposable.

of answers resulting from the queries to any kd-tree in the set of kd-trees can be used for constructing the answer to the single static tree.

Fortunately, the requirement holds for many range query data structures and also for general search query data structures. For example, in Figure 10.6, a rectangular range query is answered by the union of rectangular range queries. With respect to the binary decomposition, we have to combine the answer of $O(\log n)$ substructures. Therefore, the running time of the query will rise within a factor of $O(\log n)$.

### 10.1.4. Generic Delete Operations for a Kd-Tree by the Half-Size Rule

Reconstruction and rebalancing in a full dynamic kd-tree may also be required because of delete operations. For a generic implementation of delete operations, we follow a similar idea as presented in the previous section. We discuss the following idea, which goes back to [van Leeuwen and Maurer 80]. For a while, we perform simple delete operations—that is, we do not delete an element physically; we only mark an element as deleted. This weak delete operation is denoted as WeakDelete, and for a while, the number of *dead* elements in the tree will not

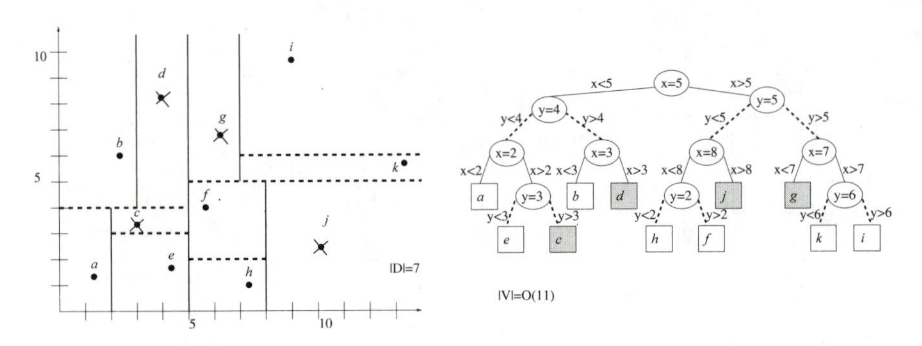


**Figure 10.7.** A set of objects are weakly deleted from the static kd-tree.

affect the efficiency of query operations or new delete operations. See Figure 10.7 for the application of a sequence of WeakDelete operations.

We want to rebuild the data structure if the actual number of objects and the total number of objects (the actual number plus the number of weakly deleted objects) differ too much. As a good rule of thumb, the set of actual numbers should be at least as big as the set of weakly deleted objects. Therefore, we reconstruct the static data structure if the number of weakly deleted objects equals or exceeds the number of real objects. This method can be denoted as the *half-size rule*. Obviously, a query operation can easily filter out the needed objects.

We can spread the cost for the reconstruction of $V$ by $\frac{1}{2}|V|$ WeakDelete operations. Additionally, we have to account for the WeakDelete operation itself; see Section 10.3.2 for details.

Up to now, we have considered insert and delete operations independently. Obviously, a sequence of operations will contain both insert and delete operations (and query operations also). In this case, the idea of using a binary decomposition for handling insertions and applying the half-size rule for a set of WeakDelete operations must interact. [Overmars and van Leeuwen 81c] showed that for a special class of decomposable search queries, one can handle delete operations directly within the binary decomposition.

## 10.1.5. Half-Size Rule and Binary Decomposition of a Kd-Tree

Finally, we should be able to combine the methods of the preceding sections. Performing a WeakDelete operation of an element $p$ in a single kd-tree is not difficult. The weakly deleted object has to be find and marked.

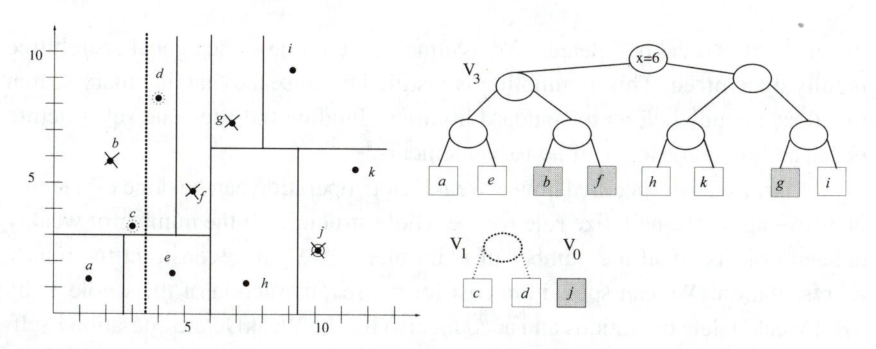


**Figure 10.8.** Some WeakDelete operations in the binary decomposition of a static kd-tree.

In the binary decomposition, we first have to find the corresponding structure $V_j$ with $p \in V_j$. This search problem is solved with a standard one-dimensional balanced search tree; see Figure 2.7. Every object is represented by a unique ID, so that we can compare two objects efficiently. Every element in the search tree has a pointer to its current substructure. For example, let us assume that the elements in Figure 10.8 can be identified by $\text{ID}(a) = 1, \text{ID}(b) = 2, \ldots, \text{ID}(k) = 11$. The corresponding balanced binary search tree is shown in Figure 10.9. For every ID, there is a pointer to the corresponding substructure. In the substructure,

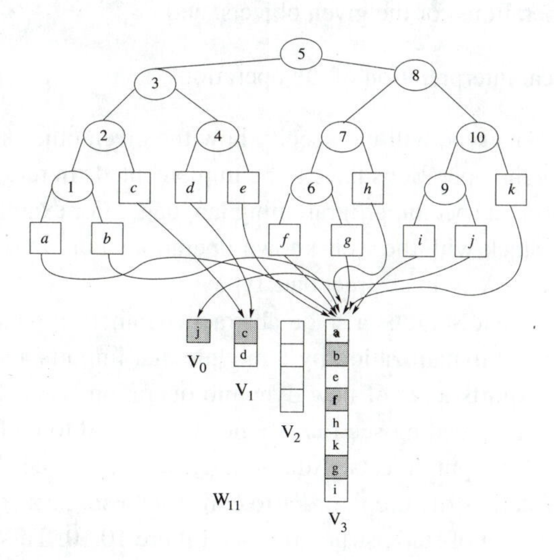


**Figure 10.9.** The balanced binary search tree for 11 elements. The entries have a pointer to the corresponding substructure.

the element is weakly deleted. We assume that the one-dimensional search tree is fully dynamized. This assumption is justified because a dynamic binary search tree does already belong to standard libraries. Finding the adequate substructure needs additional $O(\log n)$ time for $n$ elements.

So far, we have specified how a WeakDelete operation can be done efficiently. Now, we apply the half-size rule for the whole structure. If the number of weakly deleted objects equal the number of real objects, we will reconstruct the binary representation. We can spread the cost for the reconstruction of the whole $V$ by $\frac{1}{2}|V|$ WeakDelete operations and account also for the WeakDelete operation itself. The use of the binary tree gives a summand of $O(\log n)$; see Section 10.3.3 for details.

## 10.2. Model of the Dynamization

In this section, we will formalize the presented ideas in a very general setting. Thus, we need a data structure description independent from a special problem. We use the notion of an abstract (geometric) data type. An abstract data type Static is defined by the following:

- a set of objects,
- a set of operations for the given objects, and
- a semantical interpretation of the operations.

The abstract data type will not specify how the given objects are represented internally or how the operations have to be implemented. In turn, the description is independent from a specific programming language. For example, the standard description of a Stack with the well-known operations Pop, Push, Top, and New over a set of objects $X$ is an abstract data type.

Let us assume that Static is a static abstract (geometric) data type. We want to define the generic dynamization by a module that imports a set of operations from Static and exports a set of new dynamic operations. As shown in the kd-tree example in the preceding section, we need a method to build a single static structure for a set of input objects. Additionally, we have to be able to collect all elements from a static structure in order to construct a single static data structure from the input of a set of static structures; see Figure 10.10. This also means that we need a method for deleting a given static structure. Furthermore, since we have a geometric data structure, we can assume that there is a query operation on

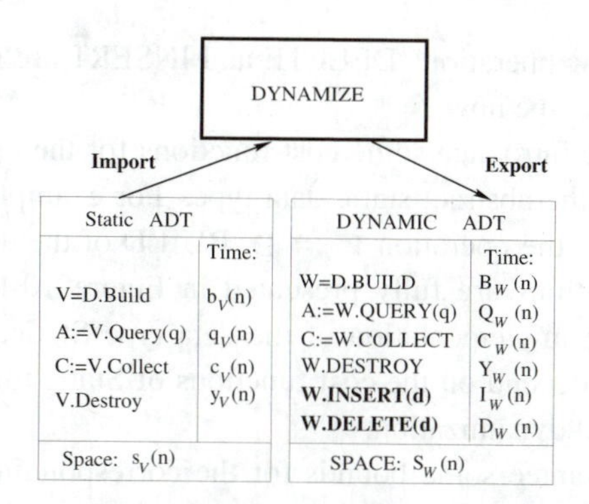


**Figure 10.10.** Dynamization in a generic sense. (Redrawn and adapted courtesy of Rolf Klein.)

the set of stored geometric objects D in Static. That is, for a query object $q$, the answer is a subset of D. Note that the answer might be empty.

Altogether, we can assume that the following operations are given.

| | |
|---|---|
| $V$ := D . Build: | Build the structure $V$ of type Static with all data objects in the set D. |
| $A$ := $V$ . Query($q$): | Return the answer $A$ (objects of D) to a query to $V$ with query object $q$. |
| $C$ := $V$ . Collect: | Collect all data objects of $V$ in $C$ (objects of D represented in $V$). |
| $V$ . Destroy: | Delete the complete data structure $V$ from storage. |

The dynamization module should export a dynamic abstract (geometric) data type DYNAMIC with the same set of operations. Since we export a dynamic data structure, additional operations for insert and delete are made available.

| | |
|---|---|
| $W$ := D . BUILD: | Build the structure $W$ of type DYNAMIC with data objects of the set D. |
| $A$ := $W$ . QUERY($q$): | Return the answer $A$ (objects of D) to a query to $W$ with query object $q$. |
| $C$ := $W$ . COLLECT: | Collect all data objects of $W$ in $C$ (objects of D represented in $W$). |
| $W$ . DESTROY: | Delete the complete data structure $W$ from storage. |
| $W$ . INSERT(d): | Insert object d into $W$. |
| $W$ . DELETE(d): | Delete d from $W$. |

Note that the *new* operations DELETE and INSERT are necessary, since we have a dynamic data type now.

Additionally, we introduce some cost functions for the operations of the abstract dynamic and the abstract static data type. For example, let $b_V(n)$ denote the time function for the operation $V$ := D.BUILD of the data type Static with $|D| = n$. The notions are fully presented in Figure 10.10. The cost functions depend on the implementation of the data type Static. The cost function of DYNAMIC will depend on the cost functions of Static together with the efficiency of the general dynamization.

In order to guarantee some bounds for the corresponding cost functions of DYNAMIC, the cost functions of Static have to behave well and run within certain bounds. Additionally, for the proofs of the time bounds, we need some kind of monotonic behavior of the cost functions. The functions should not oscillate. Altogether, we define the following simple requirements to the static data structure and its cost functions.

1. The operations Query and Destroy become more time-consuming if the object set grows. This means that the functions $q_V(n)$ and $y_V(n)$ are non-decreasing in $n$. Example functions: 1, $\log n$, $\sqrt{n}$, $n$, $n \log n$, $n^2$, $2^n$.

2. The Build operation and the needed space for $n$ objects lies in $\Omega(n)$. This means that the functions $b_V(n)$ and $s_V(n)$ are at least linear in $n$. Example functions: $n$, $n \log n$, $n^2$, $2^n$.

3. All operations do not change their behavior significantly if the set of object grows only by a constant factor. This means that all functions $f \in \{q_V, b_V, c_V, s_V, y_V\}$ are somehow smooth. That is, there is a constant $K$, so that $f(cn) \leq K\ f(n)$ for a constant $c > 1$. Example functions: 1, $\sqrt{n}$, $n$, $n^2$, and also $\log n$ with $n > 1$, as well as the products of these functions, but not $2^n$!

4. Collecting all elements needs at most twice the rebuilding time. This means that we require $c_V(n) \leq 2 \cdot b_V(n)$.

5. Moreover, we assume that the Query operation is decomposable; i.e., for a decomposition $V = V_1 \cup V_2 \cup \cdots \cup V_j$ of the data set $V$, the results of the single operations $V_i$.Query($q$) lead to the solution of $V$.Query(d). This holds for many kinds of search queries.

We consider an amortized cost model. Let Work be an arbitrary operation with cost function h. For a sequence of $t$ different operations, let Work be applied

$k$ times. That is, among the $t$ operations there are only $k$ Work operations. If

$$\frac{\text{total cost of } k \text{ Work operationen}}{k} \leq \overline{\mathrm{h}}(t)$$

holds for a cost function $\overline{\mathrm{h}}$, we say that the operation Work is performed in *amortized time* $\overline{\mathrm{h}}(t)$.

Note that this is not an expected value and that $\overline{\mathrm{h}}$ is a function of $t$, i.e., the length of the operation sequence. The current data set may have a number of elements $n \leq t$. The worst-case cost of all operations should be given by functions in $n$.

# 10.3. Amortized Insert and Delete

## 10.3.1. Amortized Insert: Binary Structure

We want to implement the operation $W$.INSERT(d) efficiently. First, we have to compute $W$. Therefore we distribute the $n$ data objects of the static structure $V$ among several structures $V_i$. If a new element has to be inserted, we hope that only a single structure $V_i$ may be concerned. Let

$$n = a_l 2^l + a_{l-1} 2^{l-1} + \ldots + a_1 2 + a_0 \text{ mit } a_i \in \{0, 1\}.$$

Then $a_l a_{l-1} \ldots a_1 a_0$ is the *binary representation* of $n$. For every $a_i = 1$, we build a structure $V_i$ that has $2^i$ elements. The collection of these structures is a representation of $W$ that is called the *binary decomposition* (see Figure 10.4). To build up the binary decomposition $W$, we proceed as follows. In Algorithm 10.1 we use array entries $V[i]$ to represent $V_i$.

The following is the BUILD(D) sketch.

- Compute the *binary representation* of $n = |\mathrm{D}|$.
- Decompose D into sets $\mathrm{D}_i$ with $|\mathrm{D}_i| = 2^i$ w.r.t. the representation of $n$.
- Compute $V_i := \mathrm{D}_i$ . Build for every $\mathrm{D}_i$.
- Return a pointer to an array of pointers to $V_i$.

Note that in Algorithm 10.1, the set $Bin$ represents the binary decomposition of $n$ by a sorted set of indices and provides access to the $O(\log n)$ filled components of the array $V[]$. For example, for $n = 1010011$, we will have

---

$W$ D . BinDecomp (D is a set of objects)

---

```
n := |D|
Bin := n. BinRep
V := new array(Bin . Last)
BinCopy := Bin . BinCopy
while BinCopy ≠ ∅ do
        i := BinCopy . First
        BinCopy . DeleteFirst
        Subset := D . First(2^i)
        V[i] := Subset . Build
        D := D . DeleteFirst(Subset)
end while
return W = (V[], Bin)
```

---

**Algorithm 10.1.** Computing the binary decomposition of a set of objects D.

$Bin = \{0, 1, 4, 6\}$. In this case, we have $Bin$ . Last $= 6$. Additionally, as mentioned earlier we make use of self-explanatory list and array manipulation. For example, $L$. First($n$) collects the first $n$ elements of list $L$, whereas $L$. Delete First($n$) deletes the first $n$ elements.

In principle, the binary decomposition $W$ can be constructed as quick as the corresponding structure $V$.

**Lemma 10.1.**

$$\mathrm{B}_W(n) \in O(b_V(n)).$$

*Proof:* Computing the binary representation of $n$ and the decomposition into $D_i$ can be done in linear time $O(n)$.

The operation $\mathrm{D}_i$ . Build needs $\mathrm{b}_V(2^i)$ time. We have $i \leq l = \lfloor \log n \rfloor$, and therefore we conclude:

$$\begin{aligned}\sum_{i=0}^{\lfloor \log n \rfloor} \mathrm{b}_V(2^i) &= \sum_{i=0}^{\lfloor \log n \rfloor} 2^i \frac{\mathrm{b}_V(2^i)}{2^i} \leq \sum_{i=0}^{\lfloor \log n \rfloor} 2^i \frac{\mathrm{b}_V(n)}{n} \\ &\leq 2^{\log n} \frac{\mathrm{b}_V(n)}{n} \in O(\mathrm{b}_V(n)).\end{aligned}$$

We used the fact that $\frac{\mathrm{b}_V(n)}{n}$ increases monotonically.

Altogether, we have

$$\mathrm{b}_W(n) \in O(n + \mathrm{b}_V(n)) = O(\mathrm{b}_V(n)),$$

since $\mathrm{b}_V(n)$ is at least linear. □

---

$A$ $W$.COLLECT($W$) ($W = (V[\,], Bin)$ represents the binary decomposition)

---

```
V[] := W.DataArray
Bin := W.BinRep
A := new list
while Bin ≠ ∅ do
        i := Bin.First
        A.ListAdd(V[i].Collect)
        Bin.DeleteFirst
end while
return A
```

**Algorithm 10.2.** Collecting all elements of $W$.

---

$A$ $W$.QUERY($q$) ($W = (V[\,], Bin)$ represents the binary decomposition, $q$ is a query)

---

```
V[] := W.DataArray
Bin := W.BinRep
A := new list
while Bin ≠ ∅ do
        i := Bin.First
        A.ListAdd(V[i].Query(q))
        Bin.DeleteFirst
end while
return A
```

**Algorithm 10.3.** Computing the query for the binary decomposition $W$.

---

$W$.DESTROY ($W = (V[\,], Bin)$ represents the binary decomposition)

---

```
V[] := W.DataArray
Bin := W.BinRep
while Bin ≠ ∅ do
        i := Bin.First
        V[i].Destroy
        Bin.DeleteFirst
end while
V[].Deallocate
Bin.Deallocate
W.Deallocate
```

**Algorithm 10.4.** Destroying the structure $W$.

Similar results hold for some other operations. The operations are designed in a straightforward manner, and we omit the sketches here; Algorithm 10.2, Algorithm 10.3, and Algorithm 10.4 are self-explanatory. The insert operation and its analysis will be explained in more detail now.

First, we can prove

$$\mathrm{C}_W(n) \leq \log n \;\; \mathrm{c}_V(n),$$

simply by applying $V_i$ . Collect for at most $\log n$ structures $V_i$; see Algorithm 10.2.

By the same argument, we have

$$\mathrm{Y}_W(n) \leq \log n \;\; \mathrm{y}_V(n).$$

Additionally,

$$\mathrm{Q}_W(n) \leq \log n \; \mathrm{q}_V(n)$$

holds if we assume that the query is decomposable; see Algorithm 10.3.

It is easy to see that

$$\mathrm{S}_W(n) \leq \sum_{i=0}^{\lfloor \log n \rfloor} \mathrm{s}_V(2^i) \in O(\mathrm{s}_V(n)),$$

which means that we do not need more space. Finally, we have to analyze the insert operation $\mathrm{I}_W(n)$ in the amortized cost model.

As we have seen from Figure 10.4, sometimes the whole structure of $W_n$ is destroyed when $W_{n+1}$ has to be built up. In the situation of Figure 10.4, we would like to perform the following construction steps:

$$\begin{array}{l} \mathrm{D}_0 := V_0\,.\,\mathrm{Collect}; \mathrm{D}_1 := V_1\,.\,\mathrm{Collect}; \mathrm{D} := \mathrm{D}_0 \cup \mathrm{D}_1 \cup \{\mathrm{d}\}; \\ V_2 := \mathrm{D}\,.\,\mathrm{Build}; \end{array}$$

In general, we have to build up $V_j$ and extract and erase $V_{j-1}, V_{j-2}, \ldots, V_0$ only if $a_i = 1$ holds for $i = 0, 1 \ldots, j-1$ and $a_j = 0$ holds (for the binary representation of the current $n$). Algorithm 10.5 also has to update the index set $\boldsymbol{Bin}$ adequately. For example, for $n = 1010011$, we have $\boldsymbol{Bin}\{0, 1, 4, 6\}$. Inserting a new element gives $\boldsymbol{Bin} = \{2, 4, 6\}$.

In this special situation, we can prove

$$\begin{array}{rcl} \mathrm{I}_W(n) & \leq & \left( \sum_{i=0}^{j-1} \mathrm{c}_V(2^i) \right) + K \cdot j + \mathrm{b}_V(2^j) \\ & \leq & \mathrm{c}_V(2^j) + K \cdot j + \mathrm{b}_V(2^j) \\ & \in & O\left(\mathrm{b}_V(2^j)\right). \end{array}$$

where $K$ is a constant. The effort $K \cdot j$ stands for combining the collected elements. The rest stems from the properties of the cost functions; see the third function property listed in Section 10.2.

For a long sequence of insertions, many of the insert operations are performed without extreme reconstructions. The effort for all $W$. INSERT(d) is amortized over time. For the $W$. INSERT(d) operation, we can prove

$$\overline{\mathrm{I}_W}(t) \in O\left(\frac{\log t}{t}\, \mathrm{b}_V(t)\right).$$

The result stems from summing up the reconstruction cost for all $V_i$ structures. $V_0$ is reconstructed after every insertion, $V_1$ is reconstructed after every second insertion, and generally $V_i$ is reconstructed after every $2^i$ insertions. For a sequence of $n$ insertions, the total insertion cost is

$$\sum_{i=1}^{n} \mathrm{I}_W(n) = \sum_{j=0}^{\log n} \frac{n}{2^j}\, \mathrm{b}_V(2^j) \in O(\log n \cdot \mathrm{b}_V(n)).$$

The average over $n$ gives the result above. Note that apart from insertions, there are only queries in $t$. Queries do not change the size of the data set, and so we can replace $t$ by the number of insertions $n$ here.

Altogether, the results are presented in the following theorem.

**Theorem 10.2.** *A static abstract data type* Static *as presented in Figure 10.10 can be dynamized by means of the binary decomposition in a dynamic abstract data type* DYNAMIC *so that the operation* $W$. INSERT(d) *is performed in amortized time*

$$\overline{\mathrm{I}_W}(t) \in O\left(\frac{\log t}{t}\, \mathrm{b}_V(t)\right),$$

*where t denotes the length of the operation sequence.*

*Let n be the size of the data set. For the remaining operations* BUILD, QUERY, DESTROY, *and* COLLECT *and the needed storage, we will achieve*

$$\begin{aligned}
\mathrm{S}_W(n) &\in O(\log n\ \mathrm{s}_V(n)),\\
\mathrm{C}_W(n) &\in O(\log n\ \mathrm{c}_V(n)),\\
\mathrm{Y}_W(n) &\in O(\log n\ \mathrm{y}_V(n)),\\
\mathrm{B}_W(n) &\in O(\mathrm{b}_V(n)),\\
\mathrm{Q}_W(n) &\in O(\log n\ \mathrm{q}_V(n)).
\end{aligned}$$

$W$.INSERT($d$) ($W = (V[], Bin)$ represents the binary decomposition, $d$ is an element)

```
W[] := W.DataArray
Bin := W.BinRep
j := 0
D = new list
while Bin ≠ ∅ and j = First(Bin) do
        D.ListAdd(V[j].Collect)
        V[j] := Nil
        Bin.DeleteFirst
        j := j + 1
end while
D.ListAdd({d})
V[j] := D.Build
Bin.AddFirst(j)
```

**Algorithm 10.5.** Inserting an element into the binary decomposition.

### 10.3.2. Amortized Delete: Half-Size Rule

Assume that we have not implemented the INSERT operation yet. If we have to delete an object, we cannot predict its location in $V$ in advance. Additionally, deletion may cause fundamental reconstruction and is much more difficult than a simple insertion.

For many data structures, it is easier to mark an object as deleted. Physically, the object remains in the structure but no longer belongs to the data set $D$. Such weakly deleted objects may have some influence on the running time of all operations. Therefore, from time to time, we would like to reconstruct the data structure for the actual data set $D$.

First, for Static we introduce an additional operation $V$ . WeakDelete(d) with cost function $\text{wd}_V(n)$. We want to construct a *strong* delete function DELETE with an acceptable amortized time bound for DYNAMIC. The WeakDelete operation strongly depends on the corresponding data structure.

We will use $V$. WeakDelete($d$) until $D$ has only the half-size of $V$. Then we erase $V$ and build a new structure $V$ out of $D$. Practically, we can make use of two counters, ActElements and WeakDeleteElements. After every WeakDelete operation, we decrease ActElements and increase WeakDeleteElements. If the counters have the same size, we collect all *real* elements and rebuild the data structure from scratch. Since the algorithm is trivial, we do not present it in pseudocode. Furthermore, we integrate the DELETE operation into Algorithm 10.6 in the next section.

The cost of the *half-size rule* is amortized over the preceding WeakDelete operations as follows. We can spread the cost for the reconstruction of $V$ by $\frac{1}{2}|V|$ WeakDelete operations. Additionally, we have to account for the WeakDelete operation itself. This means that the amortized cost of the DELETE operation is given by at most $\frac{b_V(t)}{t} + \mathrm{wd}_V(t)$ for a sequence of $t$ operations. Since the number of all data objects is at most two times the number of real data, we can apply the third function property listed in Section 10.2: $f(2n) \in O(f(n))$. Consequently, all other operations are not affected, and the running time can be considered for the actual data set. This gives the following general result.

**Theorem 10.3.** *A static abstract data type as presented in Section 10.2 with an additional operation* $V$.WeakDelete($d$) *and with additional cost function* $WD_V(n)$ *can be dynamized by means of* occasional reconstruction *in a dynamic abstract data type* $TDyn$*, so that*

$$\begin{aligned} \mathrm{B}_W(r) &\in O(b_V(r)), \\ \mathrm{C}_W(r) &\in O(e_V(r)), \\ \mathrm{Q}_W(r) &\in O(\mathrm{q}_V(r)), \\ \mathrm{S}_W(r) &\in O(\mathrm{s}_V(r)), \\ \overline{D_W}(t) &\in O\left(\mathrm{wd}_V(t) + \frac{b_V(t)}{t}\right) \end{aligned}$$

*holds. The size of the actual data set is denoted with* $r$ *and* $t$ *denotes the length of the operation sequence.*

### 10.3.3. Amortized Insert and Amortized Delete

In the preceding sections, we discussed INSERT and DELETE separately. Now, we want to show how to combine the two approaches.

A static abstract data type with a WeakDelete implementation is given. As in Section 10.3.1, we use the binary decomposition for the insertion. The WeakDelete operation is available only for the structures $V_i$, and we have to extend it to $W$ to apply the result of Section 10.3.2. If $W$.DELETE(d) is applied, $d$ should be marked as deleted in $W$, rather than physically deleted from storage. But we do not know in which of the structures $V_i$ the element $d$ lies. Therefore, in addition to the binary decomposition, we construct a balanced *search tree* $T$ that stores this information. For every $d \in W$, there is a pointer to the structure $V_i$ with $d \in V_i$ an example is shown in Figure 10.9.

We can assume that $W$ contains the array $V[]$, the index set $Bin$, and the search tree $T$. The additional cost of the search tree $T$ is covered as follows.

- QUERY operations are not involved and work as indicated in Section 10.3.1.
- Analogously, $T$ has no influence on collecting all elements. Note that only actual elements are collected.
- Algorithm 10.4 for the DESTROY operation must be extended for destroying $T$. Since $T$ has a linear size, the running time of $\mathrm{Y}(W)$ is not concerned.
- For the BUILD operation, we need to build up the search tree also. This operation needs $O(r)$ time, since the order of the elements in the search tree $T$ is unimported. Fortunately, the cost $O(r)$ is subsumed by $O(\mathrm{b}_V(r))$, because the number of data objects is a lower bound for $\mathrm{b}_V(r)$.

The corresponding algorithm for DESTROY, BUILD, QUERY, and COLLECT can be easily extended, and we omit the simple pseudocode extension here. We turn to the extended INSERT and DELETE operations.

For $W$.DELETE(d), there is an additional $O(\log n)$ for searching the corresponding $V_i$ and for marking $d$ as deleted in $V_i$; see Algorithm 10.6.

The INSERT operation needs additional $O(\log 2r)$ time for inserting the new element $d$ into the tree $T$. Creating a link to its substructure $V_j$ is done in constant time. The costs of these operations are subsumed by the amortized cost of the insertion. Furthermore, $V_0, \ldots, V_{j-1}$ has to be erased, and the corresponding objects should point to $V_j$ afterwards. This can be efficiently realized by collecting

---

$W$.DELETE($d$) ($W = (V[], Bin, T)$ represents the binary decomposition with search tree $T$, and $d$ is an element)

---

```
V[] := W.DataArray
Bin := W.BinRep
T := W.SearchTree
if W.ActElements < W.WeakDeleteElements + 1 then
        D := W.COLLECT
        W.DESTROY
        W := BUILD(D)
else
        W.WeakDeleteElements := W.WeakDeleteElements + 1
        V[i] := T.SearchTreeQuery(d)
        V[i].WeakDelete(d)
end if
```

---

**Algorithm 10.6.** The extended DELETE operation incorporates the half-size rule and the WeakDelete operation.

$W$. INSERT($d$) ($W = (V[], Bin, T)$ represents the binary decomposition with search tree $T$, and $d$ is an element)

```
V[] := W.DataArray
Bin := W.BinRep
T := W.SearchTree
j := 0
D := new list
P := new list
while Bin ≠ ∅ and j = First(Bin) do
        D.ListAdd(V[j].Collect)
        V[j] := Nil
        Bin.DeleteFirst
        j := j + 1
end while
D.Pointer(V[j])
D.ListAdd({d})
V[j] := D.Build
Bin.AddFirst(j)
T.Insert(d)
W.ActElements := W.ActElements +1
```

**Algorithm 10.7.** The extended INSERT operation.

the pointers of all elements in $V_0, \ldots, V_{j-1}, \{d\}$. We assume that the pointers belong to the elements. That is, the information will be stored within the elements; see Algorithm 10.7. We collect the pointers and change them to "$V_j$". This operation is already covered by time $O(\mathrm{b}_V(2^j))$ for constructing $V_j$. Additionally, the INSERT operation has to update the counter $W$. ActElements. The extended INSERT operation is implemented in Algorithm 10.7.

Altogether we conclude the following.

**Theorem 10.4.** *A static abstract data type* Static *as presented in Figure 10.10 with an additional operation* $V$. WeakDelete(d) *and with additional cost function* $\mathrm{wd}_V(n)$ *can be dynamized by means of* binary decomposition, search tree $T$, *and the* half-size rule *in a dynamic abstract data type* DYNAMIC *so that the amortized time for insertion reads*

$$\overline{I_W}(t) \in O\left(\log t \frac{\mathrm{b}_V(t)}{t}\right),$$

*and the amortized time for deletion reads*

$$\overline{D_W}(t) \in O\left(\log t + \mathrm{wd}_V(t) + \frac{\mathrm{b}_V(t)}{t}\right).$$

*For the remaining operations and storage, we obtain the following time functions:*

$$\begin{aligned}
\mathrm{B}_W(r) &\in O(\mathrm{b}_V(r)),\\
\mathrm{C}_W(r) &\in O(\log r \; \mathrm{c}_V(r)),\\
\mathrm{Y}_W(r) &\in O(\log r \; \mathrm{y}_V(r)),\\
\mathrm{Q}_W(r) &\in O(\log r \; \mathrm{q}_V(r)),\\
\mathrm{S}_W(r) &\in O(\mathrm{s}_V(r)).
\end{aligned}$$

*The size of the actual data set is denoted with $r$, whereas $t$ denotes the length of the operation sequence.*

## 10.4. Dynamization for Worst-Case Performance

In the previous section, you saw that it is easy to amortize the *cost* of Insert and Delete analytically over time. The main idea for the construction of the dynamic data structure was given by the binary decomposition of $W$, which has fundamental changes from time to time, but the corresponding costs were amortized. Now, we are looking for the worst-case cost of Insert and Delete. The idea is to distribute the construction of $V_j$ *itself* over time; i.e., the structure $V_j$ should be finished if $V_{j-1}, V_{j-2}, \ldots V_0$ has to be erased.

More precisely, we want to demonstrate that the given ideas can also be adapted for running under a cost model that measures every single operation in the worst case. The main ideas stem from [Overmars and van Leeuwen 81a, Overmars and van Leeuwen 81d].

For convenience, we consider the insert operation first. After a sequence of insert operations, we have to reconstruct the binary decomposition. Obviously, if this happens, the corresponding insert operation takes a lot of effort. Therefore, we will try to build up the new structure stepwise until the structure is already completed in the case of the last insertion.

The main idea is that every $V_i$ is split into at most four structures: $V_{i1}$, $V_{i2}$, $V_{i3}$, and $V_i{}^*$ with $2^i$ objects, respectively. $V_i{}^*$ is under construction. The general rule is as follows. If $V_{i1}$ is full and we insert objects into $V_{i2}$ for the first time, we start inserting the objects of $V_{i1}$ and $V_{i2}$ into $V_{i+1}{}^*$, a structure of size $2^{i+1}$. This is spread over the next $2^{i+1}$ insertions into $V_{i2}$ and $V_{i3}$. Then, $V_{i+1}{}^*$ is full and replaces $V_{i+11}$ in the next insertion step. The second rule is that we successively replace $V_{i_1}$, $V_{i_2}$, or $V_{i_3}$ with $V_i{}^*$, if $V_i{}^*$ is full. The replacement takes constant time.

For convenience, we consider a sequence of insertions into an empty structure. The steps can be seen in Figure 10.11.

1. The first object $a$ is inserted into $V_{01}$. The next object $b$ is inserted into $V_{02}$. At the same time, application of the simple rule above requires that the first element of $V_{01} = a$ is inserted into $V_1{}^*$.

2. $c$ is inserted into $V_{03}$. At the same time, $V_{02} = b$ is inserted into $V_1{}^*$, and $V_1{}^*$ is finished for the first time. That is, we successively build up $V_1{}^*$ from the elements of $V_{01}$ and $V_{02}$ after two insertions.

3. If the next element $d$ has to be inserted, $V_{01}$, $V_{02}$, and $V_{03}$ are already full. Now, $V_{11}$ is replaced by $V_1{}^*$ and $V_{01}$ is replaced by $V_{03}$, and the new element $d$ is inserted into $V_{03}$. This again means that we start to build a new $V_1{}^*$ with elements of $V_{01}$ and $V_{02}$, and $c$ is inserted into $V_1{}^*$.

4. The next element $e$ is inserted into $V_{03}$, and at the same time, $d$ is inserted into $V_1{}^*$, which is completed then.

5. If the next element $f$ is inserted, $V_{12}$ is replaced by $V_1{}^*$. With respect to the above rule, the first part of $V_2{}^*$ is constructed with the first element of $V_{11}$, which is $a$.

6. The element $g$ is inserted into $V_{03}$, $V_1{}^*$ is completed, and $V_2{}^*$ has half-size.

7. $V_1{}^*$ replaces $V_{13}$, if $h$ is inserted. Again, we start with a new construction of $V_1{}^*$ by inserting $V_{01} = g$.

8. If $i$ is inserted into $V_{03}$, the structure $V_2{}^*$ is completed after the $2^2$ insertions, namely of $f$, $g$, $h$, and $i$. Again, $V_1{}^*$ is completed also.

9. In the next step, $V_2{}^*$ replaces $V_{21}$. Additionally, $V_{13}$ replaces $V_{11}$ and $V_1{}^*$ is moved to $V_{12}$. Therefore, we also start to build up $V_1{}^*$ again from $V_{11}$ and $V_{12}$.

The structure $V_{i+1}{}^*$ is completed after $2^{i+1}$ insert operations. Altogether, for every single insert operation, we invest

$$\frac{\mathrm{b}_V(2^{i+1})}{2^{i+1}}$$

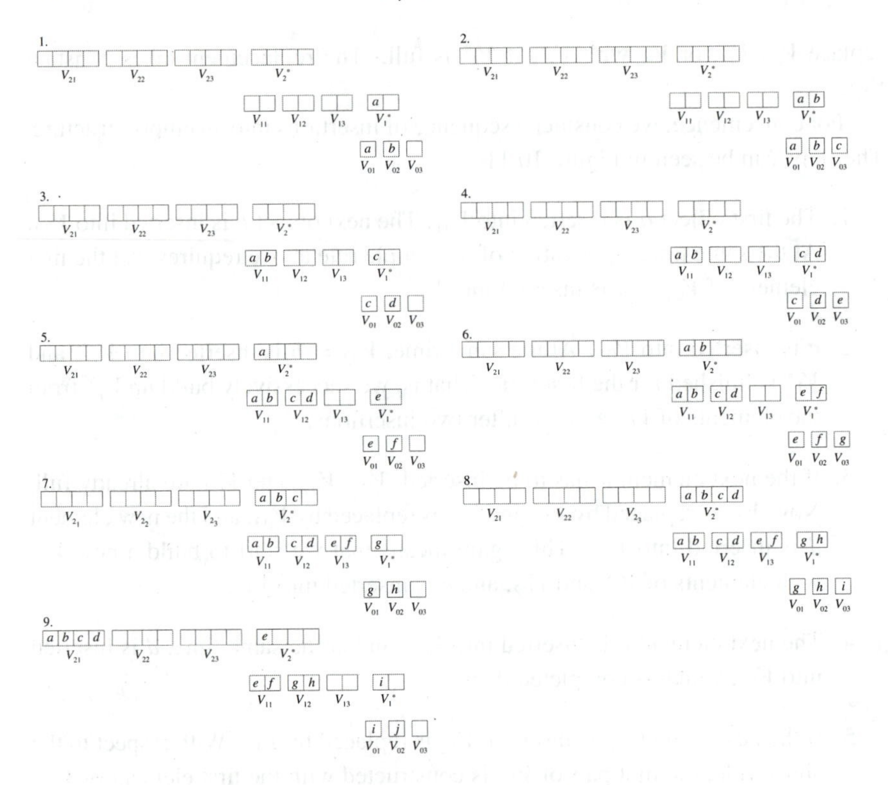


**Figure 10.11.** Spreading reconstruction over time.

time for the construction of $V_{i+1}{}^*$. Of course, we still need to sum up the cost for all structures $V_i{}^*$. The cost can be analyzed analogously to the amortized cost in Section 10.3.1. The only difference to the amortized cost model is that the work has to be partially done if the insert operations occurs.

The structures $V_{i1}$, $V_{i2}$ and $V_{i3}$ are either full or empty. For the query operation, all full structures $V_{i1}$, $V_{i2}$, and $V_{i3}$ are in use.

We have to take care that the Build and Collect operations can be split into small steps; otherwise, we cannot distribute the work adequately.

A worst-case–sensitive delete operation for the binary decomposition can be implemented as follows. We start to reconstruct a better copy $\overline{V}$ of each of the single structures $V$ if the number of all (real data and weakly deleted elements) objects $n$ in $V$ equals approximately $\frac{3}{2}$ times the number of the real data objects in $V$. If the number of all (real data and weakly deleted elements) objects $n$ of $V$ rise to approximately two times the number of the real data objects, $\overline{V}$ should should be ready to replace $V$.

The structures $V_{i1}$, $V_{i2}$, and $V_{i3}$ may have weakly deleted elements. For a single WeakDelete, we will find the corresponding substructure by a balanced search tree, as shown in the previous section. As explained earlier, we start to reconstruct $V_{ij}$ in a copy $\overline{V_{ij}}$ if $\frac{1}{4}$ of the structure $V_{ij}$ is non-real. If the half-size of the structure $V_{ij}$ becomes non-real, we replace $V_{ij}$ by $\overline{V_{ij}}$. We have to take care that $\overline{V_{ij}}$ is fully completed if the half-size limit is reached. The worst case is that after the start of constructing $\overline{V_{ij}}$, only WeakDelete operations occur. In any case, at least $\frac{1}{4}|V_{ij}|$ WeakDelete operations can be used to build $\overline{V_{ij}}$. Therefore, the cost $\mathrm{b}_V(V_{ij})$ is spread over $\frac{1}{4}|V_{ij}|$ WeakDelete operations, and each operation is charged with cost

$$4\frac{\mathrm{b}_V(V_{ij})}{|V_{ij}|}.$$

The main difference from the amortized cost model is that the work has to be partially done if the WeakDelete operation occurs.

There is one special problem here. Some of the real objects of $V_{ij}$ that were already inserted in the copy $\overline{V_{ij}}$ may be weakly deleted before $\overline{V_{ij}}$ is ready. During the reconstruction of $V_{ij}$ in $\overline{V_{ij}}$, we cannot perform a WeakDelete on the copy $\overline{V_{ij}}$. Therefore, we store the WeakDelete operations on $V_{ij}$ in a queue and apply them afterwards to $\overline{V_{ij}}$. Thus, if $V_{ij}$ is replaced by $\overline{V_{ij}}$, it may have some weakly deleted objects. But it can be guaranteed by simple arithmetic that less than 25 per cent of the new structure is non-real. There is also a special problem with $V_i^*$. We cannot weakly delete elements from $V_{i*}$, because we still have to construct it. Therefore, for $V_i^*$, we also store the WeakDelete operations on $V_{i-11}$ and $V_{i-12}$ in a queue and apply them afterwards to $V_i^*$.

We refer the interested reader to [Klein 05] and [Overmars 83] for more information.

**Theorem 10.5.** *A static abstract data type* Static, *as presented in Figure 10.10 with an additional operation* $V$. WeakDelete$(d)$ *and with additional cost function* $\mathrm{wd}_V(n)$, *can be dynamized in a dynamic abstract data type* DYNAMIC *so that*

$$\begin{array}{ll} \text{BUILD(D)} & \in O(\mathrm{b}_V(n)) \\ W.\text{QUERY}(q) & \in O(\log n \cdot \mathrm{q}_V(n)) \\ W.\text{INSERT(d)} & \in O\left(\frac{\log n}{n}\,\mathrm{b}_V(n)\right) \\ W.\text{DELETE(d)} & \in O\left(\log n + \mathrm{wd}_V(n) + \frac{\mathrm{b}_V(n)}{n}\right) \\ \text{SPACE} & O(\mathrm{s}_V(n)). \end{array}$$

*Here, n denotes the number actual data objects.*

## 10.5. Application to Search Query Data Structures

For convenience, we take some simple examples from Chapter 2 and apply Theorem 10.4, thus implementing amortized insert and delete for some static data structures.

We consider data structures for orthogonal range queries, namely windowing queries and stabbing queries. The kd-tree example in Section 10.1 belongs to the point set windowing problem with a rectangular box range query in 2D. For convenience, we concentrate on BUILD, QUERY, INSERT, DELETE, and storage space. In the following, the size of the actual data set is denoted with $n$, whereas $t$ denotes the length of the operation sequence.

**Windowing search queries.** Here, we present point set windowing queries and 2D windowing queries.

The following is a $d$-dimensional (point/axis-parallel box) windowing query.

- Input: A set $S$ of points in dimension $d$
- Query: An axis-parallel box $B$ in dimension $d$
- Output: All points of $S$ in $B$
- **Data structure: $d$-dimensional range trees**
- $\mathrm{b}_V(n) \in O(n \log^{(d-1)} n)$
- $\mathrm{s}_V(n) \in O(n \log^{(d-1)} n)$
- $\mathrm{q}_V(n) \in O(k + \log^d n)$
- $\mathrm{wd}_v(n) \in O(\log^{(d-1)} n)$
- $\mathrm{B}_W(n) \in O(n \log^{(d-1)} n)$
- $\mathrm{S}_W(n) \in O(n \log^{(d-1)} n)$
- $\mathrm{Q}_V(n) \in O(\log^{(d+1)} n)$
- $\overline{I_W}(t) \in O(\log^d t)$
- $\overline{D_W}(t) \in O(\log^d t)$

The following is a 2D (axis-parallel box/axis-parallel box) windowing query.

- Input: A set $S$ of axis-parallel boxes in 2D
- Query: An axis-parallel box $B$ in 2D
- Output: All boxes in $S$ with elements in $B$
- Three different search queries **data structures**; see Figure 2.9:
  - Query box $B$ inside: 2D **segment trees**; see below
  - Vertices inside $B$: 2D **range trees**; see above
  - Crossing segments: **interval tree with** 2D **range tree**; see Section 2.6
  - $b_V(n) \in O(n \log n)$
  - $s_V(n) \in O(n \log n)$
  - $q_V(n) \in O(\log^2 n + k)$
  - $wd_v(n) \in O(\log^2 n)$
  - $B_W(n) \in O(n \log n)$
  - $S_W(n) \in O(n \log n)$
  - $Q_V(n) \in O(\log^3 n + k)$
  - $\overline{I_W}(t) \in O(\log^2 t)$
  - $\overline{D_W}(t) \in O(\log^2 t)$

**Stabbing search queries.** Here, we present several stabbing search queries. The following is a 2D (line segment/line) stabbing query.

- Input: A set $S$ of line segments in 2D
- Query: A vertical line $l$
- Output: All segments of $S$ crossed by $l$
- **Data structure: segment tree**
- $b_V(n) \in O(n \log n)$

- $\mathrm{s}_V(n) \in O(n \log n)$
- $\mathrm{q}_V(n) \in O(k + \log n)$
- $\mathrm{wd}_v(n) \in O(\log n)$
- $\mathrm{B}_W(n) \in O(n \log n)$
- $\mathrm{S}_W(n) \in O(n \log n)$
- $\mathrm{Q}_V(n) \in O(\log^2 n + k)$
- $\overline{I_W}(t) \in O(\log^2 t)$
- $\overline{D_W}(t) \in O(\log^2 t)$

The following is a $d$-dimensional (axis-parallel box/point) stabbing query.

- Input: A set $S$ of $d$-dimensional axis-parallel boxes
- Query: A point $q$ in dimension $d$
- Output: All boxes of $B$ that contain $q$
- **Data structure: multi-level segment trees**
- $\mathrm{b}_V(n) \in O(n \log^d n)$
- $\mathrm{s}_V(n) \in O(n \log^d n)$
- $\mathrm{q}_V(n) \in O(k + \log^d n)$
- $\mathrm{wd}_v(n) \in O(\log^d n)$
- $\mathrm{B}_W(n) \in O(n \log^d n)$
- $\mathrm{S}_W(n) \in O(n \log^d n)$
- $\mathrm{Q}_V(n) \in O(\log^{(d+1)} n + k)$
- $\overline{I_W}(t) \in O(\log^{d+1} t)$
- $\overline{D_W}(t) \in O(\log^{d+1} t)$

# Bibliography

[Abellanas et al. 01a] Manuel Abellanas, Ferran Hurtado, Christian Icking, Rolf Klein, Elmar Langetepe, Lihong Ma, Belén Palop, and Vera Sacristán. "The Farthest Color Voronoi Diagram and Related Problems." In *Abstracts 17th European Workshop Comput. Geom.*, pp. 113–116. Freie Universität Berlin, 2001.

[Abellanas et al. 01b] Manuel Abellanas, Ferran Hurtado, Christian Icking, Rolf Klein, Elmar Langetepe, Lihong Ma, Belén Palop, and Vera Sacristán. "Smallest Color-Spanning Objects." In *Proc. 9th Annu. European Sympos. Algorithms*, Lecture Notes Comput. Sci., 2161, pp. 278–289. Springer-Verlag, 2001.

[Adamson and Alexa 03] Anders Adamson and Marc Alexa. "Approximating and Intersecting Surfaces from Points." In *Proc. Eurographics Symp. on Geometry Processing*, pp. 230–239, 2003.

[Adamson and Alexa 04] Anders Adamson and Marc Alexa. "Approximating Bounded, Non-orientable Surfaces from Points." In *Proc. Shape Modeling International*, pp. 243–252, 2004.

[Agarwal et al. 97] Pankaj K. Agarwal, Leonidas J. Guibas, T. M. Murali, and Jeffrey Scott Vitter. "Cylindrical Static and Kinetic Binary Space Partitions." In *Proc. 13th Annu. ACM Sympos. Comput. Geom.*, pp. 39–48, 1997.

[Agarwal et al. 98] Pankaj K. Agarwal, Jeff Erickson, and Leonidas J. Guibas. "Kinetic BSPs for Intersecting Segments and Disjoint Triangles." In *Proc. 9th ACM-SIAM Sympos. Discrete Algorithms*, pp. 107–116, 1998.

[Aho et al. 74] A. V. Aho, J. E. Hopcroft, and J. D. Ullman. *The Design and Analysis of Computer Algorithms.* Reading, MA: Addison-Wesley, 1974.

[Alexa et al. 03] M. Alexa, J. Behr, Daniel Cohen-Or, S. Fleishman, D. Levin, and C. T. Silva. "Computing and Rendering Point Set Surfaces." *IEEE Trans. on Visualization and Computer Graphics* 9:1 (2003), 3–15.

[Alt and Schwarzkopf 95] Helmut Alt and Otfried Schwarzkopf. "The Voronoi Diagram of Curved Objects." In *Proc. 11th Annu. ACM Sympos. Comput. Geom.*, pp. 89–97, 1995.

[Ar et al. 00] Sigal Ar, Bernard Chazelle, and Ayellet Tal. "Self-Customized BSP Trees for Collision Detection." *Computational Geometry: Theory and Applications* 15:1–3 (2000), 91–102.

[Ar et al. 02] Sigal Ar, Gil Montag, and Ayellet Tal. "Deferred, Self-Organizing BSP Trees." In *Eurographics*, pp. 269–278, 2002.

[Arvo and Kirk 87] James Arvo and David B. Kirk. "Fast Ray Tracing by Ray Classification." In *Computer Graphics (SIGGRAPH '87 Proceedings)*, 21, edited by Maureen C. Stone, 21, pp. 55–64, 1987.

[Arvo and Kirk 89] J. Arvo and D. Kirk. "A Survey of Ray Tracing Acceleration Techniques." In *An Introduction to Ray Tracing*, edited by A. Glassner, pp. 201–262. San Diego, CA: Academic Press, 1989.

[Arya et al. 98] S. Arya, D. M. Mount, N. S. Netanyahu, R. Silverman, and A. Wu. "An Optimal Algorithm for Approximate Nearest Neighbor Searching in Fixed Dimensions." *J. ACM* 45 (1998), 891–923.

[Aurenhammer and Klein 00] Franz Aurenhammer and Rolf Klein. "Voronoi Diagrams." In *Handbook of Computational Geometry*, edited by Jörg-Rüdiger Sack and Jorge Urrutia, pp. 201–290. Amsterdam: Elsevier Science Publishers B.V. North-Holland, 2000.

[Aurenhammer 91] F. Aurenhammer. "Voronoi Diagrams: A Survey of a Fundamental Geometric Data Structure." *ACM Comput. Surv.* 23:3 (1991), 345–405.

[Avis and Horton 85] David Avis and Joe Horton. "Remarks on the Sphere of Influence Graph." *Discrete Geometry and Convexity, Annals of the New York Academy of Sciences* 440 (1985), 323–327.

[Avnaim et al. 97] Francis Avnaim, Jean-Daniel Boissonnat, Olivier Devillers, Franco P. Preparata, and Mariette Yvinec. "Evaluating Signs of Determinants using Single-Precision Arithmetic." *Algorithmica* 17:2 (1997), 111–132.

[Baker et al. 99] R. Baker, M. Boilen, M. T. Goodrich, R. Tamassia, and B. A. Stibel. "Testers and Visualizers for Teaching Data Structures." In *Proc. of ACM SIGCSE '99*, 1999.

[Bala et al. 03] Kavita Bala, Bruce Walter, and Donald P. Greenberg. "Combining Edges and Points for Interactive High-Quality Rendering." *ACM Transactions on Graphics (SIGGRAPH 2003)* 22:3 (2003), 631–640.

[Balmelli et al. 99] Laurent Balmelli, Jelena Kovacevic, and Martin Vetterli. "Quadtrees for Embedded Surface Visualization: Constraints and Efficient Data Structures." In *Proc. of IEEE International Conference on Image Processing (ICIP)*, pp. 487–491, 1999.

[Balmelli et al. 01] Laurent Balmelli, Thmoas Liebling, and Martin Vetterli. "Computational Analysis of 4-8 Meshes with Application to Surface Simplification using global Error." In *Proc. of the 13th Canadian Conference on Computational Geometry (CCCG)*, 2001.

[Barequet et al. 96] Gill Barequet, Bernard Chazelle, Leonidas J. Guibas, Joseph S. B. Mitchell, and Ayellet Tal. "BOXTREE: A Hierarchical Representation for Surfaces in 3D." *Computer Graphics Forum* 15:3 (1996), C387–C396, C484.

[Barzel et al. 96] Ronen Barzel, John Hughes, and Daniel N. Wood. "Plausible Motion Simulation for Computer Graphics Animation." In *Proceedings of the Eurographics Workshop Computer Animation and Simulation*, edited by R. Boulic and G. Hégron, pp. 183–197. Springer, 1996.

[Basch et al. 97] J. Basch, L. J. Guibas, C. Silverstein, and L. Zhang. "A Practical Evaluation of Kinetic Data Structures." In *Proc. 13th Annu. ACM Sympos. Comput. Geom.*, pp. 388–390, 1997.

[Batut et al. 00] C. Batut, D. Bernadi, H. Cohen, and M. Olivier. "User's guide to PARI/GP." 2000. Available from World Wide Web (http://www.gn-50uma.de/ftp/pari-2.1/manuals/users.pdf).

[Beckmann et al. 90] N. Beckmann, H.-P. Kriegel, R. Schneider, and B. Seeger. "The R*-Tree: An Efficient and Robust Access Method for Points and Rectangles." In *Proc. ACM SIGMOD Conf. on Management of Data*, pp. 322–331, 1990.

[Benouamer et al. 93] M. Benouamer, P. Jaillon, D. Michelucci, and J.-M. Moreau. "A Lazy Solution to Imprecision in Computational Geometry." In *Proc. 5th Canad. Conf. Comput. Geom.*, pp. 73–78, 1993.

[Bentley and Ottmann 79] J. L. Bentley and T. A. Ottmann. "Algorithms for Reporting and Counting Geometric Intersections." *Transactions on Computing* 28:9 (1979), 643–647.

[Bentley et al. 80] Jon Louis Bentley, Bruce W. Weide, and Andrew C. Yao. "Optimal Expected-Time Algorithms for Closest Point Problems." *ACM Transactions on Mathematical Software* 6:4 (1980), 563–580.

[Bentley 77] J. L. Bentley. "Solutions to Klee's Rectangle Problems." Technical report, Carnegie-Mellon Univ., Pittsburgh, PA, 1977.

[Bentley 79] J. L. Bentley. "Decomposable Searching Problems." *Inform. Process. Lett.* 8 (1979), 244–251.

[Bernal 92] J. Bernal. "Bibliographic Notes on Voronoi Diagrams." Technical report, National Institute of Standards and Technology, Gaithersburg, MD 20899, 1992.

[Bhattacharya et al. 81] Binay K. Bhattacharya, Ronald S. Poulsen, and Godfried T. Toussaint. "Application of Proximity Graphs to Editing Nearest Neighbor Decision Rule." In *International Symposium on Information Theory*. Santa Monica, 1981.

[Boissonnat and Cazals 00] Jean-Daniel Boissonnat and Frédéric Cazals. "Smooth Surface Reconstruction via Natural Neighbour Interpolation of Distance Functions." In *Proc. 16th Annu. ACM Sympos. Comput. Geom.*, pp. 223–232, 2000.

[Boissonnat and Cazals 01] J.-D. Boissonnat and F. Cazals. "Natural Neighbor Coordinates of Points on a Surface." *Comput. Geom. Theory Appl.* 19 (2001), 155–173.

[Boissonnat and Teillaud 93] Jean-Daniel Boissonnat and Monique Teillaud. "On the Randomized Construction of the Delaunay Tree." *Theoret. Comput. Sci.* 112 (1993), 339–354.

[Borgefors 84] G. Borgefors. "Distance Transformations in Arbitrary Dimensions." In *Computer. Vision, Graphics, Image Processing*, 27, pp. 321–345, 1984.

[Boyer et al. 00] Elizabeth D. Boyer, L. Lister, and B. Shader. "Sphere-of-Influence Graphs Using the Sup-Norm." *Mathematical and Computer Modelling* 32:10 (2000), 1071–1082.

[Bremer et al. 02] Peer-Timo Bremer, Serban D. Porumbescu, Falko Kuester, Bernd Hamann, Kenneth I. Joy, and Kwan-Liu Ma. "Virtual Clay Modeling using Adaptive Distance Fields." In *Proceedings of the 2002 International Conference on Imaging Science, Systems, and Technology (CISST 2002)*, edited by H. R. Arambnia et al. Athens, Georgia, 2002.

[Brown 79] K. Q. Brown. "Voronoi Diagrams from Convex Hulls." *Inform. Process. Lett.* 9:5 (1979), 223–228.

[Burnikel et al. 94] C. Burnikel, K. Mehlhorn, and S. Schirra. "On Degeneracy in Geometric Computations." In *Proc. 5th ACM-SIAM Sympos. Discrete Algorithms*, pp. 16–23, 1994.

[Burnikel et al. 95] Christoph Burnikel, Jochen Könnemann, Kurt Mehlhorn, Stefan Näher, Stefan Schirra, and Christian Uhrig. "Exact Geometric Computation in LEDA." In *Proc. 11th Annu. ACM Sympos. Comput. Geom.*, pp. C18–C19, 1995.

[Burnikel et al. 99] C. Burnikel, R. Fleischer, K. Mehlhorn, and S. Schirra. "Efficient Exact Geometric Computation Made Easy." In *Proc. 15th Annu. ACM Sympos. Comput. Geom.*, pp. 341–350, 1999.

[Burnikel et al. 00] C. Burnikel, R. Fleischer, K. Mehlhorn, and S. Schirra. "A Strong and Easily Computable Separation Bound for Arithmetic Expressions Involving Radicals." *Algorithmica* 27:1 (2000), 87–99.

[Burnikel et al. 01] C. Burnikel, S. Funke, K. Mehlhorn, S. Schirra, and S. Schmitt. "A Separation Bound for Real Algebraic Expressions." In *Proc. 9th Annu. European Sympos. Algorithms*, Lecture Notes Comput. Sci., 2161, edited by Friedhelm Meyer auf der Heide, pp. 254–265. Springer-Verlag, 2001.

[Chew 89] L. P. Chew. "Constrained Delaunay Triangulations." *Algorithmica* 4 (1989), 97–108.

[Chew 93] L. P. Chew. "Guaranteed-Quality Mesh Generation for Curved Surfaces." In *Proc. 9th Annu. ACM Sympos. Comput. Geom.*, pp. 274–280, 1993.

[Chin 92] Norman Chin. "Partitioning a 3D Convex Polygon with an Arbitrary Plane." In *Graphics Gems III*, edited by David Kirk, chapter V.2, pp. 219–222. Academic Press, 1992.

[Clarkson 92] K. L. Clarkson. "Safe and Effective Determinant Evaluation." In *Proc. 33rd Annu. IEEE Sympos. Found. Comput. Sci.*, pp. 387–395, 1992.

[Cleveland and Loader 95] W. S. Cleveland and C. L. Loader. "Smoothing by Local Regression: Principles and Methods." In *Statistical Theory and Computational Aspects of Smoothing*, edited by W. Haerdle and M. G. Schimek, pp. 10–49. New York: Springer, 1995.

[Cohen-Or et al. 98] Daniel Cohen-Or, Amira Solomovici, and David Levin. "Three-Dimensional Distance Field Metamorphosis." *ACM Transactions on Graphics* 17:2 (1998), 116–141.

[Cole 86] R. Cole. "Searching and Storing Similar Lists." *J. Algorithms* 7 (1986), 202–220.

[Comba 99] João Luiz Dihl Comba. "Kinetic Vertical Decomposition Trees." PhD dissertation, Stanford University, 1999. Available from World Wide Web (http://graphics.stanford.edu/~comba/kvd/kvd.html).

[Coren and Girgus 78] S. Coren and J. S. Girgus. *Seeing is Deceiving: The Psychology of Visual Illusions.* Lawrence Erlbaum Associates, 1978.

[Cover and Hart 67] T.M. Cover and P.E. Hart. "Nearest Neighbor Pattern Classification." *IEEE Transactions on Information Theory* IT-13:1 (1967), 21–27.

[Dantzig 63] G. B. Dantzig. *Linear Programming and Extensions.* Princeton, NJ: Princeton University Press, 1963.

[Davenport et al. 88] J. H. Davenport, Y. Siret, and E. Tournier. *Computer Algebra: Systems and Algorithms for Algebraic Computation.* Acacdemic Press, 1988.

[de Berg et al. 00] Mark de Berg, Marc van Kreveld, Mark Overmars, and Otfried Schwarzkopf. *Computational Geometry: Algorithms and Applications*, Second edition. Berlin, Germany: Springer-Verlag, 2000.

[de Berg et al. 01] Mark de Berg, João Comba, and L. J. Guibas. "A Segment-Tree Based Kinetic BSP." In *7th Annual Symposium on Computational Geometry (SOCG)*, 2001.

[de Berg 95] Mark de Berg. "Linear Size Binary Space Partitions for Fat Objects." In *Proc. 3rd Annu. European Sympos. Algorithms*, Lecture Notes Comput. Sci., 979, pp. 252–263. Springer-Verlag, 1995.

[de Berg 00] Mark de Berg. "Linear Size Binary Space Partitions for Uncluttered Scenes." *Algorithmica* 28 (2000), 353–366.

[de Rezende and Jacometti 93a] P. de Rezende and W. Jacometti. "GeoLab: An Environment for Development of Algorithms in Computational Geometry." In *Proc. 5th Canad. Conf. Comput. Geom.*, pp. 175–180, 1993.

[de Rezende and Jacometti 93b] P. J. de Rezende and W. R. Jacometti. "Animation of Geometric Algorithms using GeoLab." In *Proc. 9th Annu. ACM Sympos. Comput. Geom.*, pp. 401–402, 1993.

[Dehne and Noltemeier 85] F. Dehne and H. Noltemeier. "A Computational Geometry Approach to Clustering Problems." In *Proc. 1st Annu. ACM Sympos. Comput. Geom.*, pp. 245–250, 1985.

[Dekker 71] T. J. Dekker. "A Floating-point Technique for Extending the Available Precision." *Numerische Mathematik* 18 (1971), 224–242.

[Deussen et al. 00] Oliver Deussen, Stefan Hiller, Cornelius van Overveld, and Thomas Strothotte. "Floating Points: A Method for Computing Stipple Drawings." *Computer Graphics Forum* 19:3.

[Devillers 02] Olivier Devillers. "The Delaunay Hierarchy." *Internat. J. Found. Comput. Sci.* 13 (2002), 163–180.

[Devroye et al. 98] Luc Devroye, Ernst Peter Mücke, and Binhai Zhu. "A Note on Point Location in Delaunay Triangulations of Random Points." *Algorithmica* 22 (1998), 477–482.

[Devroye et al. 04] L. Devroye, C. Lemaire, and J. M. Moreau. "Expected Time Analysis for Delaunay Point Location." *Comput. Geom. Theory Appl.* 22 (2004), 61–89.

[Dewdney and Vranch 77] A. K. Dewdney and J. K. Vranch. "A Convex Partition of $R^3$ with Applications to Crum's Problem and Knuth's Post-Office Problem." *Utilitas Math.* 12 (1977), 193–199.

[Djidjev and Lingas 91] H. Djidjev and A. Lingas. "On Computing the Voronoi Diagram for Restricted Planar Figures." In *Proc. 2nd Workshop Algorithms Data Struct.*, Lecture Notes Comput. Sci., 519, pp. 54–64. Springer-Verlag, 1991.

[Dobkin and Laszlo 89] D. P. Dobkin and M. J. Laszlo. "Primitives for the Manipulation of Three-Dimensional Subdivisions." *Algorithmica* 4 (1989), 3–32.

[Dobkin and Lipton 76] D. P. Dobkin and R. J. Lipton. "Multidimensional Searching Problems." *SIAM J. Comput.* 5 (1976), 181–186.

[Du et al. 99] Q. Du, V. Faber, and M. Gunzburger. "Centroidal Voronoi Ressellations: Applications and Algorithms." *SIAM Reviews* 41 (1999), 637–676.

[Dwyer 91] R. A. Dwyer. "Higher-Dimensional Voronoi Diagrams in Linear Expected Time." *Discrete Comput. Geom.* 6 (1991), 343–367.

[Dwyer 95] Rex A. Dwyer. "The Expected Size of the Sphere-of-Influence Graph." *Computational Geometry: Theory and Applications* 5:3 (1995), 155–164.

[Edelsbrunner and Maurer 81] H. Edelsbrunner and H. A. Maurer. "On the Intersection of Orthogonal Objects." *Inform. Process. Lett.* 13 (1981), 177–181.

[Edelsbrunner and Mücke 90] H. Edelsbrunner and E. P. Mücke. "Simulation of Simplicity: A Technique to Cope with Degenerate Cases in Geometric Algorithms." *ACM Trans. Graph.* 9:1 (1990), 66–104.

[Edelsbrunner and Shah 96] H. Edelsbrunner and N. R. Shah. "Incremental Topological Flipping Works for Regular Triangulations." *Algorithmica* 15 (1996), 223–241.

[Edelsbrunner and Waupotitsch 86] H. Edelsbrunner and R. Waupotitsch. "Computing a Ham-Sandwich Cut in Two Dimensions." *J. Symbolic Comput.* 2 (1986), 171–178.

[Edelsbrunner et al. 84] H. Edelsbrunner, Leonidas J. Guibas, and J. Stolfi. "Optimal Point Location in Monotone Subdivisions." Technical Report 2, DEC/SRC, 1984.

[Edelsbrunner et al. 89] Herbert Edelsbrunner, Günter Rote, and Emo Welzl. "Testing the Necklace Condition for Shortest Tours and Optimal Factors in the Plane." *Theoretical Computer Science* 66:2 (1989), 157–180.

[Edelsbrunner 80] H. Edelsbrunner. "Dynamic Data Structures for Orthogonal Intersection Queries." Report F59, Inst. Informationsverarb., Tech. Univ. Graz, Graz, Austria, 1980.

[Edelsbrunner 87] H. Edelsbrunner. *Algorithms in Combinatorial Geometry*, EATCS Monographs on Theoretical Computer Science, 10. Heidelberg, West Germany: Springer-Verlag, 1987.

[Ehmann and Lin 01] Stephan A. Ehmann and Ming C. Lin. "Accurate and Fast Proximity Queries Between Polyhedra Using Convex Surface Decomposition." In *Computer Graphics Forum*, 20, 20, pp. 500–510, 2001.

[Elassal and Caruso 84] Atef A. Elassal and Vincent M. Caruso. "USGS Digital Cartographic Data Standards—Digital Elevation Models." Technical Report Geological Survey Circular 895-B, US Geological Survey, 1984.

[Emiris and Canny 92] I. Emiris and J. Canny. "An Efficient Approach to Removing Geometric Degeneracies." In *Proc. 8th Annu. ACM Sympos. Comput. Geom.*, pp. 74–82, 1992.

[Emiris et al. 97] I. Z. Emiris, J. F. Canny, and R. Seidel. "Efficient Perturbations for Handling Geometric Degeneracies." *Algorithmica* 19:1–2 (1997), 219–242.

[Fabri et al. 00] A. Fabri, G.-J. Giezeman, L. Kettner, S. Schirra, and S. Schönherr. "On the Design of CGAL a Computational Geometry Algorithms Library." *Softw. – Pract. Exp.* 30:11 (2000), 1167–1202.

[Field 86] D. A. Field. "Implementing Watson's Algorithm in Three Dimensions." In *Proc. 2nd Annu. ACM Sympos. Comput. Geom.*, pp. 246–259, 1986.

[Fortune and Milenkovic 91] S. Fortune and V. Milenkovic. "Numerical Stability of Algorithms for Line Arrangements." In *Proc. 7th Annu. ACM Sympos. Comput. Geom.*, pp. 334–341, 1991.

[Fortune and Van Wyk 93] S. Fortune and C. J. Van Wyk. "Efficient Exact Arithmetic for Computational Geometry." In *Proc. 9th Annu. ACM Sympos. Comput. Geom.*, pp. 163–172, 1993.

[Fortune 87] S. J. Fortune. "A Sweepline Algorithm for Voronoi Diagrams." *Algorithmica* 2 (1987), 153–174.

[Fortune 89] S. Fortune. "Stable Maintenance of Point Set Triangulations in Two Dimensions." In *Proc. 30th Annu. IEEE Sympos. Found. Comput. Sci.*, pp. 494–505, 1989.

[Fortune 92a] S. Fortune. "Numerical Stability of Algorithms for 2-D Delaunay Triangulations and Voronoi Diagrams." In *Proc. 8th Annu. ACM Sympos. Comput. Geom.*, pp. 83–92, 1992.

[Fortune 92b] S. Fortune. "Voronoi Diagrams and Delaunay Triangulations." In *Computing in Euclidean Geometry*, Lecture Notes Series on Computing, 1, edited by D.-Z. Du and F. K. Hwang, pp. 193–233. Singapore: World Scientific, 1992.

[Fortune 96] S. Fortune. "Robustness Issues in Geometric Algorithms." In *Applied Computational Geometry: Towards Geometric Engineering*, Lecture Notes Comput. Sci., 1148, edited by M. C. Lin and D. Manocha, pp. 9–14. Springer-Verlag, 1996.

[Frisken et al. 00] Sarah F. Frisken, Ronald N. Perry, Alyn P. Rockwood, and Thouis R. Jones. "Adaptively Sampled Distance Fields: A General Representation of Shape for Computer Graphics." In *Siggraph 2000, Computer Graphics Proceedings, Annual Conference Series*, edited by Kurt Akeley, pp. 249–254. ACM Press / ACM SIGGRAPH / Addison Wesley Longman, 2000.

[Fuchs et al. 80] H. Fuchs, Z. M. Kedem, and B. F. Naylor. "On Visible Surface Generation by a Priori Tree Structures." In *Computer Graphics (SIGGRAPH '80 Proceedings)*, 14, 14, pp. 124–133, 1980.

[Fussell and Subramanian 88] Donald Fussell and K. R. Subramanian. "Fast Ray Tracing Using K-D Trees." Technical Report TR-88-07, U. of Texas, Austin, Dept. Of Computer Science, 1988.

[Gelfand et al. 98] Natasha Gelfand, Michael T. Goodrich, and Roberto Tamassia. "Teaching Data Structure Design Patterns." In *Proc. ACM Symp. Computer Science Education*, 1998.

[Gibbons 85] A. Gibbons. *Algorithmic Graph Theory*. Cambridge: Cambridge University Press, 1985.

[Glassner 89] Andrew S. Glassner, editor. *An Introduction to Ray Tracing*. Academic Press, 1989.

[Goldberg 91] D. Goldberg. "What Every Computer Scientist Should Know About Floating-Point Arithmetic." *ACM Comput. Surv.* 23:1 (1991), 5–48.

[Goldsmith and Salmon 87] Jeffrey Goldsmith and John Salmon. "Automatic Creation of Object Hierarchies for Ray Tracing." *IEEE Computer Graphics and Applications* 7:5 (1987), 14–20.

[Gomez et al. 97] Francisco Gomez, Suneeta Ramaswami, and Godfried T. Toussaint. "On Removing Non-degeneracy Assumptions in Computational Geometry." In *CIAC '97: Proceedings of the Third Italian Conference on Algorithms and Complexity*, pp. 86–99. London, UK: Springer-Verlag, 1997.

[Gomez et al. 01] Francisco Gomez, Ferran Hurtado, Toni Sellares, and Godfried Toussaint. "On Degeneracies Removable by Perspective Projections." *International Journal of Mathematical Algorithms* 2 (2001), 227–248.

[Goodrich and Tamassia 98] Michael T. Goodrich and Roberto Tamassia. *Data Structures and Algorithms in Java*. New York, NY: John Wiley & Sons, 1998.

[Gottschalk et al. 96] Stefan Gottschalk, Ming Lin, and Dinesh Manocha. "OBB-Tree: A Hierarchical Structure for Rapid Interference Detection." In *SIGGRAPH 96 Conference Proceedings*, edited by Holly Rushmeier, pp. 171–180. ACM SIGGRAPH, Addison Wesley, 1996.

[Gradshteyn and Ryzhik 00] I. S. Gradshteyn and I. M. Ryzhik. *Tables of Integrals, Series, and Products*, Sixth Edition. San Diego, CA: Academic Press, 2000.

[Granlund 96] Torbjörn Granlund. *GMP, The GNU Multiple Precision Arithmetic Library*, Second edition, 1996. Available from World Wide Web (http://www.swox.com/gmp/).

[Guibas and Stolfi 85] Leonidas J. Guibas and J. Stolfi. "Primitives for the Manipulation of General Subdivisions and the Computation of Voronoi Diagrams." *ACM Trans. Graph.* 4:2 (1985), 74–123.

[Guibas et al. 89] Leonidas J. Guibas, D. Salesin, and J. Stolfi. "Epsilon Geometry: Building Robust Algorithms from Imprecise Computations." In *Proc. 5th Annu. ACM Sympos. Comput. Geom.*, pp. 208–217, 1989.

[Guibas et al. 92a] L. Guibas, J. Pach, and M. Sharir. "Generalized Sphere-of-Influence Graphs in Higher Dimensions." Manuscript, Tel-Aviv University, 1992.

[Guibas et al. 92b] Leonidas J. Guibas, D. E. Knuth, and Micha Sharir. "Randomized Incremental Construction of Delaunay and Voronoi Diagrams." *Algorithmica* 7 (1992), 381–413.

[Guibas et al. 93] Leonidas J. Guibas, D. Salesin, and J. Stolfi. "Constructing Strongly Convex Approximate Hulls with Inaccurate Primitives." *Algorithmica* 9 (1993), 534–560.

[Guibas 98] L. J. Guibas. "Kinetic Data Structures—A State of the Art Report." In *Proc. Workshop Algorithmic Found. Robot.*, edited by P. K. Agarwal, L. E. Kavraki, and M. Mason, pp. 191–209. Wellesley, MA: A. K. Peters, 1998.

[Haeberli 90] Paul E. Haeberli. "Paint By Numbers: Abstract Image Representations." In *Computer Graphics (SIGGRAPH '90 Proceedings)*, edited by Forest Baskett, pp. 207–214, 1990.

[Hamacher 95] Horst W. Hamacher. *Mathematische Lösungsverfahren für planare Standortprobleme*. Wiesbaden: Verlag Vieweg, 1995.

[Härdle 90] W. Härdle. *Applied Nonparametric Regression*, Econometric Society Monograph, 19. New York: Cambridge University Press, 1990.

[Hausner 01] Alejo Hausner. "Simulating Decorative Mosaics." In *SIGGRAPH 2001, Computer Graphics Proceedings, Annual Conference Series*, edited by Eugene Fiume, pp. 573–578, 2001.

[Higham 90] Nicholas J. Higham. "Analysis of the Cholesky Decomposition of a Semi-Definite Matrix." In *Reliable Numerical Computation*, edited by M. G. Cox and S. J. Hammarling, pp. 161–185. Oxford University Press, 1990.

[Hiyoshi and Sugihara 00] Hisamoto Hiyoshi and Kokichi Sugihara. "Voronoi-Based Interpolation with Higher Continuity." In *Proc. 16th Annu. ACM Sympos. Comput. Geom.*, pp. 242–250, 2000.

[Hoff III et al. 99] Kenneth E. Hoff III, Tim Culver, John Keyser, Ming Lin, and Dinesh Manocha. "Fast Computation of Generalized Voronoi Diagrams Using Graphics Hardware." In *Proc. of the Conference on Computer Graphics (Siggraph99)*, pp. 277–286. LA, California, 1999.

[Hoppe et al. 92] Hugues Hoppe, Tony DeRose, Tom Duchamp, John McDonald, and Werner Stuetzle. "Surface Reconstruction from Unorganized Points." In *Computer Graphics (SIGGRAPH '92 Proceedings)*, 26, edited by Edwin E. Catmull, 26, pp. 71--78, 1992.

[Huang et al. 01] Jian Huang, Yan Li, Roger Crawfis, Shao Chiung Lu, and Shuh Yuan Liou. "A Complete Distance Field Representation." In *Proceedings of the conference on Visualization 2001*, pp. 247–254. IEEE Press, 2001.

[Hubbard 95] Philip M. Hubbard. "Real-Time Collision Detection and Time-Critical Computing." In *SIVE 95, The First Worjshop on Simulation and Interaction in Virtual Environments, 1*, pp. 92–96. University of Iowa, Iowa City, Iowa: Informal Proceedings, 1995.

[Hurtado et al. 04] Ferran Hurtado, Rolf Klein, Elmar Langetepe, and Vera Sacristán. "The Weighted Farthest Color Voronoi Diagram on Trees and Graphs." *Computational Geometry: Theory and Applications* 27 (2004), 13–26.

[Hwang 79] F. K. Hwang. "An $O(n \log n)$ Algorithm for Rectilinear Minimal Spanning Tree." *J. ACM* 26 (1979), 177–182.

[IEEE 85] *IEEE Standard for Binary Floating Point Arithmetic, ANSI/IEEE Std* 754 – 1985. New York, NY, 1985. Reprinted in SIGPLAN Notices, 22(2):9–25, 1987.

[Inagaki et al. 92] H. Inagaki, K. Sugihara, and N. Sugie. "Numerically Robust Incremental Algorithm for Constructing Three-Dimensional Voronoi Diagrams." In *Proc. 4th Canad. Conf. Comput. Geom.*, pp. 334–339, 1992.

[Jaromczyk and Toussaint 92] J. W. Jaromczyk and Godfried T. Toussaint. "Relative Neighborhood Graphs and their Relatives." *Proc. of the IEEE* 80:9 (1992), 1502–1571.

[Joe 89] B. Joe. "3-Dimensional Triangulations from Local Transformations." *SIAM J. Sci. Statist. Comput.* 10:4 (1989), 718–741.

[Joe 91a] B. Joe. "Construction of Three-Dimensional Delaunay Triangulations Using Local Transformations." *Comput. Aided Geom. Design* 8:2 (1991), 123–142.

[Joe 91b] B. Joe. "Geompack. A Software Package For The Generation Of Meshes Using Geometric Algorithms." *Advances in Engineering Software and Workstations* 13:5–6 (1991), 325–331.

[Jones and Satherley 01] M. W. Jones and R. A. Satherley. "Shape Representation using Space Filled Sub-Voxel Distance Fields." In *Proc. of the Int'l Conf. on Shape Modeling and Applications (SMI)*, edited by Bob Werner, pp. 316–325. Genova, Italy: IEEE, 2001.

[Ju et al. 02] Lili Ju, Qiang Du, and Max Gunzburger. "Probabilistic Methods for Centroidal Voronoi Tessellations and Their Parallel Implementations." *Parallel Computing* 28:10 (2002), 1477–1500.

[Kaltofen and Villard 04] Erich Kaltofen and Gilles Villard. "Computing the Sign or the Value of the Determinant of an Integer Matrix—A Complexity Survey." *J. Comput. Appl. Math.* 162:1 (2004), 133–146.

[Kamphans and Langetepe 03] Tom Kamphans and Elmar Langetepe. "The Pledge Algorithm Reconsidered under Errors in Sensors and Motion." In *Proc. of the 1th Workshop on Approximation and Online Algorithms*, Lecture Notes Comput. Sci., 2909, pp. 165–178. Berlin: Springer, 2003.

[Kamphans and Langetepe 05] Tom Kamphans and Elmar Langetepe. "Optimal Competitive Online Ray Search with an Error-Prone Robot." In *Proc. of the 4th International Workshop on Efficient and Experimental Algorithms*, 2005.

[Kao and Mount 92] T. C. Kao and D. M. Mount. "Incremental Construction and Dynamic Maintenance of Constrained Delaunay Triangulations." In *Proc. 4th Canad. Conf. Comput. Geom.*, pp. 170–175, 1992.

[Karamcheti et al. 99a] Vijay Karamcheti, Chen Li, Igor Pechtchanski, and Chee Yap. "A Core Library for Robust Numeric and Geometric Computation." In *Proc. 15th Annu. ACM Sympos. Comput. Geom.*, pp. 351–359, 1999.

[Karamcheti et al. 99b] Vijay Karamcheti, Chen Li, Igor Pechtchanski, and Chee Yap. *The CORE Library Project*, First edition, 1999. Available from World Wide Web (http://www.cs.nyu.edu/exact/core/).

[Karasick et al. 89] M. Karasick, D. Lieber, and L. R. Nackman. "Efficient Delaunay Triangulations using Rational Arithmetic." Report RC 14455, IBM T. J. Watson Res. Center, Yorktown Heights, NY, 1989.

[Katayama and Satoh 97] Norio Katayama and Shin'ichi Satoh. "The SR-Tree: An Index Structure for High-Dimensional Nearest Neighbor Queries." In *Proc. ACM SIGMOD Conf. on Management of Data*, pp. 369–380, 1997.

[Kay and Kajiya 86] Timothy L. Kay and James T. Kajiya. "Ray Tracing Complex Scenes." In *Computer Graphics (SIGGRAPH '86 Proceedings)*, edited by David C. Evans and Russell J. Athay, pp. 269–278, 1986.

[Keil and Gutwin 89] J. M. Keil and C. A. Gutwin. "The Delaunay triangulation closely approximates the complete Euclidean graph." In *Proc. 1st Workshop Algorithms Data Struct.*, Lecture Notes Comput. Sci., 382, pp. 47–56. Springer-Verlag, 1989.

[Keller-Gehrig 85] W. Keller-Gehrig. "Fast Algorithms for the Characteristic Polynomial." *Theoret. Comput. Sci.* 36 (1985), 309–317.

[Kimmel et al. 98] Ron Kimmel, Nahum Kiryati, and Alfred M. Bruckstein. "Multi-Valued Distance Maps for Motion Planning on Surfaces with Moving Obstacles." *IEEE Transactions on Robotics and Automation* 14:3 (1998), 427–436.

[Klein and Zachmann 03] Jan Klein and Gabriel Zachmann. "ADB-Trees: Controlling the Error of Time-Critical Collision Detection." In *8th International Fall Workshop Vision, Modeling, and Visualization (VMV)*, pp. 37–45. University München, Germany, 2003.

[Klein and Zachmann 04a] Jan Klein and Gabriel Zachmann. "Point Cloud Collision Detection." In *Computer Graphics forum (Proc. EUROGRAPHICS)*, 23, edited by M.-P. Cani and M. Slater, pp. 567–576. Grenoble, France, 2004.

[Klein and Zachmann 04b] Jan Klein and Gabriel Zachmann. "Point Cloud Surfaces using Geometric Proximity Graphs." *Computers & Graphics* 28:6 (2004), 839–850.

[Klein and Zachmann 05] Jan Klein and Gabriel Zachmann. "Interpolation Search for Point Cloud Intersection." In *Proc. of WSCG 2005*, pp. 163–170. University of West Bohemia, Plzen, Czech Republic, 2005.

[Klein et al. 93] Rolf Klein, Kurt Mehlhorn, and Stefan Meiser. "Randomized Incremental Construction of Abstract Voronoi Diagrams." *Comput. Geom. Theory Appl.* 3:3 (1993), 157–184.

[Klein et al. 99] Reinhard Klein, Andreas Schilling, and Wolfgang Straßer. "Reconstruction and Simplification of Surfaces from Contours." In *Pacific Graphics*, pp. 198–207. IEEE Computer Society, 1999.

[Klein 05] Rolf Klein. *Algorithmische Geometrie*. Bonn: Springer, 2005.

[Klosowski et al. 98] James T. Klosowski, Martin Held, Jospeh S. B. Mitchell, Henry Sowrizal, and Karel Zikan. "Efficient Collision Detection Using Bounding Volume Hierarchies of $k$-DOPs." *IEEE Transactions on Visualization and Computer Graphics* 4:1 (1998), 21–36.

[Knuth 81] D. E. Knuth. *Seminumerical Algorithms*, The Art of Computer Programming, 2, Second edition. Reading, MA: Addison-Wesley, 1981.

[Kruskal, Jr. 56] J. B. Kruskal, Jr. "On the shortest spanning subtree of a graph and the traveling salesman problem." *Proc. Amer. Math. Soc.* 7 (1956), 48–50.

[Larsson and Akenine-Möller 01] Thomas Larsson and Tomas Akenine-Möller. "Collision Detection for Continuously Deforming Bodies." In *Eurographics*, pp. 325–333, 2001. Short presentation.

[Larsson and Akenine-Möller 03] Thomas Larsson and Tomas Akenine-Möller. "Efficient Collision Detection for Models Deformed by Morphing." *The Visual Computer* 19:2 (2003), 164–174.

[Latombe 91] J.-C. Latombe. *Robot Motion Planning*. Boston: Kluwer Academic Publishers, 1991.

[Lawson 77] C. L. Lawson. "Software for $C^1$ Surface Interpolation." In *Math. Software III*, edited by J. R. Rice, pp. 161–194. New York, NY: Academic Press, 1977.

[Lee and Lin 86] D. T. Lee and A. K. Lin. "Generalized Delaunay Triangulation for Planar Graphs." *Discrete Comput. Geom.* 1 (1986), 201–217.

[Lee and Wong 80] D. T. Lee and C. K. Wong. "Voronoi Diagrams in $L_1$ ($L_\infty$) Metrics with 2-Dimensional Storage Applications." *SIAM J. Comput.* 9:1 (1980), 200–211.

[Lee 80] D. T. Lee. "Two-Dimensional Voronoi Diagrams in the $L_p$-Metric." *J. ACM* 27:4 (1980), 604–618.

[Lee 00] I.-K. Lee. "Curve Reconstruction from Unorganized Points." *Computer Aided Geometric Design* 17:2 (2000), 161–177.

[Leutenegger et al. 97] S. Leutenegger, J. Edgington, and M. Lopez. "STR : A Simple and Efficient Algorithm for R-Tree Packing." In *Proceedings of the 13th International Conference on Data Engineering (ICDE'97)*, pp. 497–507. Washington - Brussels - Tokyo: IEEE, 1997.

[Levin 03] David Levin. "Mesh-Independent Surface Interpolation." In *Geometric Modeling for Scientific Visualization*, edited by Hamann Brunnett and Mueller, pp. 37–49. Springer, 2003.

[Li and Chen 98] Tsai-Yen Li and Jin-Shin Chen. "Incremental 3D Collision Detection with Hierarchical Data Structures." In *Proc. VRST '98*, pp. 139–144. Taipei, Taiwan: ACM, 1998.

[Li and Yap 01a] C. Li and C. Yap. "A New Constructive Root Bound for Algebraic Expressions." In *Proc. 12th ACM-SIAM Sympos. Discrete Algorithms*, pp. 496–505, 2001.

[Li and Yap 01b] Chen Li and Chee Yap. "Recent Progress in Exact Geometric Computation." Technical Report, Dept. Comput. Sci., New York University, 2001.

[Li 01] Chen Li. "Exact Geometric Computation: Theory and Application." Ph.D. dissertation, Department of Computer Science, New York University, New York, 2001.

[Lindstrom and Pascucci 01] Peter Lindstrom and Valerio Pascucci. "Visualization of Large Terrains Made Easy." In *Proc. IEEE Visualization*. San Diego, 2001.

[Lindstrom et al. 96] Peter Lindstrom, David Koller, William Ribarsky, Larry F. Hughes, Nick Faust, and Gregory Turner. "Real-Time, Continuous Level of Detail Rendering of Height Fields." In *SIGGRAPH 96 Conference Proceedings*, edited by Holly Rushmeier, pp. 109–118. ACM SIGGRAPH, Addison Wesley, 1996.

[Liotta et al. 96] Giuseppe Liotta, Franco P. Preparata, and Roberto Tamassia. "Robust Proximity Queries in Implicit Voronoi Diagrams." Technical Report CS-96-16, Center for Geometric Computing, Comput. Sci. Dept., Brown Univ., Providence, RI, 1996.

[Lorensen and Cline 87] William E. Lorensen and Harvey E. Cline. "Marching Cubes: A High Resolution 3D Surface Construction Algorithm." In *Computer Graphics (SIGGRAPH '87 Proceedings)*, 21, edited by Maureen C. Stone, 21, pp. 163–169, 1987.

[Maurer and Ottmann 79] H. A. Maurer and T. A. Ottmann. "Dynamic Solutions of Decomposable Searching Problems." In *Discrete Structures and Algorithms*, edited by U. Pape, pp. 17–24. München, Germany: Carl Hanser Verlag, 1979.

[McCreight 80] E. M. McCreight. "Efficient Algorithms for Enumerating Intersecting Intervals and Rectangles." Report CSL-80-9, Xerox Palo Alto Res. Center, Palo Alto, CA, 1980.

[Mehlhorn and Näher 94] Kurt Mehlhorn and Stefan Näher. "The Implementation of Geometric Algorithms." In *Proc. 13th World Computer Congress IFIP94*, pp. 223–231, 1994.

[Mehlhorn and Näher 00] Kurt Mehlhorn and Stefan Näher. *LEDA: A Platform for Combinatorial and Geometric Computing*. Cambridge, UK: Cambridge University Press, 2000.

[Mehlhorn and Overmars 81] K. Mehlhorn and M. H. Overmars. "Optimal Dynamization of Decomposable Searching Problems." *Inform. Process. Lett.* 12 (1981), 93–98.

[Michael and Quint 03] T. S. Michael and Thomas Quint. "Sphere of Influence Graphs and the $L_\infty$-Metric." *Discrete Applied Mathematics* 127:3 (2003), 447 – 460.

[Michelucci and Moreau 97] D. Michelucci and J.-M. Moreau. "Lazy Arithmetic." *IEEE Transactions on Computers* 46:9 (1997), 961–975.

[Michelucci 96] D. Michelucci. "Arithmetic Isuues in Geometric Computations." In *Proc. 2nd Real Numbers and Computer Conf.*, pp. 43–69, 1996.

[Michelucci 97] D. Michelucci. "The Robustness Issue." 1997. Available from World Wide Web (http://citeseer.ist.psu.edu/361993.html).

[Mosaic a] "Mosic from a Roman bath in Bath, UK." Available from World Wide Web (http://www2.sjsu.edu/depts/jwss/bath2004/baths.html).

[Mosaic b] "Mosaic floor from a Late Roman bath, with main theme two peacocks enframing a vessel of *kantharos* shape." Available from World Wide Web (http://www.culture.gr/2/21/211/21110m/e211jm04.html).

[Müller et al. 00] Gordon Müller, Stephan Schäfer, and W. Dieter Fellner. "Automatic Creation of Object Hierarchies for Radiosity Clustering." *Computer Graphics Forum* 19:4.

[Naylor et al. 90] Bruce Naylor, John Amanatides, and William Thibault. "Merging BSP Trees Yields Polyhedral Set Operations." In *Computer Graphics (SIGGRAPH '90 Proceedings)*, 24, edited by Forest Baskett, 24, pp. 115–124, 1990.

[Naylor 96] Bruve F. Naylor. "A Tutorial on Binary Space Partitioning Trees." *ACM SIGGRAPH '96 Course Notes 29.*

[Nievergelt and Hinrichs 93] J. Nievergelt and K. H. Hinrichs. *Algorithms and Data Structures: With Applications to Graphics and Geometry*. Englewood Cliffs, NJ: Prentice Hall, 1993.

[Nievergelt and Schorn 88] J. Nievergelt and P. Schorn. "Das Rätsel der verzopften Geraden." *Informatik Spektrum* 11 (1988), 163–165.

[Nievergelt et al. 91] J. Nievergelt, P. Schorn, M. de Lorenzi, C. Ammann, and A. Brüngger. "XYZ: Software for Geometric Computation." Report 163, Institut für Theorische Informatik, ETH, Zürich, Switzerland, 1991.

[Novotni and Klein 01] M. Novotni and R. Klein. "A Geometric Approach to 3D Object Comparison." In *International Conference on Shape Modeling and Applications*, pp. 167–175, 2001.

[Okabe and Suzuki 97] A. Okabe and A. Suzuki. "Locational Optimization Problems Solved Through Voronoi Diagrams." *European J. Oper. Res.* 98:3 (1997), 445–456.

[Okabe et al. 92] Atsuyuki Okabe, Barry Boots, and Kokichi Sugihara. *Spatial Tessellations: Concepts and Applications of Voronoi Diagrams*. Chichester, UK: John Wiley & Sons, 1992.

[Osada et al. 01] R. Osada, T. Funkhouser, B. Chazelle, and D. Dobkin. "Matching 3D Models with Shape Distributions." In *Proceedings of the International Conference on Shape Modeling and Applications (SMI-01)*, edited by Bob Werner, pp. 154–166. Los Alamitos, CA: IEEE Computer Society, 2001.

[Ouchi 97] K. Ouchi. "Real/Expr: Implementation of Exact Computation." 1997. Available from World Wide Web (http://cs.nyu.edu/exact/realexpr).

[Overmars and van Leeuwen 81a] M. H. Overmars and J. van Leeuwen. "Dynamization of Decomposable Searching Problems Yielding Good Worst-Case Bounds." In *Proc. 5th GI Conf. Theoret. Comput. Sci.*, Lecture Notes Comput. Sci., 104, pp. 224–233. Springer-Verlag, 1981.

[Overmars and van Leeuwen 81b] M. H. Overmars and J. van Leeuwen. "Some Principles for Dynamizing Decomposable Searching problems." *Inform. Process. Lett.* 12 (1981), 49–54.

[Overmars and van Leeuwen 81c] M. H. Overmars and J. van Leeuwen. "Two General Methods for Dynamizing Decomposable Searching problems." *Computing* 26 (1981), 155–166.

[Overmars and van Leeuwen 81d] M. H. Overmars and J. van Leeuwen. "Worst-Case Optimal Insertion and Deletion Methods for Decomposable Searching Problems." *Inform. Process. Lett.* 12 (1981), 168–173.

[Overmars 83] M. H. Overmars. *The Design of Dynamic Data Structures*, Lecture Notes Comput. Sci., 156. Heidelberg, West Germany: Springer-Verlag, 1983.

[Overmars 96] Mark H. Overmars. "Designing the Computational Geometry Algorithms Library CGAL." In *Proc. 1st ACM Workshop on Appl. Comput. Geom.*, Lecture Notes Comput. Sci., 1148, pp. 53–58. Springer-Verlag, 1996.

[Paglieroni 92] David W. Paglieroni. "Distance Transforms: Properties and Machine Vision Applications." *CVGIP: Graphical Models and Image Processing* 54:1 (1992), 56–74.

[Palmer and Grimsdale 95] I. J. Palmer and R. L. Grimsdale. "Collision Detection for Animation using Sphere-Trees." *Computer Graphics Forum* 14:2 (1995), 105–116.

[Paterson and Yao 90] M. S. Paterson and F. F. Yao. "Efficient Binary Space Partitions for Hidden-Surface Removal and Solid Modeling." *Discrete Comput. Geom.* 5 (1990), 485–503.

[Pauly et al. 02] Mark Pauly, Markus H. Gross, and Leif Kobbelt. "Efficient Simplification of Point-Sampled Surfaces." In *IEEE Visualization 2002*, pp. 163–170, 2002.

[Pauly et al. 03] Mark Pauly, Richard Keiser, Leif P. Kobbelt, and Markus Gross. "Shape Modeling with Point-Sampled Geometry." *ACM Transactions on Graphics (SIGGRAPH 2003)* 22:3 (2003), 641–650.

[Payne and Toga 92] Bradley A. Payne and Arthur W. Toga. "Distance Field Manipulation of Surface Models." *IEEE Computer Graphics and Applications* 12:1 (1992), 65–71.

[Perry and Frisken 03] Ronald N. Perry and Sarah F. Frisken. "Method for Generating a Two-Dimensional Distance Field within a Cell Associated with a Corner of a Two-Dimensional Object." Patent application 10/396,267, Mitsubishi Electric Research Laboratories - MERL, 2003.

[Pfister et al. 00] Hanspeter Pfister, Jeroen van Baar, Matthias Zwicker, and Markus Gross. "Surfels: Surface Elements as Rendering Primitives." *ACM Transactions on Graphics (SIGGRAPH 2000)* 19:3 (2000), 335–342.

[Piper 93] B. Piper. "Properties of Local Coordinates Based on Dirichlet Tesselations." In *Geometric modelling*, Computing. Supplementum, 8, edited by G. Farin, H. Hagen, H. Noltemeier, and W. Knödel, pp. 227–240. Wien / New York: Springer, 1993.

[Preparata and Shamos 90] F. P. Preparata and M. I. Shamos. *Computational Geometry: An Introduction*, Third edition. Springer-Verlag, 1990.

[Press et al. 92] William H. Press, Saul A. Teukolsky, William T. Vetterling, and Brian P. Flannery. *Numerical Recipes in C: The Art of Scientific Computing (2nd ed.)*. Cambridge: Cambridge University Press, 1992.

[Rajan 91] V. T. Rajan. "Optimality of the Delaunay triangulation in $R^d$." In *Proc. 7th Annu. ACM Sympos. Comput. Geom.*, pp. 357–363, 1991.

[RAS] "Cassiopeia constellation from Uranometreia." Available from World Wide Web (http://www.ras.org.uk/html/library/rare.html).

[Revelles et al. 00] J. Revelles, C. Urena, and M. Lastra. "An Efficient Parametric Algorithm for Octree Traversal." In *WSCG 2000 Conference Proceedings*. University of West Bohemia, Plzen, Czech Republic, 2000.

[Rogers 63] C.A. Rogers. "Covering a Sphere with Spheres." *Mathematika* 10 (1963), 157–164.

[Roussopoulos and Leifker 85] Nick Roussopoulos and Daniel Leifker. "Direct Spatial Search on Pictorial Databases using Packed R-Trees." In *Proceedings of ACM-SIGMOD 1985 International Conference on Management of Data*, edited by Sham Navathe, pp. 17–31. New York: ACM Press, 1985.

[Rusinkiewicz and Levoy 00] Szymon Rusinkiewicz and Marc Levoy. "QSplat: A Multiresolution Point Rendering System for Large Meshes." *ACM Transactions on Graphics (SIGGRAPH 2000)* 19:3 (2000), 343–352.

[Sanchez et al. 97] J.S. Sanchez, F. Pla, and F.J. Ferri. "Prototype Selection for the Nearest Neighbour Rule through Proximity Graphs." *Pattern Recognition Letters* 18:6 (1997), 507–513.

[Santisteve 99] Francisco J. Santisteve. "Robust Geometric Computation (RGC), State of the Art." 1999. Available from World Wide Web (http://citeseer.ist.psu.edu/santisteve99robust.html).

[Sarnak and Tarjan 86] N. Sarnak and R. E. Tarjan. "Planar Point Location using Persistent Search Trees." *Commun. ACM* 29:7 (1986), 669–679.

[Sattar 04] Junaed Sattar. "The Spheres-of-Influence Graph." 2004. Available from World Wide Web (http://www.cs.mcgill.ca/~jsatta/pr507/about.html).

[Saxe and Bentley 79] J. B. Saxe and J. L. Bentley. "Transforming Static Data Structures to Dynamic Structures." In *Proc. 20th Annu. IEEE Sympos. Found. Comput. Sci.*, pp. 148–168, 1979.

[Schirra 00] Stefan Schirra. "Robustness and Precision Issues in Geometric Computation." In *Handbook of Computational Geometry*, edited by Jörg-Rüdiger Sack and Jorge Urrutia, Chapter 14, pp. 597–632. Amsterdam: Elsevier Science Publishers B.V. North-Holland, 2000.

[Schorn 90] P. Schorn. "An Object-Oriented Workbench for Experimental Geometric Computation." In *Proc. 2nd Canad. Conf. Comput. Geom.*, pp. 172–175, 1990.

[Schorn 91] P. Schorn. *Robust Algorithms in a Program Library for Geometric Computation*, Informatik-Dissertationen ETH Zürich, 32. Zürich: Verlag der Fachvereine, 1991.

[Schwarz 89] H. R. Schwarz. *Numerical Analysis: A Comprehensive Introduction.* Stuttgart: B. G. Teubner, 1989.

[Sedgewick 89] Robert Sedgewick. *Algorithms*, Second edition. Reading: Addison-Wesley, 1989.

[Seidel 88] R. Seidel. "Constrained Delaunay Triangulations and Voronoi Diagrams with Obstacles." Technical Report 260, IIG-TU Graz, Austria, 1988.

[Serpette et al. 89] B. Serpette, J. Vuillemin, and J. C. Herv'e. "BigNum: A Portable and Efficient Package for Arbitrary-Precision Arithmetic." Technical report, INRIA, 1989.

[Sethian 82] J. A. Sethian. "An Analysis of Flame Propagation." PhD diss., Dept. of Mathematics, University of California, Berkeley, 1982.

[Sethian 96] J. A. Sethian. *Level Set Methods.* Cambridge University Press, 1996.

[Sethian 99] J. A. Sethian. *Level Set Methods and Fast Marching Methods.* Cambridge University Press, 1999.

[Shamos 78] M. I. Shamos. "Computational Geometry." Ph.D. thesis, Dept. Comput. Sci., Yale Univ., New Haven, CT, 1978.

[Sharir and Agarwal 95] Micha Sharir and P. K. Agarwal. *Davenport-Schinzel Sequences and Their Geometric Applications.* New York: Cambridge University Press, 1995.

[Shewchuk 96] Jonathan R. Shewchuk. "Robust Adaptive Floating-Point Geometric Predicates." In *Proc. 12th Annu. ACM Sympos. Comput. Geom.*, pp. 141–150, 1996.

[Shewchuk 97] Jonathan Richard Shewchuk. "Adaptive Precision Floating-Point Arithmetic and Fast Robust Geometric Predicates." *Discrete Comput. Geom.* 18:3 (1997), 305–363.

[Shewchuk 99] J. Shewchuk. "Lecture Notes on Geometric Robustness." Technical report, University of California at Berkeley, Berkeley, CA, 1999.

[Sibson 80] R. Sibson. "A Vector Identity for the Dirichlet Tesselation." *Math. Proc. Camb. Phil. Soc.* 87 (1980), 151–155.

[Sibson 81] R. Sibson. "A Brief Description of Natural Neighbour Interpolation." In *Interpreting Multivariate Data*, edited by Vic Barnet, pp. 21–36. Chichester: John Wiley & Sons, 1981.

[Sugihara and Iri 89] K. Sugihara and M. Iri. "Two Design Principles of Geometric Algorithms in Finite-Precision Arithmetic." *Appl. Math. Lett.* 2:2 (1989), 203–206.

[Sugihara et al. 00] K. Sugihara, M. Iri, H. Inagaki, and T. Imai. "Topology-Oriented Implementation - An Approach to Robust Geometric Algorithms." *Algorithmica* 27:1 (2000), 5–20.

[Sugihara 94] Kokichi Sugihara. "Robust Gift-Wrapping for the Three-Dimensional Convex Hull." *J. Comput. Syst. Sci.* 49:2 (1994), 391–407.

[Sugihara 00] Kokichi Sugihara. "How to Make Geometric Algorithms Robust." *IEICE Transactions on Information and Systems* E83-D:3 (2000), 447–454.

[Tamassia and Vitter 96] R. Tamassia and J. S. Vitter. "Optimal Cooperative Search in Fractional Cascaded Data Structures." *Algorithmica* 15:2.

[Tamassia et al. 97] Roberto Tamassia, Luca Vismara, and James E. Baker. "A Case Study in Algorithm Engineering for Geometric Computing." In *Proc. Workshop on Algorithm Engineering*, pp. 136–145, 1997.

[Tanemura et al. 83] M. Tanemura, T. Ogawa, and W. Ogita. "A New Algorithm for Three-Dimensional Voronoi Tesselation." *J. Comput. Phys.* 51 (1983), 191–207.

[Torres 90] Enric Torres. "Optimization of the Binary Space Partition Algorithm (BSP) for the Visualization of Dynamic Scenes." In *Eurographics '90*, edited by C. E. Vandoni and D. A. Duce, pp. 507–518. North-Holland, 1990.

[Toussaint 88] Godfried T. Toussaint. "A Graph-Theoretical Primal Sketch." In *Computational Morphology*, edited by Godfried T. Toussaint, pp. 229–260, 1988.

[Uno and Slater 97] S. Uno and M. Slater. "The Sensitivity of Presence to Collision Response." In *Proc. of IEEE Virtual Reality Annual International Symposium (VRAIS)*, p. 95. Albuquerque, New Mexico, 1997.

[V. Barnett 94] T. Lewis V. Barnett. *Outliers in Statistical Data*. New York: John Wiley and Sons, 1994.

[Vaishnavi and Wood 80] V. K. Vaishnavi and D. Wood. "Data Structures for the Rectangle Containment and Enclosure Problems." *Comput. Graph. Image Process.* 13 (1980), 372–384.

[Vaishnavi and Wood 82] V. K. Vaishnavi and D. Wood. "Rectilinear Line Segment Intersection, Layered Segment Trees and Dynamization." *J. Algorithms* 3 (1982), 160–176.

[Vaishnavi et al. 80] V. K. Vaishnavi, H. P. Kriegel, and D. Wood. "Space and Time Optimal Algorithms for a Class of Rectangle Intersection Problems." *Inform. Sci.* 21 (1980), 59–67.

[van den Bergen 97] Gino van den Bergen. "Efficient Collision Detection of Complex Deformable Models using AABB Trees." *Journal of Graphics Tools* 2:4 (1997), 1–14.

[van Kreveld 92] M. J. van Kreveld. "New Results on Data Structures in Computational Geometry." Ph.D. dissertation, Dept. Comput. Sci., Utrecht Univ., Utrecht, Netherlands, 1992.

[van Leeuwen and Maurer 80] J. van Leeuwen and H. A. Maurer. "Dynamic Systems of Static Data-Structures." Report F42, Inst. Informationsverarb., Tech. Univ. Graz, Graz, Austria, 1980.

[van Leeuwen and Overmars 81] J. van Leeuwen and M. H. Overmars. "The Art of Dynamizing." In *Proc. 10th Internat. Sympos. Math. Found. Comput. Sci.*, Lecture Notes Comput. Sci., 118, pp. 121–131. Springer-Verlag, 1981.

[van Leeuwen and Wood 80] J. van Leeuwen and D. Wood. "Dynamization of Decomposable Searching Problems." *Inform. Process. Lett.* 10 (1980), 51–56.

[Wahl et al. 04] R. Wahl, M. Massing, P. Degener, M. Guthe, and R. Klein. "Scalable Compression and Rendering of Textured Terrain Data." In *Journal of 12th WSCG*, 12, edited by V. Skala and R. Scopigno, 12, pp. 521–528. UNION Agency - Science Press, 2004.

[Wan et al. 01] Ming Wan, Frank Dachille, and Arie Kaufman. "Distance-Field Based Skeletons for Virtual Navigation." In *Proceedings of the conference on Visualization 2001*, pp. 239–246. IEEE Press, 2001.

[Wang and Schubert 87] C. A. Wang and L. Schubert. "An Optimal Algorithm for Constructing the Delaunay Triangulation of a Set of Line Segments." In *Proc. 3rd Annu. ACM Sympos. Comput. Geom.*, pp. 223–232, 1987.

[Wang 93] C. A. Wang. "Efficiently Updating the Constrained Delaunay Triangulations." *BIT* 33 (1993), 238–252.

[Watson 81] D. F. Watson. "Computing the $n$-Dimensional Delaunay Tesselation with Applications to Voronoi Polytopes." *Comput. J.* 24:2 (1981), 167–172.

[Weghorst et al. 84] Hank Weghorst, Gary Hooper, and Donald P. Greenberg. "Improved Computational Methods for Ray Tracing." *ACM Transactions on Graphics* 3:1 (1984), 52–69.

[Wei and Levoy 00] Li-Yi Wei and Marc Levoy. "Fast Texture Synthesis Using Tree-Structured Vector Quantization." In *Siggraph 2000, Computer Graphics Proceedings, Annual Conference Series*, edited by Kurt Akeley, pp. 479–488. ACM Press / ACM SIGGRAPH / Addison Wesley Longman, 2000.

[Wendland 95] Holger Wendland. "Piecewise Polynomial, Positive Definite and Compactly Supported Radial Basis Functions of Minimal Degree." *Advances in Computational Mathematics* 4 (1995), 389–396.

[Wilhelms and Gelder 90] Jane Wilhelms and Allen Van Gelder. "Octrees for Faster Isosurface Generation Extended Abstract." In *Computer Graphics (San Diego Workshop on Volume Visualization)*, 24, 24, pp. 57–62, 1990.

[Yap and Dubé 95] C.-K. Yap and T. Dubé. "The exact computation paradigm." In *Computing in Euclidean Geometry*, Lecture Notes Series on Computing, 4, edited by D.-Z. Du and F. K. Hwang, pp. 452–492. Singapore: World Scientific, 1995.

[Yap 87a] C.-K. Yap. "An $O(n \log n)$ Algorithm for the Voronoi Diagram of a Set of Simple Curve Segments." *Discrete Comput. Geom.* 2 (1987), 365–393.

[Yap 87b] C.-K. Yap. "Symbolic Treatment of Geometric Degeneracies." In *Proc. 13th IFIP Conf. System Modelling and Optimization*, pp. 348–358, 1987.

[Yap 93] C.-K. Yap. "Towards Exact Geometric Computation." In *Proc. 5th Canad. Conf. Comput. Geom.*, pp. 405–419, 1993.

[Yap 97a] C.-K. Yap. "Towards Exact Geometric Computation." *Comput. Geom. Theory Appl.* 7:1 (1997), 3–23.

[Yap 97b] C.-K. Yap. "Robust Geometric Computation." In *Handbook of Discrete and Computational Geometry*, edited by Jacob E. Goodman and Joseph O'Rourke, Chapter 35, pp. 653–668. Boca Raton, FL: CRC Press LLC, 1997.

[Youngblut et al. 02] Christine Youngblut, Rob E. Johnson, Sarah H. Nash, Ruth A. Wienclaw, and Craig A. Will. "Different Applications of Two-Dimensional Potential Fields for Volume Modeling." Technical Report UCAM-CL-TR-541, University of Cambridge, Computer Laboratory, 15 JJ Thomson Avenue, Cambridge CB3 0FD, United Kingdom, 2002.

[Yu 92] Jiaxun Yu. "Exact Arithmetic Solid Modeling." Technical Report CSD-TR-92-037, Comput. Sci. Dept., Purdue University, 1992.

[Zachmann 97] Gabriel Zachmann. "Real-time and Exact Collision Detection for Interactive Virtual Prototyping." In *Proc. of the 1997 ASME Design Engineering Technical Conferences*. Sacramento, California, 1997. Paper no. CIE-4306.

[Zachmann 98] Gabriel Zachmann. "Rapid Collision Detection by Dynamically Aligned DOP-Trees." In *Proc. of IEEE Virtual Reality Annual International Symposium; VRAIS '98*, pp. 90–97. Atlanta, Georgia, 1998.

[Zachmann 02] Gabriel Zachmann. "Minimal Hierarchical Collision Detection." In *Proc. ACM Symposium on Virtual Reality Software and Technology (VRST)*, pp. 121–128. Hong Kong, China, 2002.

[Zhu and Mirzaian 91] B. Zhu and A. Mirzaian. "Sorting Does Not Always Help in Computational Geometry." In *Proc. 3rd Canad. Conf. Comput. Geom.*, pp. 239–242, 1991.

[Zwicker et al. 02] Matthias Zwicker, Hanspeter Pfister, Jeroen van Baar, and Markus Gross. "EWA Splatting." *IEEE Trans. on Visualization and Computer Graphics* 8:3 (2002), 223–238.

# Index